# Mathematical Adventures for Students and Amateurs

Edited by

David F. Hayes and Tatiana Shubin
*San Jose State University*

With the assistance of

Gerald L. Alexanderson and Peter Ross
*Santa Clara University*



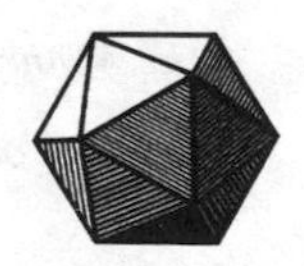

*Published and Distributed by*
The Mathematical Association of America

*Library of Congress Catalog Card Number 2004104647*

ISBN 0-88385-548-8

*Printed in the United States of America*

Current Printing (last digit):
10 9 8 7 6 5 4 3 2 1

SPECTRUM SERIES

The Spectrum Series of the Mathematical Association of America was so named to reflect its purpose: to publish a broad range of books including biographies, accessible expositions of old or new mathematical ideas, reprints and revisions of excellent out-of-print books, popular works, and other monographs of high interest that will appeal to a broad range of readers, including students and teachers of mathematics, mathematical amateurs, and researchers.

*777 Mathematical Conversation Starters,* by John de Pillis

*All the Math That's Fit to Print,* by Keith Devlin

*Carl Friedrich Gauss: Titan of Science,* by G. Waldo Dunnington, with additional material by Jeremy Gray and Fritz-Egbert Dohse

*The Changing Space of Geometry,* edited by Chris Pritchard

*Circles: A Mathematical View,* by Dan Pedoe

*Complex Numbers and Geometry,* by Liang-shin Hahn

*Cryptology,* by Albrecht Beutelspacher

*Five Hundred Mathematical Challenges,* Edward J. Barbeau, Murray S. Klamkin, and William O. J. Moser

*From Zero to Infinity,* by Constance Reid

*The Golden Section,* by Hans Walser. Translated from the original German by Peter Hilton, with the assistance of Jean Pedersen.

*I Want to Be a Mathematician,* by Paul R. Halmos

*Journey into Geometries,* by Marta Sved

*JULIA: a life in mathematics,* by Constance Reid

*The Lighter Side of Mathematics: Proceedings of the Eugène Strens Memorial Conference on Recreational Mathematics & Its History,* edited by Richard K. Guy and Robert E. Woodrow

*Lure of the Integers,* by Joe Roberts

*Magic Tricks, Card Shuffling, and Dynamic Computer Memories: The Mathematics of the Perfect Shuffle,* by S. Brent Morris

*The Math Chat Book,* by Frank Morgan

*Mathematical Adventures for Students and Amateurs,* edited by David F. Hayes and Tatiana Shubin. With the assistance of Gerald L. Alexanderson and Peter Ross

*Mathematical Apocrypha,* by Steven G. Krantz

*Mathematical Carnival,* by Martin Gardner

*Mathematical Circles Vol I: In Mathematical Circles Quadrants I, II, III, IV,* by Howard W. Eves

*Mathematical Circles Vol II: Mathematical Circles Revisited and Mathematical Circles Squared,* by Howard W. Eves

*Mathematical Circles Vol III: Mathematical Circles Adieu and Return to Mathematical Circles,* by Howard W. Eves

*Mathematical Circus,* by Martin Gardner

*Mathematical Cranks,* by Underwood Dudley

*Mathematical Evolutions,* edited by Abe Shenitzer and John Stillwell

*Mathematical Fallacies, Flaws, and Flimflam,* by Edward J. Barbeau

*Mathematical Magic Show,* by Martin Gardner

*Mathematical Reminiscences,* by Howard Eves

*Mathematical Treks: From Surreal Numbers to Magic Circles,* by Ivars Peterson
*Mathematics: Queen and Servant of Science,* by E.T. Bell
*Memorabilia Mathematica,* by Robert Edouard Moritz
*Musings of the Masters: An Anthology of Mathematical Reflections,* edited by Raymond G. Ayoub
*New Mathematical Diversions,* by Martin Gardner
*Non-Euclidean Geometry,* by H. S. M. Coxeter
*Numerical Methods That Work,* by Forman Acton
*Numerology or What Pythagoras Wrought,* by Underwood Dudley
*Out of the Mouths of Mathematicians,* by Rosemary Schmalz
*Penrose Tiles to Trapdoor Ciphers ... and the Return of Dr. Matrix,* by Martin Gardner
*Polyominoes,* by George Martin
*Power Play,* by Edward J. Barbeau
*The Random Walks of George Pólya,* by Gerald L. Alexanderson
*Remarkable Mathematicians, from Euler to von Neumann,* Ioan James
*The Search for E.T. Bell, also known as John Taine,* by Constance Reid
*Shaping Space,* edited by Marjorie Senechal and George Fleck
*Sherlock Holmes in Babylon and Other Tales of Mathematical History,* edited by Marlow Anderson, Victor Katz, and Robin Wilson
*Student Research Projects in Calculus,* by Marcus Cohen, Arthur Knoebel, Edward D. Gaughan, Douglas S. Kurtz, and David Pengelley
*Symmetry,* by Hans Walser. Translated from the original German by Peter Hilton, with the assistance of Jean Pedersen.
*The Trisectors,* by Underwood Dudley
*Twenty Years Before the Blackboard,* by Michael Stueben with Diane Sandford
*The Words of Mathematics,* by Steven Schwartzman

MAA Service Center
P.O. Box 91112
Washington, DC 20090-1112
800-331-1622 FAX 301-206-9789

# Preface

During the past few decades much has been done to improve mathematics teaching in the schools, but with the exception of a variety of problem competitions, activities to inspire interested and gifted middle school or high school students have not matched the efforts of the 1960s. An oft-heard comment is that highly gifted students are so well motivated, they really don't need much help. From my experience I am convinced this is not the case. Very talented students can become bored if there is nothing to stimulate their mathematical development, and they can drop out of school or, perhaps more commonly, major in some related field like engineering. In either case they're largely lost to mathematics. Problem competitions are important, of course, in motivating many students but timed tests appeal to only a part of the population of very talented students and they represent only one aspect of mathematics.

Stimulated in 1998 by a presentation by Zvezdelina Stankova and Paul Zeitz on plans for the Bay Area Mathematical Olympiad (BAMO) in the San Francisco Bay Area (a project promoting local mathematics competitions and "Math Circles," discussion sessions on high school or college campuses), Tatiana Shubin of San Jose State University and Peter Ross of Santa Clara University came up with the idea of sponsoring a series of talks aimed at bright secondary school students and their teachers and given by mathematicians who are not only admired for their research but are also renowned as inspiring speakers. The planning group was expanded to include Glenn Appleby of Santa Clara and David Hayes of San Jose State. They decided to call the program BAMA (Bay Area Mathematical Adventures). Fortunately, since our institutions are in the center of "Silicon Valley," we were convinced that a potential audience for these talks existed.

The program was launched on the Santa Clara campus on September 30 of that year by Ron Graham—at that time still at Bell Labs—who talked about the mathematics of juggling, filling to capacity (and slightly beyond) an auditorium that seats 250, not only with enthusiastic mathematics students and teachers but with jugglers as well. Subsequent talks, without jugglers, have continued to draw large audiences that include parents and even members of the general public. Funded by the departments at the two institutions, we were able to launch the program, which is still thriving six years later. Further assistance has been provided by Brian Conrey in his role as Director of the American Institute of Mathematics in Palo Alto, and by a parent who was so impressed by the quality of the lectures that he anonymously made a significant contribution to the financial support of the program. Professor Hendrik W. Lenstra, Jr., of the University of California, Berkeley, and the Universitiet Leiden, and a speaker early in the BAMA program, has been especially helpful in identifying additional speakers and has advocated starting up a similar program in The Netherlands.

When Don Albers at the MAA heard of the program he suggested that we collect some of these talks, not only to preserve them but also to allow people elsewhere to refer able students to them and perhaps inspire similar efforts in other regions.

Clearly, what you will find here is not a collection of transcripts of the talks—instead you will find the content of the talks presented in, we hope, easily readable form. We also point out that the speakers have different styles and were speaking on widely disparate topics in mathematics. No attempt has been made to make them uniform in style, as would befit chapters in a book devoted to a single topic.

We hope that readers will find much to inspire them in these articles, which range from classical geometry to very recent work in number theory using elliptic curves, from the purest mathematics to applications to the space program. And we hope that groups elsewhere will be moved to try something in their own geographical areas, something to supplement the problem competitions, to encourage and nurture the kind of talent we need in mathematics to make the next moves forward, prove the big theorems, solve the important problems, and inspire the next generation and the generation after that.

Gerald L. Alexanderson
January, 2004

# Contents

# Introduction

BAMA, the Bay Area Mathematical Adventures, is a series of talks for young people; its goal is to expose high school age students to the beauties of mathematics. What sparks an interest in mathematics? Is it an intriguing problem from a favorite teacher or a chance look into a book that catches our attention? Whatever it is, wouldn't it be nice to be able to make that potential moment of epiphany available on a regular basis? That is what BAMA is attempting to do and this book is a partial record of the form that attempt has taken.

From a more prosaic point of view this book is a collection of some of the BAMA talks on mathematics for bright middle and high school students, hosted alternately by Santa Clara University and San Jose State University. Many prominent mathematicians have agreed to expend their time and efforts to make these "adventures" possible. One of the goals is to provide students who have an interest in mathematics with the opportunity to expand and deepen that interest. Another goal is to furnish that unexpected spark in some of those who just happen to attend, say to accompany a friend or because their teacher encouraged them.

The first BAMA lecture, "Juggling Permutations of the Integers," was given by Ron Graham. The lecture was accompanied by breath-taking demonstrations of juggling by the lecturer and attracted not only students interested in mathematics but a large contingent of jugglers, both professional and amateur. It was a great beginning for the BAMA series.

But one lecture, no matter how good, does not a series make. BAMA has been sustained through subsequent adventures by such mathematical luminaries as Carl Pomerance, Hendrik W. Lenstra, Jr., Joseph Gallian, Robert Osserman, Sherman Stein, Robin Wilson, Persi Diaconis, Donald Saari, and many others as you will see in reading through the table of contents. With a format of three talks each Fall and three each Spring, the series is still going strong and attracting good audiences.

A measure of the success of BAMA is that it has consistently drawn audiences with significant percentages of sixth, seventh, and eighth graders as well as a good, though varying, number of high school students, parents, and teachers. While we're speaking of parents, the first time Hendrik Lenstra's mother ever heard him give a complete mathematics lecture was at his BAMA talk. There are also a fair number of other interested adults who attend the BAMA lectures on a regular basis.

Turning the lively and often highly interactive BAMA events into written text would be nearly impossible, so this book consists of a set of chapters, each based on the author's lecture in the series. In some cases the material here is rather more loosely based on the original delivery than in others, but all are presented by authors courageous and generous enough to put into writing material intended originally for a teenage audience. This book represents an effort to spread to a wider audience some of the enthusiasm for mathematics the series has helped to generate locally. After all, why should teens in the south San Francisco Bay Area have all the fun?

Of course, not every one of the speakers was able to prepare a written version of his or her lecture; the act of writing differs greatly from the act of presenting a lecture to a live (and often lively) audience, and writing takes a great deal of time. Still we hope that the reader will find in this collection some of the flavor of these adventures and come away with a sense of the range and depth of the presentations.

The chapters of the book vary in different dimensions: by theme, approach to the material, and in the expected level of mathematical maturity.

The book opens with an article that takes us beyond our own planet; in this article Carl Pomerance reminds us of the wonderful universality of mathematics. He shows us that mathematics is a marvelous tool that can be used for preserving and retrieving a wealth of information in a concise form, understandable for any intelligent being. Closer to home but still in space Helen Moore tackles NASA's problem of making sure the shuttle boosters were round to motivate the theory of curves of constant width; to the surprise of the students, the circle is not the only representative of this type of curve. Sheldon Axler brings us back to earth with an examination of the effects of computers on mathematics; he asks if computers have made mathematicians obsolete and concludes with a refreshing 'far from it.' Along the way we are shown many interesting examples of computers helping mathematicians.

In the next article of this opening section we are led through a mathematical detective story by Joe Gallian, who gives a brilliant example of mathematical sleuthing in the area of coding applied to real life—can you predict your driver's license number? The curious properties of the digit 1 as the lead digit in measured data are developed beautifully by Paul Zeitz into the theoretic underpinnings which show why this is really as it should be.

Thematically the largest group of articles is devoted to number theory and algebraic geometry. Karl Rubin, Ed Schaefer, and Peter Stevenhagen in their articles consider three similar Diophantine equations that correspond to three elliptic curves, and use them as starting points to demonstrate methods, current status, conjectures, and problems in number theory.

Articles by Joe Buhler & Ron Graham, and Art Benjamin & Jennifer Quinn each belong to that nebulous borderland between number theory and combinatorics. These papers describe new techniques or theories developed by the authors, and apply them to real life—especially to the real life of jugglers—or to number-theoretical problems. Susan Holmes considers the surprises lurking in probability; in this highly interactive article readers who perform the experiments will be richly rewarded.

Not surprisingly in a series for teens, geometry is the main topic in a number of the articles. These begin with Don Chakerian demonstrating how an ancient technique of solving algebraic equations (going back to the Babylonians) leads to Newton's method of approximating the zeros of non-linear functions, Pascal's theorem on the "mystic hexagon," projective geometry, and abstract algebra. Zvezdelina Stankova explores how Euclidian problems are sometimes better solved by using projective geometry, and then moves on to looking at problems in planar geometry as projections of three-dimensional objects. This later view generates even more elegant solutions.

Jean Pedersen's approach to a difficult problem in 3-dimensional combinatorial geometry demonstrates that, although mathematics may be the product of pure thought, a hands on approach can be very helpful. She brings to it a wealth of concepts such as convexity, duality, and symmetry groups, and at the same time provides the reader with a very practical and original way of constructing models of various polyhedra that turns out to be surprisingly useful in studying their properties. Next, we return to real life again as Dmitry Fuchs' article on caustics finds numerous examples in our normal surroundings; in this he even calls on the work of Pablo Picasso to provide examples. Richard Scott puts triangles on curved surfaces, at first on friendly surfaces, such as the plane and the sphere, but then surprises with results on surfaces of non-positive curvature.

Sherman Stein's paper belongs to the classical genre of viewing elementary mathematics from a higher point of view, or, if you wish, of "ancient math revisited." Dedicated to studying one of Archimedes' works, it may give a reader some inkling as to why Archimedes is considered one of the greatest mathematicians of all time. Bob Osserman examines the geometric problems rooted in constructing "good" geographical maps. Can you preserve the shape? Can you preserve the relative sizes? Can you do both at once? After we find the answers to these and other questions we embark on an elegant mini-tour of the geometric aspects of complex analysis.

Finally, in an account of the mathematics of Charles Dodgson, also known as Lewis Carroll, the audience happily joined Alice in Wonderland. Robin Wilson presents a play that comes from the same source of inspiration as his lecture—this is a form that may best capture the interactive flavor of his original talk.

The readers may well ask why there is no algebra in the table of contents. Surely algebra is a true companion of geometry—a full partner in the body of mathematics. Despite the absence of any specific section devoted to it, algebra permeates and serves as a background in many of the articles, most notably Scott's. This lecture had a very broad appeal—from the high school students intrigued by voluminous objects of no volume to the well-known mathematician in the audience who found a new fascination in the correspondence between strings of symbols and geometry.

Of course, the division of the contents into parts is approximate, at best. Our subject is like a fractal in which each piece seems to contain reference to the whole and may lead one to wander further into the rest of that body of knowledge we call mathematics. This collection of seemingly unrelated lectures provides evidence of this wonderful interconnectedness in mathematics. This is not only found inside many of the articles, but also in relation of articles showing similar topics from different points of view. For all of that, each talk stood alone, not building on any of the others. So these articles may be read independently of each other, in whatever order strikes the reader's fancy.

That the demands on the reader vary greatly between articles is an accurate reflection of the talks themselves. Each article contains something of interest and, we hope, inspiration, for mathematically-advanced middle and high school students. Everything in this book can be understood by bright high school students, and those who love to think at least as much as they like to be entertained.

David F. Hayes and Tatiana Shubin

January, 2004

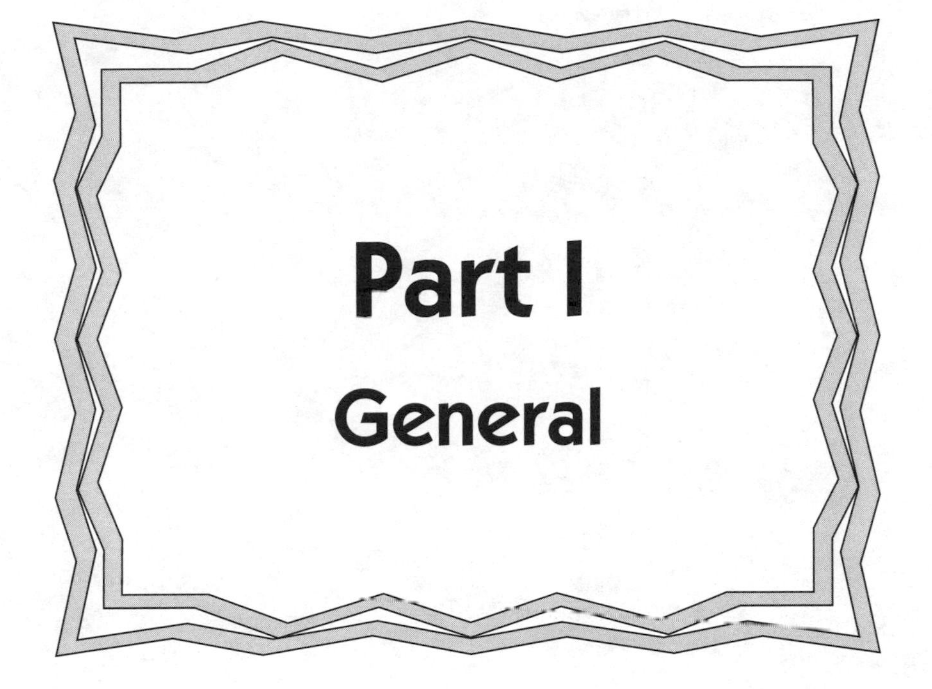

# Part I

## General

# 1

# Prime Numbers and the Search for Extraterrestrial Intelligence

Carl Pomerance
*Dartmouth College*

Apart from tabloid devotees, no one knows for sure if we human beings are alone in the universe as an intelligent species or if there are others, perhaps many others. Given the vast distances involved, it is reasonable to guess that if there is ever contact between humans and intelligent aliens, it will be through the radio spectrum. Fair enough, but how possibly might we actually transmit useful information to each other? English is becoming the "universal" language on our planet at this point, but it is ridiculous to assume that at first contact our alien correspondents would communicate with us in English, or any other Earth language.

The problem of accurate communication even between humans is quite daunting. Our history is marked with episodes of misunderstanding often caused by words or nuances that are misinterpreted. Despite the tragic loss of languages spoken by small and isolated groups of people, one might argue that it is a good thing that we are moving towards just a few principal languages.[1]

We also have some experiences communicating with other species, such as our pets. A dog owner can sometimes tell from the tenor of a bark or a whimper what's up with Fido. And sometimes Fido hears and obeys his owner's command. We have also tried to teach apes to speak in sign language, with limited success, and we have tried to interpret the whistles and clicks of dolphins and other whales. While this research is exciting, we still have not found a Rosetta stone that allows for free communication across species. It may be that as technological beings, we just don't have enough in common with other Earth species.

Which brings us again to possible extraterrestrial civilizations. Could there possibly be a message they could send us that we would be able to understand? Could we likewise send a message that an alien recipient could decipher?

In 1960, a German/Dutch mathematician and philosopher, Hans Freudenthal, suggested a language he called *Lincos* to use for possible extraterrestrial communication. It was sort of a pidgin

[1]Some have argued for an adaptation of Esperanto as a more logical language that is easily learned, but it has yet to catch on in a meaningful way.

of Latin and logic with short words using the roman alphabet together with logical symbols.[2] One could well argue that such a language is not transparent enough.

A decade later, the American astronomer Frank Drake came up with a brilliant scheme for extraterrestrial communication involving prime numbers. To get an idea of this, say you were to receive a message of dots and dashes:

——•———•—•—•———•—•—•———•———————————————————••—••——•—•—•

and to distinguish it from noise, after a short pause, it repeats, and then again, and again, and so on. The repetitive nature would certainly be an eye opener. But what possibly could it mean?

You begin analyzing. The "message" has 55 symbols. Hmmm. The number 55 can be factored into primes as $5 \times 11$ or $11 \times 5$. This suggests putting the symbols into a rectangular array. Let's try $11 \times 5$:

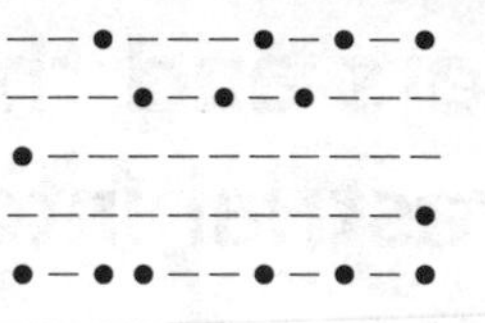

No, this does not seem to help, it still looks like random noise. Let's try the other way, $5 \times 11$:

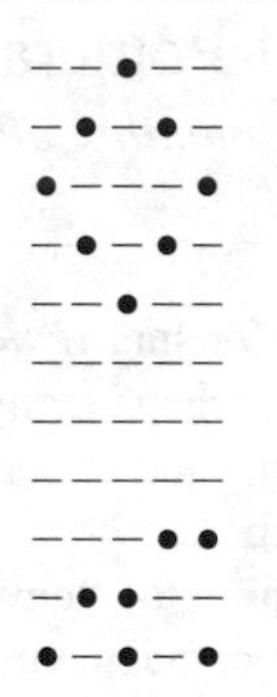

Wait, this no longer looks so random. Clearly there is some sort of diamond shape at the top. Could the aliens be baseball fans? Probably not, it is likely just a pleasing symbol to get our attention, But what is this strange staircase design at the bottom?

With just 55 pixels it is hard to communicate very much, but an inspired guess might be that the staircase design is counting in binary from 1 to 5. Think of a dot as a 1 and a dash as a 0, and read the bottoms of the 5 columns from left to right: 001, 010, 011, 100, 101. Yes, these are the base-two representations of the numbers $1, 2, 3, 4, 5$.

It is easy to imagine that using a longer repetitive message, one might easily send more detailed images. Even images of photographic quality. And by then sending different pictures, it is conceivable that one could transmit a movie in this way. Which one would you send? My favorite: *Attack of the Killer Tomatoes*. Well, maybe not, we don't want to give the aliens any ideas.

The key thing is that each repetitive string should have length $pq$ where $p, q$ are primes, so there are at most two essentially different ways of putting the string into a rectangular array.[3]

In the novel *Contact* by Carl Sagan, this is exactly the method that the aliens choose to communicate with us earthlings. Currently, SETI (Search for ExtraTerrestrial Intelligence) is systematically searching for messages from outer space, enlisting the aid of people all over the world who have spare computer time.[4]

[2] See `http://www.geocities.com/monicavdv/informatie-hans/Lincos.html`.

[3] If $p = q$ there is just one way, so maybe this is a better scheme for sending messages.

[4] See `http://setiathome.ssl.berkeley.edu/`.

**Figure 1.1.** Left: The Arecibo message. Right: The Arecibo radio telescope. Courtesy: Seth Shostak/SETI Institute.

They haven't found anything too remarkable as yet, but in some sense we have only just begun our search. In the meantime, have we sent any such messages specifically for alien consumption? Yes, we have. When the Arecibo radio telescope was inaugurated in Puerto Rico, one of its first jobs was to send a repetitive message into space. This message had length 1679, which factors as $23 \times 73$. Arranged the long way from top to bottom, the picture contains an amazing wealth of information: the numbers from 1 to 10, the atomic numbers of hydrogen, carbon, nitrogen, oxygen, and phosphorous (the principal elements on which most life on Earth is based), the formulas for sugars and bases in nucleotides of DNA, the number of nucleotides in DNA, a picture of a double helix of DNA, a picture of a person, the human population on Earth, the height of a human, a schematic showing our solar system with Earth in a distinguished position, a schematic of the radiotelescope at Arecibo, and the diameter of the telescope. Measures such as the telescope diameter and the height of a person were calibrated using the wavelength of the transmission of the message, namely 12.6 centimeters. Perhaps a bit crowded to my taste, but it was merely a demonstration of an idea. The Arecibo message was beamed to the globular cluster M13 which is about 25,000 light years away. Maybe in 50,000 years we will get a reply!

It is interesting that prime numbers would have an application that makes communication transparently simple. Especially so, since many cryptographic systems, designed to *hide* messages from unauthorized recipients, are also based on prime numbers. And it was not so long ago that prime numbers were merely the province of the curious, with no discernible applications at all.

What are the chances that there is an alien species in existence close enough for us to contact? Of course this question is very difficult to answer. Those who think there are other civilizations out there point to the enormous number of stars in the universe. Even if only $1/10$ of them have planetary systems, and even if only $1/10$ of these have a planet such as Earth that can support life as we know it, and even if only 1/10 of these actually have spawned life, and even if only 1/10 of these have life that has evolved into intelligent creatures, that still leaves a tremendous number of intelligent aliens out there. On the minus side, we might argue that human civilization has only existed for a very brief time on a cosmic scale, and the technological aspect of our civilization only for a few hundred years at most. Moreover, there are indications that we will not be around very long! We use our technology to foul our own planet and make it less habitable, maybe even uninhabitable. Even though we have clear archaeological and historical records of earlier human civilizations that vanished because of climate changes, we continue to burn fossil fuels as if there were no tomorrow, literally. And when the most advanced and richest country, which produces the most pollutants, is called on the question, its leaders claim that it would hurt their economy to significantly cut back on emissions, and this greenhouse thing really hasn't been proved anyway. So, if one were to extrapolate to alien civilizations in outer space, one might conceivably guess that if they are at all like us, they won't be around very long either. So, while there may have been many intelligent civilizations, the chance that one exists at the same brief time as ours, and also within shouting distance, may be very slim indeed.[5]

Nevertheless it is fun to contemplate such contact. And the exercise may help us to be a bit more introspective and responsible as stewards of our own lovely planet.

For further reading on prime numbers:

R. Crandall and C. Pomerance, *Prime numbers: a computational perspective*, Springer-Verlag, New York, 2001.

P. Ribenboim, *The little book of bigger primes,* second edition, Springer, New York, 2004.

For further reading on the likelihood of extraterrestrial civilizations, and how we might detect and communicate with them:

A. G. W. Cameron, Editor, *Interstellar communication*, Benjamin, New York, 1963.

H. Freudenthal, *Lincos: design of a language for cosmic intercourse,* North-Holland, Amsterdam, 1960.

S. A. Kaplan, Editor, *Extraterrestrial civilizations; problems of interstellar communication,* (translated from the Russian edition of 1969), Israel Program for Scientific Translations, Jerusalem, 1971.

C. Sagan and F. Drake, The search for extraterrestrial intelligence, *Scientific American*, May, 1975, pp. 80–89.

I. S. Shklovskii and C. Sagan, *Intelligent life in the universe*, Emerson-Adams Press, 1998.

W. Sullivan, *We are not alone: the continuing search for extraterrestrial intelligence*, Plume, 1994.

S. Webb, *If the universe is teeming with aliens ... where is everybody? Fifty solutions to the Fermi Paradox and the problem of extraterrestrial life*, Copernicus, New York, 2002.

Finally, a link to the SETI Institute: `http://www.seti-inst.edu/`.

[5]The thought of multiplying various probabilities, such as we suggested above to assess the likely number of advanced civilizations, and also to assess the longevity of a civilization, is due to Frank Drake. Drake's equation in fact presents a formula that gives the number of advanced civilizations, once all of these probabilities are estimated. On the longevity question, Drake was led to muse "Is there intelligent life on Earth?"

# Space Shuttle Geometry

Helen Moore
*American Institute of Mathematics*

How can mathematical definitions and theorems help prevent accidents? Read on, and find out. For the exploration in section 4, you will need something with a straightedge (for example, a ruler), a compass, some paper and a pencil.

This talk was inspired by a talk of Robert Bryant (of Duke University) on the same topic.

## 1 Introduction

The knowledge of mathematical theorems and properties turns out to be useful in many unexpected situations. Perhaps even more important than knowing lots of mathematics, is knowing enough mathematics to realize when you should talk to someone who knows more. In this example, the definition of a circle plays a key role. Do you remember the definition of a circle?[1]

## 2 Space Shuttle Background

The U.S. space shuttle program originated in the early 1970's, as a way to save money while continuing to fly space missions with astronauts on board. Instead of building rockets that could be flown only once, reusable space shuttles were developed. Like airplanes, they would refuel and fly again, but because of the extreme stresses they would undergo, they would need months of repairs and careful testing between flights. The first space shuttle was launched in 1981, and by the year 2000, over one hundred successful flights had taken place.

On January 28, 1986, the space shuttle Challenger exploded 72 seconds after lift-off. All seven crew members were killed, including Ron McNair, the second African-American in space, Judy Resnik, the second American woman in space, and Christa McAuliffe, who was on board as part of the new Teacher in Space Program. The shuttle and the attached boosters and fuel tank were destroyed. The shuttle program was halted for over two years, while the accident was investigated and preventative recommendations were implemented.

**Figure 2.1.** A space shuttle launch

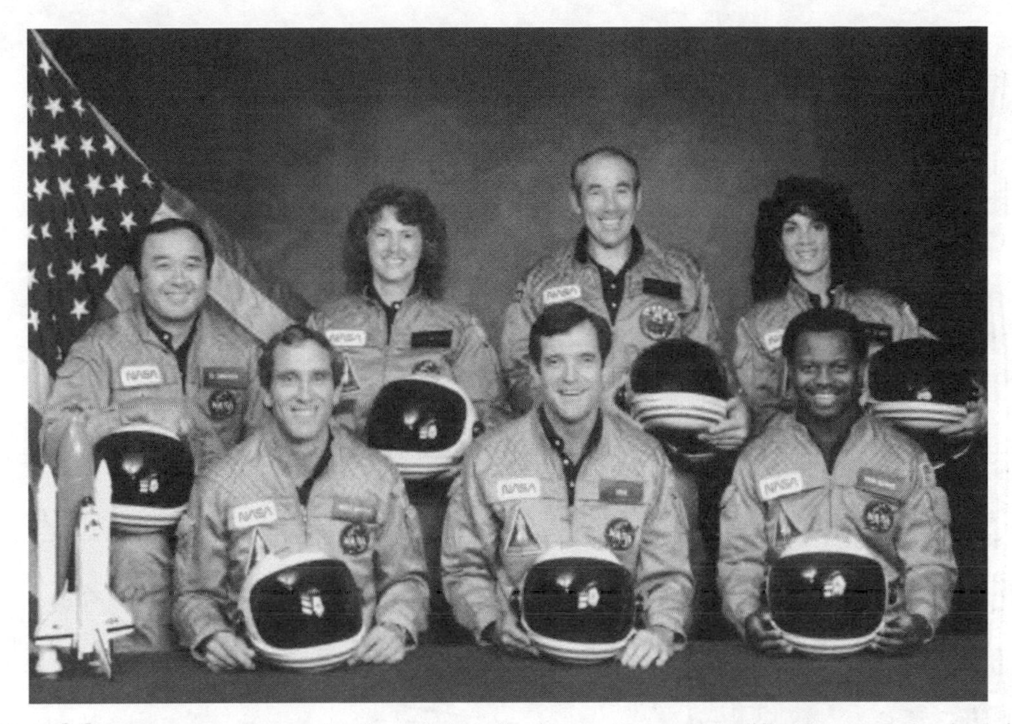

**Figure 2.2.** Back row (left to right): Ellison S. Onizuka, Sharon Christa McAuliffe, Gregory B. Jarvis, Judith A. Resnik. Front row: Michael J. Smith, Francis R. (Dick) Scobee, Ronald E. McNair.

The president of the United States appointed a committee to investigate the causes of the accident. Two of the committee members had been astronauts themselves: Neil Armstrong was the first person to set foot on the moon; Sally Ride was a physicist and the first American woman in space. Another physicist appointed to the committee was Richard Feynman, who was a professor at Caltech. Feynman wrote about his experience on the committee in his humorous and candid book *"What Do You Care What Other People Think?"* Feynman's book includes many details of the investigation of the accident.

## 3 The Cause of the Accident

The cause of the accident was determined to be the failure of a seal on one of the boosters attached to the shuttle. This seal failure led to a fuel leak, which led to the explosion. The committee decided there were three main contributing factors to the seal failure: poor booster joint design, slow o-ring response, and out-of-roundness.

### 3.1 Booster Joint Design

During shuttle lift-off, there are two solid rocket boosters and a fuel tank that are attached to the shuttle at first, and drop off after the fuel inside them is used up. The boosters are each made up of several cylindrical metal pieces that are put together. Figure 2.3 shows how two pieces of a booster fit together at a place called a "joint." The large forces inside the booster during lift-off cause the joint to bulge outward. (Because a joint is weaker than the other parts of the booster, it will give way first.) After the accident, the joints were redesigned to keep them from bulging outward so much. This helps prevent the joint seal from breaking.

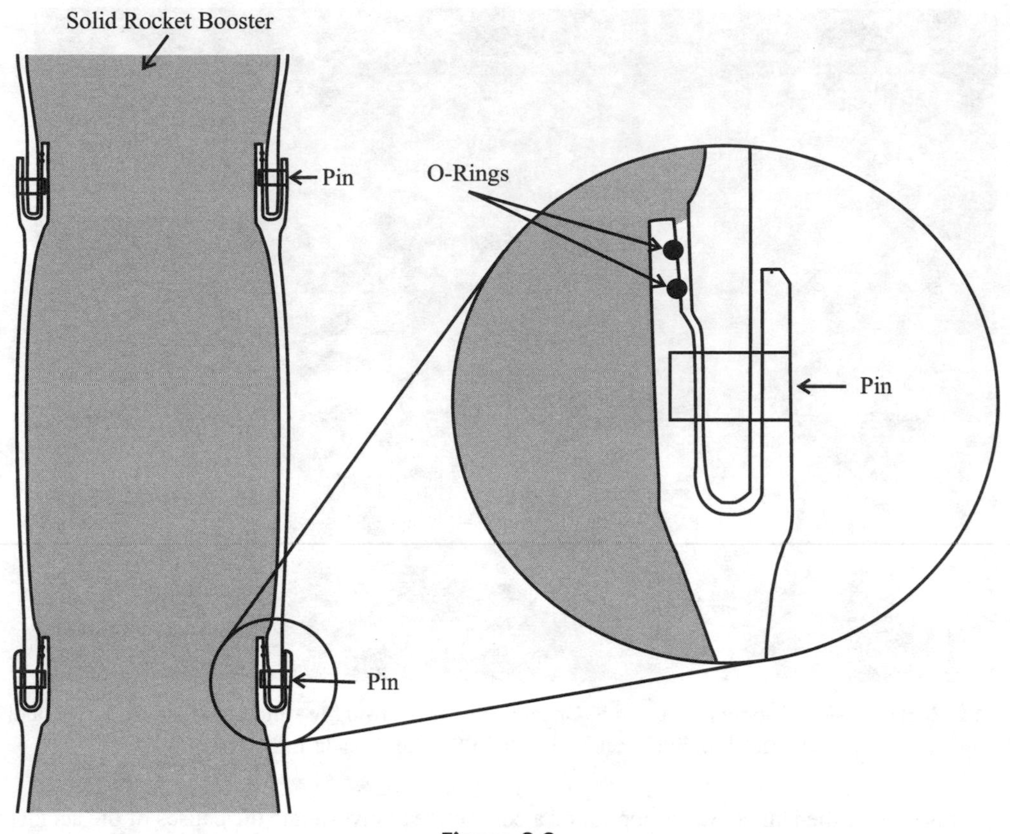


Figure 2.3.

## 3.2 O-ring Response Time

The o-rings are thin rubber rings that act like washers on a faucet. When they are in place between the metal pieces of the booster, they are compressed, like a sponge. But if a gap is formed between pieces of the booster, then the o-rings are supposed to spring back quickly to their full size, which would prevent any fuel from leaking out of the booster. The morning of the Challenger accident, the temperature was below freezing at the site of the launch. It turns out that o-rings don't spring back to full size as quickly when they are that cold. Feynman gave an impressive demonstration of this at a public press conference. He used a metal clamp to compress part of an o-ring, and he put the clamp and the o-ring piece in a glass of ice water. After a few minutes, Feynman pulled the o-ring piece out of the water, and removed the clamp. It was easy to see that the piece of o-ring took several seconds to spring back to its normal shape. During lift-off, however, it was estimated that such a lengthy delay in springing back to shape could be disastrous. It was determined that lift-offs should never again occur under such cold conditions.

## 3.3 Out-of-roundness

After a lift-off from Kennedy Space Center, in Cape Canaveral, Florida, it takes about five minutes for all of the fuel in the shuttle boosters to be used. The empty boosters are then released from the shuttle, and fall into the Atlantic Ocean. They are recovered, taken apart, and the pieces reconditioned. Then each part is refilled with solid fuel, and examined to make sure that it is very round. If the pieces are not round enough, then they have to be reshaped, since the metal pieces have to fit together precisely to prevent fuel leaks. NASA's test for roundness at the time of the Challenger lift-off was the following: measure the longest distance (we will call this a "diameter") across the cross-section at three different places. If the three diameters were all the same, then the booster was declared round.

What do you think of their test? How many diameters would need to be the same in order to guarantee that the booster is round? On a piece of paper, try to draw a curve that has that number of diameters all the same, but isn't a circle. If you have trouble, get someone else to try it with you. You can make a game: if you give them a number, say, $n$, they can try to draw a curve (we'll allow corners and straight pieces) that has $n$ diameters all the same, but isn't a circle. Who do you think will win the game? See Figure 2.4 for some attempts to play this game.

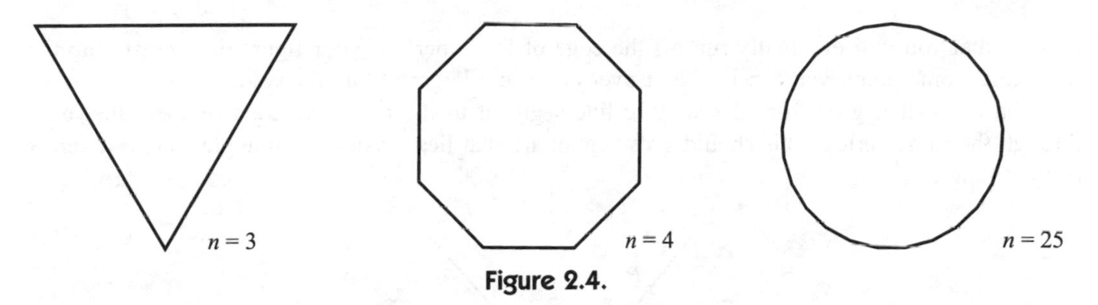


**Figure 2.4.**

What if you could measure *all* of the diameters, and they were all the same? Would the curve have to be a circle? Try measuring the diameters of the figure shown below (Figure 2.5). It has all of the diameters the same, but it is clearly not a circle! We will call it a "fake circle" or a curve of constant width. It turns out that there are infinitely many fake circles that can be drawn. In the next section, you will learn how to make them yourself.

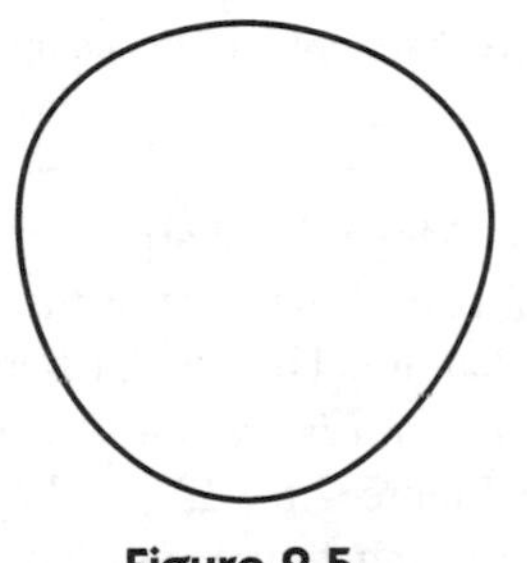

**Figure 2.5.**

## 4 Watch Out! That's a Fake Circle

How do you construct a fake circle? Start by drawing three straight line segments that form a triangle. It doesn't matter what kind of triangle is formed. It could be equilateral (all sides having the same length), isosceles (exactly two of the sides having the same length), or scalene (none of the sides having the same length). The line segments should actually be extended beyond the corners, or vertices, of the triangle. (See Figure 2.6.)

To draw the first arc of the fake circle, take a compass, put the sharp point on a corner, or vertex, of the triangle, and put the pencil point on one of the nearest extended segments, reasonably far outside of the triangle. You should make it far enough outside the triangle that you'll be able to continue the figure without ever crashing into a side of the triangle, but not so far outside the

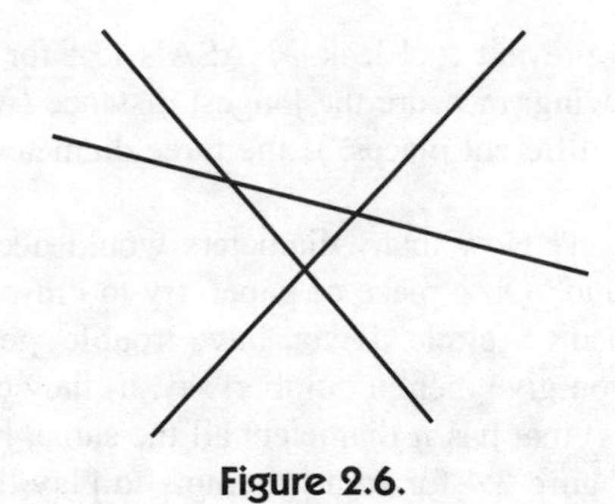

**Figure 2.6.**

triangle that you will eventually run off the edge of the paper! (If your figure does crash into the triangle at some point, you can just start over and use a larger radius for your first arc.) Draw an arc of a circle that goes from the starting line segment to the other extended segment that goes through the same vertex. This should give you an arc that lies outside the triangle, across a vertex of the triangle.

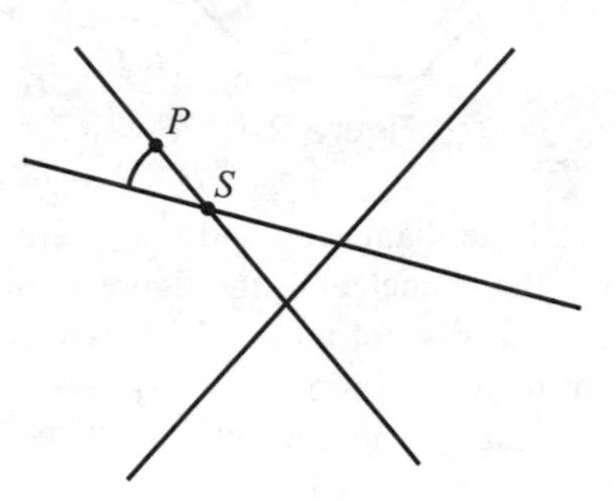


**Figure 2.7.** $S$ is where you should put the sharp point of the compass; $P$ is where you should put the pencil point at first.

Now you will draw an adjacent arc. Move the sharp point of the compass to a different vertex of the triangle. Put the pencil point on one of the ends of the first arc you drew, by opening up the compass until the pencil point meets that arc. The crucial step here is to start with the pencil point on the correct end of the arc: *the correct end is the one that makes the entire compass lie directly over one of the line segments.* (See Figure 2.8.) If you begin with part of your compass over blank space on the paper, then the resulting figure will have corners. Once you have positioned the compass over a line segment, go ahead and start drawing an arc in the new region, stopping when you hit the next line segment.

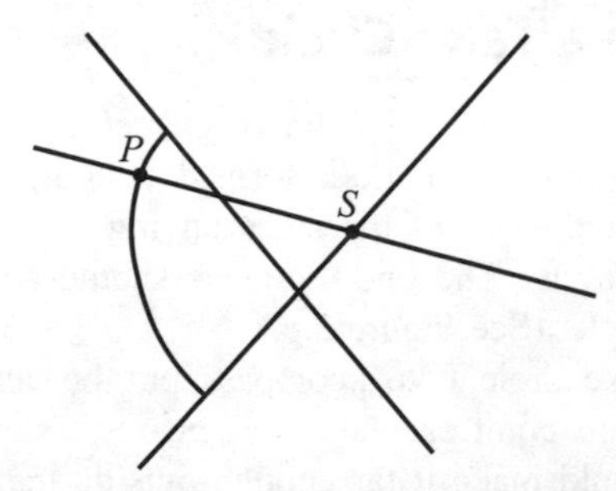


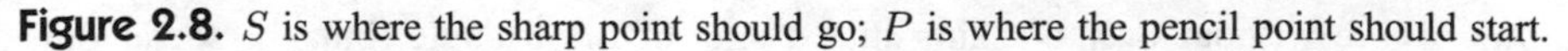

**Figure 2.8.** $S$ is where the sharp point should go; $P$ is where the pencil point should start.

For the third arc, start with your pencil point where it stopped when you finished the second arc. But move the sharp point to a different vertex than the two you have already used. Make sure that once again the entire compass initially lies directly over one of the line segments. If it doesn't, then you've placed the sharp point of the compass on the wrong vertex, and should move it. When placing the sharp point on a new vertex, you will need to change the radius of the compass. If everything is lined up, then go ahead and draw an arc in the next region, stopping when you encounter the next line segment.

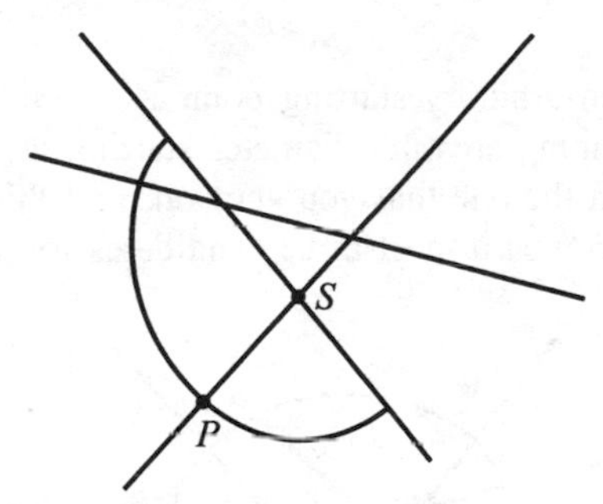


**Figure 2.9.** Third arc.

For the fourth arc, keep the pencil point where it ended for the third arc, but move the sharp point of the compass back to the first vertex you used. Now draw an arc to the next line segment. (See Figure 2.10a.) Can you figure out where to put the sharp point of the compass for the fifth and sixth arcs?[2] If you draw it correctly, your figure shouldn't have any corners. (See Figures 2.10b and 2.10c.)

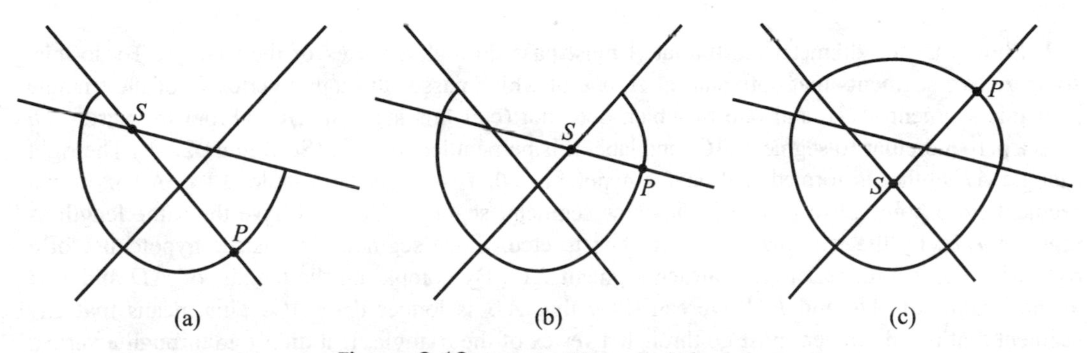


**Figure 2.10.** Fourth, fifth, and sixth arcs.

Try experimenting with triangles of different shapes (thin, thick, tiny, large, isosceles, equilateral, etc.). What effect does this have on the resulting curve of constant width? What kind of triangles lead to curves that look very much like circles?[3] Are there other properties besides the shape and relative size of thc triangle you can use to affect the shape of the resulting curve?

## 5 How Do You Know For SURE That the Curve Has Constant Width?

If this is the question that's on your mind now, then you should consider a career in mathematics or science! It is an excellent question, and we will use mathematics to give a proof that the figure has constant width. When I say proof here, I mean a convincing argument, assuming that you know some geometric facts about circles. Why is a proof important? For one thing, it would have shown NASA that their test for roundness would fail, even if they could take perfect measurements! We describe this kind of situation by saying that the theory behind their test wasn't sound. Of course, measurements can never be made perfectly precise (and for similar reasons, we can never count to infinity). But if the theory is sound, then measurements can be made as accurately as possible, and if everything is within certain small error sizes, then it can all work out okay. In the remainder of this section, we will go through the main parts of the proof that the fake circle we constructed has constant width.

**Proof in 4 steps: 1)** Given an arbitrary starting point $A$ in one of the six arcs on the curve, figure out in which of the remaining arcs the diameter starting at $A$ will land. It should always be the arc directly opposite from the one that you start on, and this can be justified with a little thought. (See Figure 2.11 for a comparison of three candidates for a diameter.)

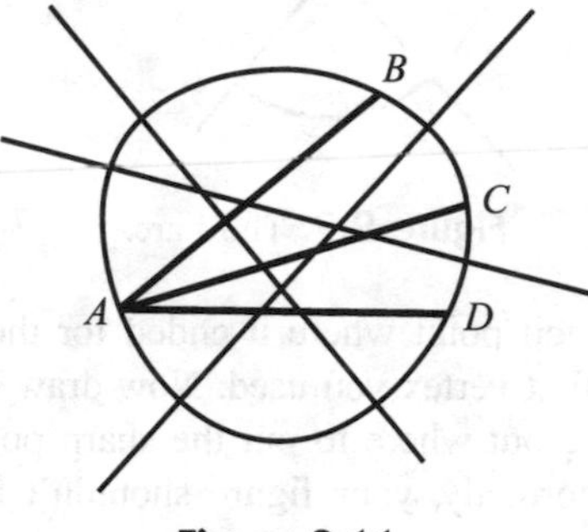


Figure 2.11.

**2)** Show that the diameter starting at $A$ must pass through a vertex of the triangle. To do this, compare two segments that both start at $A$, one of which passes through a vertex $D$ of the triangle (call this segment $AB$), and one of which does not (call this segment $AC$). From the vertex $D$, drop a perpendicular to segment $AC$, and label this perpendicular $DE$. (See Figure 2.12.) The right triangle $ADE$ that is formed will have a hypotenuse that is longer than its long leg. Also, draw a segment from point $D$ to point $C$. This new segment, segment $DC$ will have the same length as segment $DB$, by the way the figure was constructed. Since segment $DC$ is the hypotenuse of a right triangle, it will be longer than leg segment $EC$. By comparing the lengths of $AD$ and $DB$ to the lengths of $AE$ and $EC$, we can show that $AB$ is longer than $AC$. This means that any segment that is a diameter, must go through a vertex of the triangle. If it didn't go through a vertex, then it wouldn't be as long as a segment that did, and therefore it couldn't be a diameter.

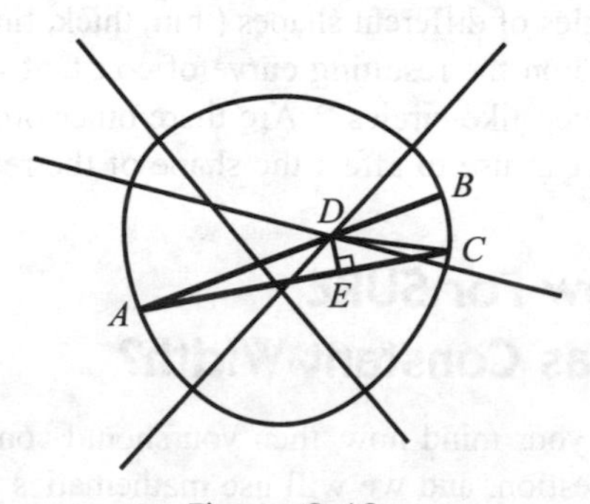


Figure 2.12.

**3)** Show that all the diameters passing through this vertex will have the same length. This uses the fact that the fake circle was constructed from pieces of circles, and so segments which are radii of the same circle will have the same length.

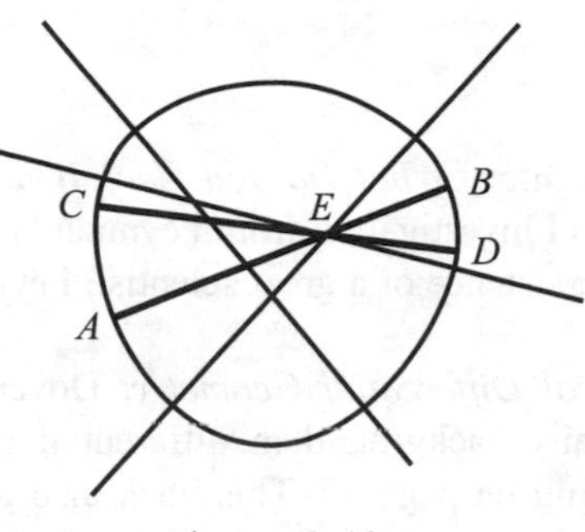


Figure 2.13.

**4)** Show that diameters that originate in a neighboring region will have the same length as diameters originating in the region where $A$ is located. To do this, just compare the diameter in a region with the diameter on the boundary of a region, and then compare the boundary diameter with a diameter in the next neighboring region. You can show that diameters in adjacent regions will have the same length, so then all of the diameters have the same length.

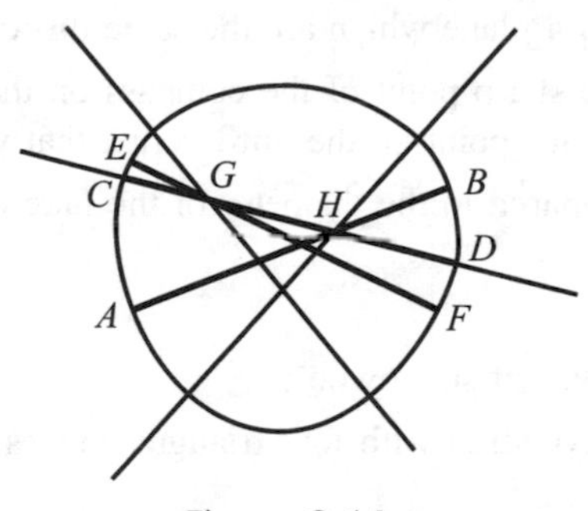


Figure 2.14.

## 6 Questions You Are Probably Asking

Having a scientific, inquisitive mind, you might be wondering what test for roundness NASA is using now! NASA currently uses the following technique to test for roundness: they take a hollow cylindrical tube, and place it around the booster component being tested. They measure the distance from the outer tube to the booster at various points. If the distance is the same at *every* point, then they know for sure that the booster is round. By putting a tube on the outside, they are establishing a fixed center point (the center point of the tube), without ever having to disturb the solid propellant in the interior of the booster. Can you prove that if all of the distances are the same, then the booster is round?[4] Is it ever possible for the distances to be different, but for the booster to still be round? Suppose this happened. What would be a good way to adjust the tube in order to get some more measurements?

Here are some other questions for you: Are there other ways to make fake circles? What if you started with a quadrilateral (a figure with 4 sides) inside? A pentagon (5 sides)? A hexagon (6 sides)? A heptagon (7 sides)? Others?[5] Explore!

Is there a shape like a beach ball that has all its diameters the same but isn't perfectly spherical (round)? How would you try to make such an object? What kind of shape would you start with inside?[6] Can you generalize this for higher-dimensional objects?

## 7 References

Richard Feynman and Ralph Leighton, *What Do You Care What Other People Think?*, Bantam Books, 1988. (Tales of the panel investigation from Feynman's perspective. Highly entertaining, and shows the curiosity and persistence of a great scientist. Feynman wasn't afraid to ask "silly" or "troublesome" questions.)

Dirk Struik, *Lectures on Classical Differential Geometry*, Dover Publications, 1961. (A bit advanced, but you don't need any background in differential geometry to read the section on curves of constant width starting on page 50. This book also gives excellent further references for the mathematics of these curves.)

`http://science.ksc.nasa.gov/shuttle/missions/51-l/mission-51-l.html` (A web site with details of the Challenger accident, including links to explore some things in more depth.)

## 8 Answers and Hints to Footnoted Questions

1) A circle is the set of points in a plane which are the same distance from a fixed (center) point.

2) To draw the fifth arc, put the sharp point of the compass on the second vertex that you used. To draw the sixth arc, put the sharp point on the third vertex that you used.

3) The smaller the triangle (compared to the diameter of the fake circle), the more the curve will resemble a circle.

4) Yes.

5) Hint: Some of these will work, but some won't.

6) Hint: Try a tetrahedron (a polyhedron with four triangular faces).

# 3

# Mathematicians versus the Silicon Age: Who Wins?

Sheldon Axler
*San Francisco State University*

Mathematicians invented computers. Computers then changed the way mathematics progresses. Have computers made mathematicians, and the techniques they developed over the past few thousand years, obsolete? Are mathematicians merely trying to preserve their jobs by sticking with outmoded requirements, such as insisting on rigorous proofs even in the face of overwhelming numeric evidence? This possibility was envisioned by Alan Turing, a mathematician who helped create the dawn of the computer age (and who helped the Allies win World War II by playing a key role in cracking the German Enigma code), when in 1947 he wrote:

> The masters are liable to get replaced because as soon as any technique becomes at all stereotyped it becomes possible to devise a system of instruction tables which will enable the electronic computer to do it for itself. It may happen however that the masters will refuse to do this. They may be unwilling to let their jobs be stolen from them in this way. In that case they would surround the whole of their work with mystery and make excuses, couched in well chosen gibberish, whenever any dangerous suggestions were made. I think that a reaction of this kind is a very real danger.

## Computer Triumphs

An early and dramatic example of the intrusion of computers into serious mathematics came with the resolution of the four-color conjecture. The four-color conjecture arose in 1852 as the question of whether every conceivable map of the world can be colored using only four colors so that no two adjacent countries have the same color. By 1922 this had been proved to be true for all maps that have at most 25 countries. The four-color conjecture stumped mathematicians until 1976, when Kenneth Appel and Wolfgang Haken of the University of Illinois announced that the four-color conjecture had become the four-color theorem. Press attention focused on Appel and Haken's massive use of computers to show that four colors always suffice.

Computer power has grown tremendously since 1976 when an expensive mainframe computer helped crack the four-color conjecture. Today for about a thousand dollars you can buy a fast personal computer. For a few hundred dollars more, you can add to it mathematical processing software that will put in your hands easily usable computational tools that allow you to answer questions that stumped mathematical giants of past centuries.

The two best known mathematical processing software programs are *Mathematica* and *Maple*. Both of these programs can effortlessly do almost everything you learned in high school mathematics, and they "know" number theory, calculus, differential equations, linear algebra, and much, much more. They come with many hundreds of built-in functions, extensive features for manipulating these functions, and a high-level computer language that allows you easily to create functions and procedures of your own. The two programs are roughly equivalent, and mathematicians are probably about evenly divided as to which they prefer. My personal preference is for *Mathematica*, so in the following when I write something like "my computer says" I am referring to an ordinary personal computer running *Mathematica*.

Let's look at some of the capabilities of *Mathematica* and *Maple*. At the lowest level, these programs can do arithmetic. But they surpass calculators in that they can perform exact arithmetic regardless of the number of digits involved. For example, choosing some random fractions and asking my computer to add them, I instantly see that

$$\frac{1874891749763074}{8872485740930397728975} + \frac{28352787245465764327 49}{543636634687529859284 5297}$$

equals

$$\frac{252578960392272841971636678037028055936 5253}{482340828951249608384828498505700330698338 0575},$$

where the answer has automatically been displayed in reduced form with no common factors in the numerator and denominator. Instead of having the answer displayed as a fraction, I could have requested it to be displayed as a decimal.

As another example, my computer tells me instantly that $3^{101}$ equals

$$1546132562196033993109383389296863818106322566003.$$

As a final example of arithmetic, when I ask my computer to compute 1000!, it instantly gives all 2568 digits correctly. This last example particularly amuses me because when I was an undergraduate, before software such as *Mathematica* and *Maple* existed, I spent several weeks writing a computer program to compute 1000! exactly.

If we turn our attention to algebra, we will not be surprised to find that *Mathematica* can solve large systems of linear equations and that it knows the quadratic formula. Hardly any mathematicians know the cubic formula, but *Mathematica* does, telling me that one of the solutions of the equation

$$x^3 + 5x - 7 = 0$$

is

$$-5\left(\frac{2}{3(63+\sqrt{5469})}\right)^{1/3} + \frac{\left(\frac{63+\sqrt{5469}}{2}\right)^{1/3}}{3^{2/3}}.$$

*Mathematica* also knows the quartic formula for finding the roots of fourth-degree polynomials. Alas, for fifth-degree and higher polynomials, *Mathematica* can usually provide only decimal approximations to the roots (accurate to thousands of decimal places, if so requested by the user), not exact roots. However, no person or computer could do better, because mathematicians have proved that no formula exists for finding roots of polynomials of degree higher than four.

Could you factor the polynomial

$$x^8 + 30x^7 + 233x^6 + 192x^5 + 141x^4 + 99x^3 + 163x^2 + 61x + 72$$

in less than a day? My computer instantly factors the polynomial above as

$$(x^3 + 17x^2 + 5x + 8)(x^5 + 13x^4 + 7x^3 + 2x + 9).$$

You should have learned in high school that

$$1 + 2 + 3 + \cdots + n = \frac{n(n+1)}{2}$$

for every positive integer $n$. *Mathematica* knows that, and it also tells me that

$$1^8 + 2^8 + 3^8 + \cdots + n^8$$

equals

$$\frac{n(n+1)(2n+1)(5n^6 + 15n^5 + 5n^4 - 15n^3 - n^2 + 9n - 3)}{90}.$$

Instead of asking for the sum of eighth powers, I could have asked for the sum of eightieth or eight-hundredth powers and received equally quick answers from my computer.

## Mersenne Primes

An integer greater than 1 is called *prime* if it has no divisors other than 1 and itself. For example, 13 is prime but 21, which is divisible by 3 and 7, is not. The first ten primes are 2, 3, 5, 7, 11, 13, 17, 19, 23, 29. The number 1 is excluded, by definition, from the list of primes because otherwise positive integers would not have unique factorizations into products of primes (here we regard two factorizations as the same if they differ only in the order of the factors).

Computers allow us to calculate with prime numbers in ways that would take mathematicians lifetimes without machines. My computer quickly tells me that the one-millionth prime is 15485863 and the one-billionth prime is 22801763489. When I asked my computer for the one-trillionth prime, I had to wait almost seven minutes before getting the answer: the one-trillionth prime is 29996224275833. The ancient Greeks had proved that there are an infinite number of primes, so there is no largest prime number.

A *Mersenne prime* is a prime number of the form

$$2^n - 1,$$

where $n$ is a positive integer. Mersenne primes are named in honor of the French monk Marin Mersenne, who in 1644 claimed that the only values of $n$ for which $2^n - 1$ is prime are

$$n = 2, 3, 5, 7, 13, 17, 19, 31, 67, 127, 257.$$

For the first four values specified by Mersenne, we have

$$\begin{aligned} 2^2 - 1 &= 3 \\ 2^3 - 1 &= 7 \\ 2^5 - 1 &= 31 \\ 2^7 - 1 &= 127, \end{aligned}$$

all of which are obviously prime.

The next three values specified by Mersenne are not so obvious:

$$2^{13} - 1 = 8191$$
$$2^{17} - 1 = 131071$$
$$2^{19} - 1 = 524287.$$

However, the first of these had been shown to be prime in 1456, and the next two had been shown to be prime in 1588.

The next value specified by Mersenne,

$$2^{31} - 1 = 2147483647,$$

was shown to be prime by Euler in 1772.

The next value specified by Mersenne,

$$2^{67} - 1 = 147573952589676412927,$$

resisted all attacks until 1903, when the American mathematician Frank Cole discovered that

$$2^{67} - 1 = 193707721 \times 761838257287.$$

Thus $2^{67} - 1$ is the product of two large prime numbers, but it is not itself prime. Mersenne was wrong!

*Mathematica* gives the factorization above of $2^{67} - 1$ in under a second. The point here is that a problem that had stymied mathematicians for over two centuries can now be answered quickly on an ordinary home computer. My computer also instantly disposes of the last two values specified by Mersenne, telling me that $2^{127} - 1$ is prime (as had Mersenne claimed) but that $2^{257} - 1$ is not.

At the time of this writing (Spring 2002), 39 Mersenne primes are known. The largest is

$$2^{13466917} - 1,$$

which was proved to be prime in 2001. In addition to being the largest known Mersenne prime, this number is also currently the largest known prime number of any type. It was discovered by the Great Internet Mersenne Prime Search, a remarkable network of tens of thousands of ordinary computers working in their idle time to find additional Mersenne primes. You can join the Great Internet Mersenne Prime Search, and perhaps have your computer be the discoverer of the next Mersenne prime, by downloading the free software from `http://www.mersenne.org/prime.htm`.[1]

How many Mersenne primes await discovery? Most mathematicians strongly suspect that there are infinitely many Mersenne primes, but no one has been able to prove that. The brief survey of Mersenne primes given above looks like a victory for computers in the battle between mathematicians and machines.

## Perfect Numbers

A positive integer greater than 1 is called *perfect* if the sum of its divisors, including 1 but not including itself, equals the number. For example, 6 is a perfect number because its divisors are 1, 2, 3 and $6 = 1 + 2 + 3$. The next perfect number is 28, with $28 = 1 + 2 + 4 + 7 + 14$. The next

[1] On November 17, 2003 Michael Shafer's computer found the 40th known Mersenne prime, $2^{20,996,011} - 1$. This number has 6,320,430 decimal digits, surpassing the 39th Mersenne prime by over 2 million digits.

two perfect numbers are 496 and 8128. After that there is a big jump, as the next perfect number is 33550336.

Let's factor the first five perfect numbers to see if we can discover a pattern:

$$\begin{array}{ccccc} 6 & = & 2 \times 3 & = & 2 \times (2^2 - 1) \\ 28 & = & 2^2 \times 7 & = & 2^2 \times (2^3 - 1) \\ 496 & = & 2^4 \times 31 & = & 2^4 \times (2^5 - 1) \\ 8128 & = & 2^6 \times 127 & = & 2^6 \times (2^7 - 1) \\ 33550336 & = & 2^{12} \times 8191 & = & 2^{12} \times (2^{13} - 1) \end{array}$$

The numbers in the last column above should look familiar from the previous section. As we see from the factorizations above, each of the first five perfect numbers is of the form

$$2^{n-1} \times (2^n - 1),$$

where $2^n - 1$ is a Mersenne prime.

An old theorem states that an even number is perfect if and only if it is of the form $2^{n-1} \times (2^n - 1)$, where $2^n - 1$ is a Mersenne prime. So the pattern noted above is not a coincidence. In fact, we can use the theorem just stated to find the next even perfect number: the next Mersenne prime after $2^{13} - 1$ is $2^{17} - 1$, and thus $2^{16} \times (2^{17} - 1)$, which equals 8589869056, is a perfect number.

Note that the theorem mentioned in the paragraph above characterizes only the even perfect numbers. A longstanding unanswered question is whether there are any odd perfect numbers. No one has ever found an odd perfect number, but no one has been able to prove that none exist.

Thus I was surprised several years ago, when I was an Associate Editor of the *American Mathematical Monthly*, to receive a paper submitted for publication that had as its main result the following statement:

For any odd number, the sum of its divisors (not counting itself) is less than the number.

Now if the sum of the divisors (not counting itself) of each odd number is less than the number, then it cannot equal the number, and hence the number cannot be perfect. So if the statement above is correct, then there would be no odd perfect numbers. Curiously, the paper did not point out that the claimed result would answer a famous previously unanswered question; in fact, the paper did not even mention perfect numbers.

Perfect numbers are far from my area of expertise, but I knew enough to realize that the claimed result in the submitted paper would, if correct, solve a big problem and make the author famous. Could the claimed result possibly be true? I experimented by checking the odd numbers less than 50. The table below shows the results, where I have not bothered to list the prime numbers, because the sum of the divisors of each prime number, not counting itself, obviously equals 1.

| $n$ | **sum of divisors of $n$, not counting $n$** |
|---|---|
| 9 | 4 |
| 15 | 9 |
| 21 | 11 |
| 25 | 6 |
| 27 | 13 |
| 33 | 15 |
| 35 | 13 |
| 39 | 17 |
| 45 | 33 |
| 49 | 8 |

As we see from the table above, for each odd number less than 50, the sum of the divisors of the number is less than the number, and in fact usually considerably so. At that point I turned to my computer and asked it to check all odd numbers less than 1000. The computer found one counterexample to the claimed result. Specifically, the sum of the divisors of 945, not counting 945, equals 975.

So the claimed result in the paper was false, and the alleged proof must contain an error. I had trouble following the proof, but at one point it seemed to me that the author was implicitly assuming that the odd number under consideration has no repeated prime factors. For example,

$$945 = 3^3 \times 5 \times 7,$$

so 945 has a repeated prime factor of 3. Could it be true for each odd number with no repeated prime factors that the sum of the divisors of the number (not counting itself) is less than the number?

To answer this question, I turned back to my computer and asked it to find all the odd numbers less than 7500 whose divisors (not counting the number itself) add up to more than the number. Here is the list the computer provided, along with the prime factorizations of those numbers (also provided by the computer):

$$\begin{aligned}
945 &= 3^3 \times 5 \times 7\\
1575 &= 3^2 \times 5^2 \times 7\\
2205 &= 3^2 \times 5 \times 7^2\\
2835 &= 3^4 \times 5 \times 7\\
3465 &= 3^2 \times 5 \times 7 \times 11\\
4095 &= 3^2 \times 5 \times 7 \times 13\\
4725 &= 3^3 \times 5^2 \times 7\\
5355 &= 3^2 \times 5 \times 7 \times 17\\
5775 &= 3 \times 5^2 \times 7 \times 11\\
5985 &= 3^2 \times 5 \times 7 \times 19\\
6435 &= 3^2 \times 5 \times 11 \times 13\\
6615 &= 3^3 \times 5 \times 7^2\\
6825 &= 3 \times 5^2 \times 7 \times 13\\
7245 &= 3^2 \times 5 \times 7 \times 23\\
7425 &= 3^3 \times 5^2 \times 11
\end{aligned}$$

As we see from the table above, for each odd number less than 7500 whose divisors (not counting the number itself) add up to more than the number, the number in question has a prime factorization that repeats some prime factor. Thinking that this result might be true in general, I extended the range of my computer's search, asking it to check the odd numbers less than 15000. My computer found fifteen odd numbers between 7500 and 15000 whose divisors (not counting the number) add up to more than the number. When the computer factored those numbers I saw that they, too, have repeated prime factors.

By then I thought I had good numeric evidence to suggest strongly that repeated prime factors play a key role in this question. But I asked my computer to search once more, doubling my previous search range to the odd numbers between 15000 and 30000. Alas, my computer found six odd numbers between 15000 and 30000 that smashed my hopes. The smallest of these numbers is

15015. The sum of the divisors of 15015 (not countintg 15015) equals 17241, which obviously is larger than 15015. But

$$15015 = 3 \times 5 \times 7 \times 11 \times 13,$$

so 15015 has no repeated prime factors.

The computer was killing all my conjectures. Hoping to salvage something from this excursion, I looked at the prime factorizations of all the odd numbers less than 30000 whose divisors (not counting the number itself) add up to more than the number. Nothing jumped out at me except that each of these numbers has 3 as a factor. Could it be that for every odd number not divisible by 3, the sum of the divisors (not counting the number itself) is less than the number? My feeling was that surely this question should have a negative answer. To find an example showing this, I asked my computer to find an odd number not divisible by 3 whose divisors (not counting the number itself) add up to more than the number.

First I told my computer to test the odd numbers less than a million. It found no odd numbers not divisible by 3 whose divisors (not counting the number itself) add up to more than the number. So I extended the range of the test up to ten-million, but after working for an hour the computer stated that there are no examples in that range either. Finally I cxtended the range of the test up to one-hundred-million. Checking the odd numbers in that range took my computer fourteen hours, and again it told me that there are no examples of the kind I sought.

Either my intuition about this question was wrong or an example was too hard for a computer to find. So I began thinking about it. Suppose that $p > 5$ is a prime number large enough such that

$$\frac{1}{5} + \frac{1}{7} + \frac{1}{11} + \cdots + \frac{1}{p} > 1.$$

In other words, suppose that the sum of the reciprocals of the primes from 5 to $p$ is greater than 1. Rewrite the left side of the incquality above, putting everything over the common denominator $5 \times 7 \times 11 \times \cdots \times p$. The numerator will be a sum of terms, the first of which is $7 \times 11 \times \cdots \times p$, the second of which is $5 \times 11 \times \cdots \times p$, on so on. Because the fraction is greater than 1, the sum of the terms in the numerator is greater than the denominator of $5 \times 7 \times 11 \times \cdots \times p$. But each of the terms in the numerator is a divisor of $5 \times 7 \times 11 \times \cdots \times p$. Thus the sum of all the divisors of $5 \times 7 \times 11 \times \cdots \times p$, not counting this number itself (it does not appear in the numerator), must be greater than $5 \times 7 \times 11 \times \cdots \times p$. Because $5 \times 7 \times 11 \times \cdots \times p$ is not divisible by 3, we will have found our example.

Now we just need to find a prime number $p > 5$ such that

$$\frac{1}{5} + \frac{1}{7} + \frac{1}{11} + \cdots + \frac{1}{p} > 1.$$

This is a problem that the computer can quickly do. My computer tells me that $p = 109$ is the smallest prime number larger than 5 that makes the inequality above true. Thus $5 \times 7 \times 11 \times \cdots \times 109$, which (according to my computer) equals

$$46622499469642489363046026978677968279166205,$$

is not divisible by 3 but the sum of its divisors (not counting the number itself) is larger than the number. This number is much larger than one-hundred-million, which is why my earlier search did not find it.

In fact 46622499469642489363046026978677968279166205 is not the smallest number with the desired property, but because the smallest such number is larger than one-hundred-million, a brute-force computer search would take too long. So for this problem a mathematician, with some small help from a computer, easily solved a problem that stumped the computer alone.

Before leaving this subject, let's consider a harder question. Suppose we want to show that there exists an odd number not divisible by 3 such that the sum of the divisors of the number (not counting the number itself) is greater than 2 times the number. Or greater than 100 times the number? Is that possible? Using the same reasoning as before, all we need to do is show that there exists a prime $p > 5$ such that

$$\frac{1}{5} + \frac{1}{7} + \frac{1}{11} + \cdots + \frac{1}{p}$$

is larger than 2, or larger than 100 for the even more difficult question. A famous theorem (proved by mathematicians, not by computers!) states that the sum of the reciprocals of all the prime numbers equals infinity. That means that the sum above can be as large as we want, provided we take $p$ large enough. My computer tells me that to get the sum above to be larger than 2, we need to take $p = 483281$ (or of course any larger value of $p$ will work). My computer cannot compute a choice of $p$ that will make the sum above larger than 100 (the smallest such $p$ will be incredibly huge), although mathematics proves that such a prime number $p$ exists. In this case mathematicians outperform the computer!

The results in this section show that even though a mathematical conjecture is true for the first few million examples, the conjecture could still be false. A computer can check more examples in a few minutes than a mathematician could check in a century, but without a proof, we cannot be sure that there is no counterexample lurking in some unchecked case. If a conjecture involves an infinite number of possible examples (as is the case with the integers), then a computer cannot check them all. It can check a large number of them, hoping to find a counterexample, in which case the conjecture is false. But if no counterexample is found, even among millions of cases, we still need a mathematician to prove that the result is true (or to find a clever way of finding a counterexample).

## Twin Primes

A twin prime is a pair of prime numbers differing by 2. The first ten twin primes are $\{3,5\}$, $\{5,7\}$, $\{11,13\}$, $\{17,19\}$, $\{29,31\}$, $\{41,43\}$, $\{59,61\}$, $\{71,73\}$, $\{101,103\}$, $\{107,109\}$. Although humans have known for over two thousand years that there are an infinite number of primes, we do not know whether there are an infinite number of twin primes.

As I write this paragraph in Spring 2002, the largest known twin prime is

$$\{318032361 \times 2^{107001} - 1,\ 318032361 \times 2^{107001} + 1\}.$$

But the record for the largest twin prime has been broken twelve times in the last three years, so there will likely be a new champion twin prime soon.[2] A list of the twenty largest twin primes currently known is kept up to date on the web at `http://www.utm.edu/research/primes/lists/top20/twin.html`. Finding twin primes of this size without using a computer would be impossible.

Just for fun, let's list the first ten twin primes and the sum of the two numbers in each twin prime pair:

| twin prime | sum | twin prime | sum |
|---|---|---|---|
| $\{3,5\}$ | 8 | $\{41,43\}$ | 84 |
| $\{5,7\}$ | 12 | $\{59,61\}$ | 120 |
| $\{11,13\}$ | 24 | $\{71,73\}$ | 144 |
| $\{17,19\}$ | 36 | $\{101,103\}$ | 204 |
| $\{29,31\}$ | 60 | $\{107,109\}$ | 216 |

[2] Indeed, there were three larger twin primes found in 2002.

Notice anything special about the sums above? Except for the first entry of 8, all of the other sums above are divisible by 12. Could this be true for all twin primes other than $\{3,5\}$?

To answer this question, we turn again to a computer, asking it to check all twin primes less than a million. My computer says that the sums are all divisible by 12, except for $\{3,5\}$. I get the same answer after asking my computer to check all twin primes less than ten-million. This may sound convincing, but in the previous section we looked at a mathematical statement that turned out to be false even though it was true for all numbers less than one-hundred million.

This time the evidence provided by the computer points in the right direction. It really is true that for each twin prime pair except $\{3,5\}$, the sum of the two primes is divisible by 12. A computer cannot prove that, because the computer cannot check all possible twin primes. But you can prove it. I will leave this as a homework exercise for you. The proof does not require any high-powered tools, just some simple reasoning that can be discovered by anyone reading this article. Give it a try!

## The Winner

Earlier I mentioned the proof via computer of the four-color theorem as one of the great and early triumphs of computers in mathematics. We have seen examples, however, emphasizing that computers can check only a finite number of cases. Because there are an infinite number of possible maps, how could a computer prove that each of them can be colored with at most four colors so that no two adjacent countries have the same color? What happened here is that mathematicians proved a theorem showing that only certain cases had to be checked. The number of cases that must be checked, according to this theorem, is finite. This finite number of cases is too large to check by hand, so computers were called in to finish the job.

The cooperation demonstrated in the proof of the four-color theorem between mathematicians and computers may be typical of much of the future progress of mathematics. Some problems a computer can solve simply by virtue of being able to calculate very quickly—an example would be the factorization of $2^{67} - 1$, which had been falsely suspected of being a Mersenne prime. Other problems are so hard that computers can provide no help; here one thinks of the remarkable proof of Fermat's Last Theorem by Andrew Wiles, who worked completely without computer assistance.

Computers still cannot provide the creative, clever, and imaginative proofs that give mathematics much of its beauty. But mathematicians do not cling to proofs just because we can do them better than computers. *We insist upon proofs because proof is the only method we have for determining most mathematical truths.*

A computer can check a hundred-million examples, provide evidence, carry out numeric experiments, draw fantastic pictures, and sometimes find counterexamples. Mathematicians need to take advantage of this wonderful tool. Working with computers, we can explore and discover at a pace that would have been inconceivable before the silicon age. Machines will not replace mathematicians, but computers will help mathematicians do mathematics better. Mathematics will emerge as the big winner in this collaboration between mathematicians and computers.

# 4

# Breaking Driver's License Codes

Joseph A. Gallian
*University of Minnesota Duluth*

"You know my name
look up the number"
— John Lennon and Paul McCartney, *You know my name*, Single, B-side of *Let it be*, March 1970

Many years ago while eating my corn flakes for breakfast I noticed the UPC bar code on the bottom of the package. The thought occurred to me that there might be some elementary mathematics involved with assigning the UPC code. It turns out that the UPC code does use a check digit for error detection. (A check digit is an extra digit appended to a number for the purpose of detecting an error in electronically reading the number—see references.) That started my interest in methods for assigning identification numbers. With the help of a student, Steve Winters, over the next few years I was able to discover, by reading or by figuring them out, how check digit schemes on books, credit cards, airline tickets, car rentals, money orders and some others are calculated. Eventually, I knew enough of them to be able to give talks on how they work. After one of these talks someone asked me if I knew how Minnesota driver's license numbers were coded. I said that I did not. A few months later, after another talk on ID numbers, a person asked me about Minnesota driver's license numbers. This time I replied that I would check into it.

I began by collecting and analyzing a hundred or so samples. A typical Minnesota number has the form S530-676-465-162. After looking at just a few samples one observes that the first character of the license number matches the first character of the last name. Moreover, people with matching last names have matching first blocks (an alphabetic character followed by three digits), people with matching first names have matching second blocks. For example, the first name "Paul" always yields 676 as the second block. If "Paul" is the middle name then 676 appears in the third block. Finally, people with matching day and month of birth (but not necessarily year of birth) have the identical last three digits. So, we know in a matter of minutes from looking at samples[1] that the generic format is: last name code—first name code—middle name code—day and month

[1] The Baseball Hall of Fame New York Yankee catcher Yogi Berra, now better known for his sayings such as "It's déjà vu all over again" and "It ain't over till it's over," once said something that is applicable here: "You can see a lot just by observing."

of birth code.

That gender does not matter follows because samples reveal that the names Paul, Paula, Patrick and Patricia are all coded as 676. Samples also show that Kevin, Keith, Kent, Kelvin, Kenneth and Kerry are coded as 465. So, something these names have in common must be triggering the number 676 for the first group and 465 for the second. What is this? Well, all that the names in each group have in common are the first two characters of the names. So, we speculate that the first two characters of the first and middle names are what determines the codes for the names. George Pólya says in one of his books that the most important technique for scientific discovery is "Guess and Test." We have our guess so now we have to test. Looking over our samples we notice that Aaron is coded as 028, Adam is coded as 031 and Agnes is coded as 034. This led me to think that the code goes like this:

$$\begin{aligned}
\text{Aa} &\rightarrow 028\\
\text{Ab} &\rightarrow 029\\
\text{Ac} &\rightarrow 030\\
\text{Ad} &\rightarrow 031\\
\text{Ae} &\rightarrow 032\\
\text{Af} &\rightarrow 033\\
\text{Ag} &\rightarrow 034\\
\text{Ah} &\rightarrow 035
\end{aligned}$$

and so on. These check against my samples perfectly until we reach names that begin with Al, which should be coded as 039 if the pattern continues. Instead, we have

$$\begin{array}{lcl}
\text{Alan} & \rightarrow & 040\\
\text{Albert} & \rightarrow & 041\\
\text{Alice} & \rightarrow & 048.
\end{array}$$

Also, we have Amanda coded as 066 instead of 040 as we expected. Even though the pattern that we thought held did not, it is easy to see what happens instead.

$$\begin{aligned}
\text{Ala} &\rightarrow 040\\
\text{Alb} &\rightarrow 041\\
\text{Alc} &\rightarrow 042\\
\text{Ald} &\rightarrow 043\\
\text{Ale} &\rightarrow 044\\
\text{Alf} &\rightarrow 045\\
\text{Alg} &\rightarrow 046\\
\text{Alh} &\rightarrow 047\\
&\vdots\\
\text{Alz} &\rightarrow 065\\
\text{Am} &\rightarrow 066\\
\text{An} &\rightarrow 067
\end{aligned}$$

So, the algorithm seems to be to use the first two letters in most cases but once in a while branch to three letters, then return to two. Here is the complete data for names that begin with "A."

Minnesota code for first and middle names beginning with A except Al.

| | | | | | | | | | |
|---|---|---|---|---|---|---|---|---|---|
| | | | | A | 027 | | | | |
| Aa | 028 | Ag | 034 | Al | — | Aq | 070 | Av | 075 |
| Ab | 029 | Ah | 035 | Am | 066 | Ar | 071 | Aw | 076 |
| Ac | 030 | Ai | 036 | An | 067 | As | 072 | Ax | 077 |
| Ad | 031 | Aj | 037 | Ao | 068 | At | 073 | Ay | 078 |
| Ae | 032 | Ak | 038 | Ap | 069 | Au | 074 | Az | 079 |
| Af | 033 | | | | | | | | |

Minnesota Code for first and middle names beginning with Al.

| | | | | | | | | | |
|---|---|---|---|---|---|---|---|---|---|
| | | | | Al | 039 | | | | |
| Ala | 040 | Alg | 046 | All | 051 | Alq | 056 | Alv | 061 |
| Alb | 041 | Alh | 047 | Alm | 052 | Alr | 057 | Alw | 062 |
| Alc | 042 | Ali | 048 | Aln | 053 | Als | 058 | Alx | 063 |
| Ald | 043 | Alj | 049 | Alo | 054 | Alt | 059 | Aly | 064 |
| Ale | 044 | Alk | 050 | Alp | 055 | Alu | 060 | Alz | 065 |
| Alf | 045 | | | | | | | | |

The guess that in most cases the first two letters determine the number holds up when we examine the names that begin with J. In my samples I had a Jill with number 414, Joan with 421 and Jodi with 424. So, we predict

| | | | | | |
|---|---|---|---|---|---|
| Ji | → | 414 | Joa | → | 421 |
| Jj | → | 415 | Job | → | 422 |
| Jk | → | 416 | Joc | → | 423 |
| Jl | → | 417 | Jod | → | 424 |
| Jm | → | 418 | Joe | → | 425 |
| Jn | → | 419 | Jof | → | 426 |
| Jo | → | 420 | Jog | → | 427 |
| | | | Joh | → | 428 |

However, the three-letter pattern breaks down at John. Instead of 428 which should code names that begin with Joh, John is assigned 429. After 429 we resume the three-letter pattern again until we reach Joseph where once more the three-letter pattern fails.

| | | |
|---|---|---|
| John | → | 429 |
| Joi | → | 430 |
| Joj | → | 431 |
| Jok | → | 432 |
| Jol | → | 433 |
| Jom | → | 434 |
| Jon | → | 435 |
| Joo | → | 436 |
| Jop | → | 437 |
| Joq | → | 438 |
| Jor | → | 439 |
| Jos | → | 440 |
| Joseph | → | 441 |
| Jot | → | 442 |

We see that very common names such as John, Joseph, James, William, Robert and Mary cause us to adjust our conjecture one more time. In these situations the names are assigned numbers that are not assigned to other names that share the same first three letters. For example, Roberta is given the code for names that begin with Rob (744) but Robert is coded 745. Similarly, Wilbur is assigned 886 using the first three letters but William is assigned 887.

We conclude that first and middle names in Minnesota are coded as follows:

- most names are coded by the first two characters starting with 028 for "Aa";
- in cases where the first two characters are common such as "Al," "Ja" or "Jo" we branch to the first three characters;
- when within a sequence of three letter combinations we reach a name that is particularly common such as James or Joseph, these names will be assigned a number of their own. The number for the common names is 1 more than the one assigned to names that share the same first three characters of the name (for example, Jose is assigned 440 but Joseph 441).

Our data confirms that this is the correct algorithm. The only problem is that our data are not complete enough to reveal every possible combination of letters. If a name such as "Henry" were not among my samples I would guess that I would use only the first two characters since "He" is not common enough to warrant branching to three letters. However, your intuition can mislead you because this system was invented many years ago and names that are uncommon now may have been very common when it was invented and vice versa. For example, both Daniel and David are common now but were not common enough when this scheme was designed to cause the designers of the scheme to branch to three characters for names that begin with "Da."

Now let's look at the last three digits of the license number. We know that they represent the day and month of birth since people with the same day of birth—but not same year of birth—have the same final three digits.

To glean the pattern involved we examine dates that differ by 1. Here are some samples

January 2 → 007
January 3 → 010
January 4 → 012

January 9 → 027
January 10 → 030
January 11 → 032

February 3 → 091
February 4 → 093

In most cases consecutive dates differ by increments of 2 and 3 which alternate. However, we also have

March 19 → 207
March 20 → 227.

The biologist Thomas Huxley once lamented that the slaying of a beautiful hypothesis by an ugly fact was the great tragedy of Science. Well, the gap of 20 between March 19 and March 20 is an ugly fact that slayed my hypothesis that the increments alternate by 2s and 3s. Below are the complete data for March. Notice the anomaly at March 9 and March 20. In some months the pattern of alternating increments of 2 and 3 has no exceptions. So, as was the case with the first and middle names, we know the general pattern for the date of birth but we do not have full information. We never know when an exception to the general pattern may occur.

Minnesota code for dates in March.

March 158

| | | | | | | | |
|---|---|---|---|---|---|---|---|
| 1 | - 159 | 11 | - 187 | 21 | - 229 |
| 2 | - 162 | 12 | - 189 | 22 | - 232 |
| 3 | - 164 | 13 | - 192 | 23 | - 234 |
| 4 | - 167 | 14 | - 194 | 24 | - 237 |
| 5 | - 169 | 15 | - 197 | 25 | - 239 |
| 6 | - 172 | 16 | - 199 | 26 | - 242 |
| 7 | - 174 | 17 | - 202 | 27 | - 244 |
| 8 | - 177 | 18 | - 204 | 28 | - 247 |
| 9 | - 182 | 19 | - 207 | 29 | - 249 |
| 10 | - 184 | 20 | - 227 | 30 | - 252 |
| | | | | 31 | - 254 |

Now we move to the last name. The first character of the Minnesota license is the first character of the last name. This is followed by three digits. Fortunately, someone tipped me off that these three digits are assigned using a scheme called "Soundex" and that I could find it described in Knuth's book *The Art of Computer Programming* [7]. It turns out that the Minnesota scheme is a slight variation of the one described by Knuth. Here is the algorithm for the surname.

1. Delete all occurrences of h and w. (For example, Schworer becomes Scorer and Hughgill becomes uggill.)
2. Assign numbers to the remaining letters as follows:

   a, e, i, o, u, y $\to$ 0
   b, f, p, v $\to$ 1     l $\to$ 4
   c, g, j, k, q, s, x, z $\to$ 2     m, n $\to$ 5
   d, t $\to$ 3     r $\to$ 6
3. If two or more letters with the same numeric value are adjacent, omit all but the first. (For example, Scorer becomes Sorer and uggill becomes ugil).
4. Delete the first character of the original name if still present. (Sorer becomes orer).
5. Delete all occurrences of a, e, i, o, u and y.
6. Retain only the first three digits corresponding to the remaining letters; append trailing zeros if fewer than three letters remain; precede the three digits obtained in Step 6 by the first letter of the surname.

Here are three examples.

| | Step 1 | | Step 2 | |
|---|---|---|---|---|
| Schworer | $\to$ | Scorer | $\to$ | Scorer<br>220606 |

| Step 3 | | Step 4 | | Step 5 | | Step 6 | |
|---|---|---|---|---|---|---|---|
| $\to$ | Sorer<br>20606 | $\to$ | orer<br>0606 | $\to$ | rr<br>66 | $\to$ | S-660 |

| | Step 1 | | Step 2 | |
|---|---|---|---|---|
| Hughgill | → | uggill | → | uggill 022044 |

| Step 3 | | Step 4 | | Step 5 | | Step 6 | |
|---|---|---|---|---|---|---|---|
| → | ugil 0204 | → | ugil 0204 | → | gl 24 | → | H-240 |

| | Step 1 | | Step 2 | |
|---|---|---|---|---|
| Schmidlapper | → | Scmidlapper | → | Scmidlapper 22503401106 |

| Step 3 | | Step 4 | |
|---|---|---|---|
| → | Smidlaper 250340106 | → | midlaper 50340106 |

| Step 5 | | Step 6 | | Step 7 | |
|---|---|---|---|---|---|
| → | mdlpr 53416 | → | mdl 534 | → | S-534 |

With this piece of information in place we now pretty much know how Minnesota license numbers are coded.

It is natural to wonder what advantage such a scheme might have. The answer is that it is an error correction scheme. It is designed so that if you are close to the correct spelling of a name you will get the correct code for the name. For example, the correct spelling of my daughter's name is Kristin Gallian. The correct code for her name is G-450-478. Here are some ways that her name might be misspelled: Kristen Galliam, Kristel Gallion, Kristen Galion, Kristin Galliano, Kristin Galliani, Kristen Gallahan, Kristen Gillian, Kristin Gilliam, Kristin Gilham. But all of these are coded as G-450-478. Similarly, names such as Anderson and Andersen are coded the same, and Johnson and Jonson are coded the same. As the name "Soundex" implies it is the sound rather than the spelling that determines the code. To appreciate the beauty of this system think of telephone numbers. Suppose that they were assigned in such a way that you could dial the wrong number but have the correct phone ring as long as you were only a few digits off! Such a numbering scheme could be devised by mathematicians but there would be some disadvantages as well. In fact, the "Soundex" system has its own disadvantages. Consider two people in Minnesota with the names Josephine Lynn Gallian and Joshua Lyndon Gallahan. These two names are coded exactly the same. If people with these names happen to have matching day and month of births then their entire license numbers would be the same. This kind of situation is especially likely in the case of twins. Twins automatically have the same last name and the same birth date and often parents name twins with similar names such as "Keith" and "Kent." In these cases the algorithm produces the identical license number for both. Another instance where this kind of problem occurs is when a son is given the same name as the father such as John Allen Smith and John Allen Smith, Jr. or John Allen Smith II. If the father and son share the same birth date then the algorithm yields the same number. My name is the same as my father's and John Lennon and his son Sean had the same birth date. Computer scientists call these kinds of occurrences "collisions." I call it "collapsing" because different names "collapse" to the same code.

So, here is a review of what I have told you thus far. I can do the last names as well as first and middle names but there may be some first or middle names where I am not positive about branching to three characters or assigning a common name a number of its own. Similarly, I usually can deduce the code for a birth date missing from my samples by using my knowledge that the increments typically alternate by 2s and 3s.

At this point I was satisfied that I understood the Minnesota scheme well enough.

Now what? By this time I had begun to collect license numbers from other states. I had a sizable collection from Wisconsin since many students in my classes are from Wisconsin. I would ask people I meet from other states if I could see their licenses. I still recall the time I saw my first Missouri license. Each Fall the Math Department at my school has a party to welcome new faculty and new graduate students. At the party the Math Department secretary Jane said that she wanted to introduce me to her sister from Missouri. I said "Hello, may I see your license." The sister was a bit taken aback by my question and did not know how she should respond but Jane assured her that I was harmless and she showed it to me. I told her that her number was very interesting and that I would add it to my collection. I also wrote to some friends who teach at universities in other states and asked them to ask their students for samples. I then got the idea to simply write all 50 states and ask them how they code their license numbers. The Beatle George Harrison in his book *I, Me, Mine* titled after a song on the "Let it Be" album says that if you study any topic in great detail it becomes fascinating. I found this to be true of license numbers.

The responses I received from the states revealed that many of them use uninteresting schemes such as the social security number or a sequential number. Some employ check digits for error detection. Some keep their method of assigning license numbers confidential. Among those that informed me that their method was confidential were: Missouri, New York, Florida, Wisconsin, and New Jersey. A few states, among them Minnesota, did not respond to my letter. A letter from the Colorado Department of Revenue stated "Colorado does not code license numbers by any exact science." Eventually I knew how to code all but the last digit of Wisconsin which I suspected was a check digit. However, despite considerable effort on my part, I was not able to figure out how to compute the last digit of the Wisconsin number. I wrote the Wisconsin license office asking if they would at least confirm that the last digit was a check digit. They did confirm it but would tell me nothing further. During talks I would give on check digits I began offering a $100 reward to anyone who could tell me how the last digit of the Wisconsin number was calculated. One of my students figured out the method and collected the reward.

The most interesting responses I received were from Michigan and Maryland. Michigan wrote a nice letter explaining how to do some sample names and included a code book with full details on how to code every possible name. Instantly I realized that Michigan and Minnesota used the same scheme. Maryland also included a pamphlet for their scheme which also is the same scheme as Minnesota's. Maryland went so far as to include two sample license plates! So, Minnesota, Michigan and Maryland each use the same system but Minnesota won't reveal the scheme whereas Michigan and Maryland will. The people at the Michigan and Maryland license departments were so helpful I decided to ask about the frequency of having different people assigned the same number. I was informed that the gaps in the numbers for the birth date codes serve a practical purpose. In the event that there are two or more individuals born on the same month and date and with names so similar that the coding scheme does not distinguish between them (e.g., Jill Paula Smith and Jimmy Paul Smythe), the first person who applies for a license is assigned the number given by the algorithm while the second person is assigned the next higher number thereby using one of the numbers in the gap for birthdays. For example, if Jill Paula Smith is born on March 2 and is the first to receive the combination S530-441-675-162, as determined by the algorithm, then the next person who is supposed to be assigned that same number is assigned S530-441-675-163 instead. Once all of the higher numbers in a gap have been assigned, lower numbers are used. Thus the third applicant with a name yielding the combination S530-441-675 born on March 2 would be assigned the last three digits 161 since the gap between March 2 and March 3 was filled. As of 1984, this scheme had not yielded any duplications among 4,468,080 people in Maryland while of Michigan's 6,332,878 drivers by 1987 there were 56 who had a number not uniquely their own. In fact, Michigan had two numbers that were shared by four individuals and three that were shared by three individuals.

After studying the replies I received from the states I decided I had enough interesting material to write an article on the subject for the *Mathematics Magazine* [2]. The editor Jerry Alexanderson said he would publish it but he wanted to send it to referees for comments. The referees recommended against publication on the grounds that the subject wasn't mathematical enough. Jerry asked me to include some additional information about check digits and he would publish it "for those of us who like this sort of thing." When the article came out about a year later a person working for the MAA whose job it was to get publicity for math decided to send copies of the article to various newspapers across the country. Since I did not know that this was done I thought it was strange when I received a telephone call from the science editor at the Nashua, New Hampshire *Sunday Telegraph* asking me about how drivers licenses were coded. I told him what I knew about license numbers and other identification numbers such as the ones used on credit cards, car rentals, UPS packages, books and retail items. Modular arithmetic in particular was of great interest to him. His article appeared on March 3, 1991 under the title "Where do they get those numbers?" About three weeks later I received a phone call from an all news radio station in Washington, DC asking about Maryland driver's license numbers. I asked how they had heard about my interest in this subject and they replied that I was cited in a piece in the *Washington Post* on Maryland license numbers. Sure enough the *Post* had an article entitled "Cryptography: The Logic of Numbering Licenses." Unfortunately, the *Post* writer tried to simplify the algorithm for last names by telling readers to delete all occurrences of h, w, a, e, i, o, u and y at the outset. This results in inaccurate numbers in many cases since the vowels and y act as separators before they are deleted.

A few days after the *Post* article appeared I received a call from Jim Dawson from the *Star Tribune,* Minnesota's largest newspaper, asking me about how Minnesota coded license numbers. I told him what I have told you. A few days later his article appeared. The blurb "Mathematician finds method to the 12-digit madness on driver's license" was printed alongside the masthead on the front page. Dawson also put the story on the Associated Press wire service so that it went out to newspapers, radios and television stations nationwide. It was picked up by several newspapers. The *St. Paul Pioneer Press* ran it under the heading "License codes no match for crack mathematician"; the *Superior Telegraph* used "UMD mathematician cracks codes on driver's licenses"; the story was even picked up by the *Winnipeg Free Press.* I was interviewed on a call in radio show from St. Cloud. One Duluth TV station had me code a name for the 6 o'clock news. The amazing thing is that the piece was used as the tease at 5:58. The announcer comes on and says "Stay tuned for the 6 o'clock news but first get your driver's license." I had a call from a rock station in Duluth whose call letters are KOOL. The caller asked me if I was the guy who broke the license code. When I responded "yes" he said "kool man" and hung up. I received a call from a farmer from Southern Minnesota who wanted to congratulate me. He told me he spent many hours at his kitchen table trying to figure the scheme out. I received a letter from someone who gave me his name and birth date and wanted me to code his number for him. Many radio stations mentioned it on the news. My daughter went to high school that morning knowing nothing about the article or the news reports but one of her teacher whispered to her "I heard what your father did." She wondered to herself "What did he do now?"

A few days after the *Star Tribune* article appeared I received a phone call from Virgil Tipton, automotive editor of the *St. Louis Post Dispatch,* asking if I knew how to code Missouri license numbers. By that time I had accumulated five Missouri samples from which I could deduce that the last three digits represented the day and month of birth and the gender. The formula for a male born in month $m$ and day $d$ has the digits $63m + 2d$ whereas a female has the digits $63m + 2d + 1$ (insert a 0 on the left if the number is less than 100). I told him that the last 3 digits were all I could do with only five samples. He asked how many I would need to figure out the entire code. I said I would need a lot. He said the newspaper had 1.5 million in a computer file. I asked him to fax me hundreds of Smiths and Jones and Williams because I wanted names that had overlapping data.

The first question I must answer is: How much of the name was used to code the number? Well, I have these samples (disregarding the last three digits)

S133468582900 Arthur Edward Smith
S133468582900 Arthur Eugene Smith

S133469008575 William Carl Smith
S133469008575 William Charles Smith

From these (and others like them) we deduce that only the middle initial of the middle name is involved in the coding. Now we want to know how much the middle initial counts. We have the following samples

S133469008575 William C. Smith
S133469008580 William H. Smith
S133469008583 William K. Smith

We observe that the only difference in these names is the middle initial and the only difference in the numbers is that the second is 5 more than the first and the third is 3 more than the second. Moreover, the change in the alphabet from C to H is 5 positions and the change from H to K is 3 positions.

So, it must be the case that "A" counts for 1, "B" counts for 2, "C" counts for 3 and so on. The data also showed that having no middle name counts for 0. Next, I examined the Smiths to determine how many characters of the first name were factored into the formula. Noting that Carl and Carol were coded the same but Jon and John were not, told me that only the first three characters of the first name count. So, I looked for Smiths with matching middle initials and first names that match in the first two positions such as

S133468578242 Alice R. Smith
S133468578323 Allan R. Smith

Here the i in Alice is the ninth position of the alphabet whereas the l in Allan is in the twelfth position—a change of 3 positions. On the other hand, the change in numbers is 81. Observing that $81/3 = 27$ we determine that for the third letter of the first name "A" counts 27, "B" counts twice as much, "C" counts for $3 \cdot 27$ and so on. In general, if the third letter of the first name is in the $i$th position it contributes $27i$ to the license number. Continuing in this fashion we determine that if the second letter of the first name is in the $i$th position it contributes $27^2 i$ to the total. I was convinced that I would find that a first letter of the first name in the $i$th position would contribute $27^3 i$ but I was wrong. The correct factor is $19657 = 27^3 - 26$.

To decide how many characters of the last name are relevant I ask Tipton to fax me numbers for Anderson and Andersen, Johnson and Johnsen, Larson and Larsen. The first two pairs are coded the same but Larson and Larsen are not. So, I knew that only the first five characters of the last name count. Moreover, since the number begins with the first letter of the last name, I guessed that it was not involved in the calculation of the rest of the number. I checked that out by asking Tipton to fax me numbers for people with last names Bender and Henderson and Berry and Kerry. As I suspected, Bender and Henderson were coded the same as were Berry and Kerry. So, I knew that only the letters in positions 2, 3, 4 and 5 of the last name counted in the formula. To determine how much they counted I proceeded as before to see how much each change in position in the alphabet caused the number to change. All this amounts to is calculating the slopes of straight lines. The change in the position of the alphabet is the change in $x$ and the change in the number is the change in $y$. The values turned out as follows:

| Letter | Factor |
|---|---|
| 5 | 510355 |
| 4 | 13779585 |
| 3 | 371538441 |
| 2 | 10017758323. |

Now let's see if you understand what this means. Say two people have identical birth dates and identical names except for the second letter of the last name. For example, John C. Gierdahl and John C. Gjerdahl. Then the number for Gjerdahl is 10017758323 more than the number for Gierdahl since "j" is one position after "i."

I tried to see if I could find a simple formula relating the numbers for the last names to some polynomial in 27 but was not able to find one.

Putting everything I have said about Missouri thus far together we know how to code the first 11 digits of a Missouri license number. Converting the second, third, fourth and fifth letters of the last name, the first three of the first name and the middle initial to the corresponding positions in the alphabet we have eight numbers (a blank is assigned position 0). Let us call them $n_1, n_2, \ldots, n_8$, respectively. This gives the formula $10017758323n_1 + 371538441n_2 + 13779585n_3 + 510355n_4 + 19657n_5 + 729n_6 + 27n_7 + n_8$.

To my amazement, when I checked this formula against some data it did not yield the correct numbers. When I compared my numbers to the correct ones I saw that in each case I was off by the constant $-385829132$.

So, the correct formula must be $10017758323n_1 + 371538441n_2 + 13779585n_3 + 510355n_4 + 19657n_5 + 729n_6 + 27n_7 + n_8 - 385829132$. Here is an example. For John Winston Lennon we have $n_1 = 5$, $n_2 = 14$, $n_3 = 14$, $n_4 = 15$, $n_5 = 10$, $n_6 = 15$, $n_7 = 8$, $n_8 = 23$. So, the formula yields $10017758323 \cdot 5 + 371538441 \cdot 14 + 13779585 \cdot 14 + 510355 \cdot 15 + 19657 \cdot 10 + 729 \cdot 15 + 27 \cdot 8 + 23 - 385829132 = 055105277916$.

This formula worked correctly on every name I tried until I got to the last name Lee. This failure puzzled me. Why did it work on the names Smith, Jones, Anderson, and Larson but not on Lee? This is where watching *Sesame Street* paid off. On *Sesame Street* they show four panels. In one panel will be something like a running shoe; in a second one a dress shoe; in a third, a hiking shoe, and in a fourth a banana. Then they sing "One of these things is not like the other, one of these things is not the same." You shout "It is the banana!" Well, the same analysis applies here. I looked at all the names my formula worked correctly on to see how "Lee" was not like them. It appeared that the difference was that "Lee" had three characters whereas the others had at least four. I asked Tipton to fax me numbers for names such as Poe, Hoe, Low and sure enough each time my formula was off by the same amount. This means that for three letter last names the constant is different from the constant for names with four or more letters. For three letter last names the constant is $-385318778$.

Certain that I could now handle all cases, I went to bed. While lying there thinking about how the three letter last name was an exception it occurred to me that two letter last names might not use the same constant as three letter last names. Unfortunately, I did not have any two letter names among my data. The next day I asked Tipton to fax me numbers for the last names Po, Ho, Lo, Pu and Wu. This time the constant was $-371539194$.

After making this adjustment I was sure that I could handle every possible name so I faxed the formula off to Tipton.[2] Tipton decided to test my formula by having someone write a program to randomly check my numbers against 10000 actual numbers. I received a fax back that informed me that my formula worked correctly in 9934 cases. Once again an ugly fact slayed a beautiful hypothesis. Fortunately, the fax included the correct number alongside my number and the name of

[2]I do not know how to code a last name that consists of a single character.

the person. I instantly saw that in all 66 cases my number was exactly 26000 less than the correct number (including the 3 digits for the birth date). So, there was yet another constant involved. But what triggered it? The *Sesame Street* song applied this time too but it was "66 of these things are not like the other 9934." In each case the person's first name was a single initial: E. A. Behlman, C. A. Boschert, J. Thomas Buckner and so on. So, with this final modification, the algorithm worked on 10000 out of 10000. Tipton wrote a column that told the story of how I got the formula and stated that they tried it and it "seemed to work."

Since I figured out the Missouri code after I had written my *Mathematics Magazine* article about the other states, I wrote an article for the *UMAP Journal* about the Missouri license number [3]. In the course of writing the article I wondered, if given a Missouri number, was it possible to recover all the information that went into it. That is, could I recover the first five characters of the last name, the first three of the first name and the middle initial. It turns out that this is almost always possible. (In fact, I believe the values for the coefficients in the formula were chosen for this purpose). In my *UMAP* article I said that in "virtually all real-life situations" one could reproduce the data that had gone into the calculation. I qualified my remark with "all real-life situations" because it was not possible to recover the input data only for names that were composed of a bizarre mixture of z's and y's. For instance, both of the "names" Zzz D. Smisz (first name Zzz, middle initial D) and Aa A. Smit (first name Aa, middle initial A) are assigned the number S133464487123. Working backwards from the number S133464487123 yields Aa A. Smit as the name. Another such example is provided by the "names" Ba A. Azzxy and Zaa A. Azzyy.

After my formulas appeared in the *Post-Dispatch* an anonymous employee of the Missouri License Division sent the newspaper a copy of an in-house document detailing the procedure used to assign numbers that contained a *single* formula that allegedly works for all cases. The formula, although containing several more terms than mine, simplifies to the one I derived for four or more letters but differed from my formulas for last names with exactly 2 or 3 letters. For these cases, the Missouri License Division formula gives *incorrect* numbers.

At the same time I was working on the Missouri scheme Michael Fader, a computer science graduate student from New York, independently determined that New York used the same method as Missouri but appends the last two digits for the year of birth. Fader did not know that last names with fewer than four characters and first names that used only a single letter were exceptions.

A few years after I wrote the *UMAP* article I was startled to see in the *Duluth News Tribune* the headline

"Hello, is this Mr. Zzyzz?
Zach Zzyzz?"

I instantly thought of my statement in the *UMAP* article that one could reproduce the data that had gone into the calculation of the number except for names that were composed of a "bizarre mixture of z's and y's."

I ran over to my computer and typed in Zach Zzyzz and got the number Z269736398624. Then I entered this number in a program that was written to recover the data from the number. The output was _bb Z. Zzz (the underscore represents a blank). My reaction was "Oh, no! Zach Zzyzz is the only person on the planet for whom my reverse engineering program does not work."

For the first few years after I obtained the Missouri formula my talk ended at this point. One time, at the end of my talk at a college in Michigan, a student stood up and said something like "Professor Gallian, I want to thank you for coming here tonight to give your talk. It confirms what I have thought all along. You ivory tower professors waste your time worrying about Icelandic poetry, Hungarian folk dances, and Missouri driver's license numbers instead of trying to do something to help feed and clothe the people of the world." I tried to counter by saying that I did it for the intellectual challenge and just for the fun. I also said that making and breaking codes had important applications in the real world. But these comments did not change the student's opinion. I was unhappy that I did not have a better response to the student's criticism.

About a year or two after that unpleasant episode I received a telephone call from someone who asked if I was the person who knew how to code Minnesota driver's license numbers. He then introduced himself as a cancer researcher from the University of Minnesota Minneapolis campus and said that knowing how Minnesota coded license numbers would help him in his research and asked if I would send him the code. I said I would but was so caught off guard that I did not have the presence of mind to ask how the algorithm would help him. When I sent him the algorithm I asked him that question. He wrote back that he was doing a brain cancer study that selected people in the control group from the Minnesota driver's license database. As a first step he wanted to determine which people previously identified as having brain cancer were in the database. He then said "The folks at DPS [Department of Public Safety] are willing to tell us if an individual is in the database, but only if we already have his/her MN DL number. So the ability to generate the appropriate DL numbers solves a lot of problems fairly quickly." When I read this I thought of the student from Michigan. The lesson to be learned here is that you never can tell when some knowledge might be useful.

About a year after that I received another letter from the cancer researcher saying that he had found another use for knowing how to code license numbers.

Well that is the end of the talk except that I want to finish with a brief video. When the article in the *Star Tribune* came out, the Minneapolis public television station asked me if I would be willing to be taped for their *Almanac* program. This is a weekly political show featuring reporters who sit around a table with guests and discuss Minnesota politics. I said "Isn't that a political talk show?" The caller said "Yes, but we want to have you on." *Almanac* is carried on all the public stations in Minnesota on Friday nights and is repeated on Sundays. The night I was on, the program began with the usual theme song, then went directly to "Pomp and Circumstance" with still shots of scientists in formal attire such as Copernicus, Newton, Pasteur, Galileo, Darwin, and Einstein. The voiceover says in very serious tones "At certain points in history a great thinker will step forward and advance the cause of human civilization by solving one of the mysteries of the universe. Such a man exists today in Duluth. UMD mathematician Joe Gallian has deciphered the top secret formula used by the state of Minnesota to determine your driver's license number. Gallian's quest began one day at breakfast." Just as the narrator finishes saying this, and just after the still of Einstein, they show a freeze frame of me in a red and white striped shirt with a goofy expression on my face. At this point I stop the video.

## Bibliography


1. Gallian, Joseph A., "The mathematics of identification numbers," *College Mathematics Journal,* 22 (1991), 194–202.

2. ——, "Assigning driver's license numbers," *Mathematics Magazine,* 64 (1992), 13–22.

3. ——, "Breaking the Missouri license code," *The UMAP Journal,* 13 (1992), 37–42.

4. ——, "Error detection methods," *ACM Computing Surveys,* 28 (1996), 504–517.

5. Gallian, Joseph A., and Steven Winters, "Modular arithmetic in the marketplace," *American Mathematical Monthly,* 95 (1988), 548–551.

6. Kirtland, Joseph, *Identification Numbers and Check Digit Schemes,* MAA, Washington, DC, 2001.

7. Knuth, Donald E., *The Art of Computer Programming,* Vol. 3, Addison-Wesley, Reading, MA, 1973.

8. Lewand, Robert, *Cryptological Mathematics,* MAA, Washington, DC, 2002.

9. Pólya, George, *How to Solve It: A New Aspect of Mathematical Method,* 2nd ed., Doubleday, Garden City, NY, 1957.

# 5

# Jumping Frogs and Powers of Two

Paul Zeitz
*University of San Francisco*

## 1 Introduction

One of the wonderful things about mathematics is its rich interconnectedness. Often several seemingly unrelated ideas turn out to be different aspects of the very same beautiful whole. This article explores one such unexpected connection. We examine how close a jumping frog gets to the crack in the sidewalk, and we look at the distribution of the initial digits of the powers of two. Using simple tools like the pigeonhole principle, we will see that these two problems are more than just related—they are really just two different ways of looking at the very same thing.

## 2 A Jumping Frog on the Number Line

Imagine a frog that starts at a crack on the sidewalk, and then jumps $\alpha$ units every second. If the sidewalk cracks come every unit, will the frog ever hit a crack again?

This is a mathematical frog, of course, so its width is zero, and it jumps exactly $\alpha$ units every time, for eternity. Let's model the sidewalk with the real number line. The frog starts at 0, jumps $\alpha$ units in the positive direction each time, and the cracks are represented by the integers. Then the problem becomes

*Let $\alpha > 0$ be a fixed real number. Does there exist an integer $n$ such that $n\alpha$ is an integer?*

The answer is simple: If $n\alpha = m$ for some integer $m$, then $\alpha = m/n$. Thus the frog hits the crack if and only if $\alpha$ is a rational number. If $\alpha$ is irrational, the frog never hits the crack. But how close to the crack will the frog get? Your intuition should tell you that the frog should get "arbitrarily" close. In other words, no matter how tiny we pick $\epsilon > 0$, eventually the frog will land within $\epsilon$ units of a crack.

If this happens in $n$ jumps, say, then $n\alpha$ will be within $\epsilon$ of an integer. The magnitude of $n\alpha$ is not really important to us; instead we are only concerned with what lies to the right of the decimal point. Let us introduce the notation $\langle x \rangle$ to denote the "decimal" part of the real number $x$. For example,

$$\langle 2.1 \rangle = 0.1, \quad \langle 456 \rangle = 0, \quad \langle \pi \rangle = 0.14159\ldots.$$

Note that $0 \le \langle x \rangle < 1$. Sometimes $\langle x \rangle$ is called $x$ modulo 1, the "remainder" of $x$ upon division by 1. Using this notation, our intuition about irrational frog jumps tells us that eventually the decimal part of the jump will either get very close to zero or very close to 1, depending on whether the frog lands just to the right or just to the left of a crack. More formally, we have

**Theorem 1** *Let $\alpha > 0$ be a fixed irrational number, and let $\epsilon > 0$ be fixed. Then there exists a positive integer $n$ such that either $\langle n\alpha \rangle < \epsilon$ or $1 - \langle n\alpha \rangle < \epsilon$.*

Most people who are familiar with irrational numbers have no trouble believing Theorem 1. It seems "obvious" and somehow connected with the fact that any given irrational number can be approximated arbitrarily well by fractions with arbitrarily large numerators and denominators. For example, $\pi$ can be approximated by the sequence

$$\frac{3}{1}, \frac{31}{10}, \frac{314}{100}, \frac{3141}{1000}, \ldots.$$

It turns out that this insight is not enough.[1]

We need one subtle idea, the important **Pigeonhole Principle**, which states that

*If you put $p$ pigeons into $h$ holes and $p > h$, then at least one hole will contain at least two pigeons*

The pigeonhole principle, unlike Theorem 1, is truly obvious, yet has surprisingly wide and deep applications. We will use it to prove Theorem 1, but first we will modify our point of view.

Since we only need to keep track of the values of $\langle n\alpha \rangle$, and these lie in the interval $[0, 1)$, we can abandon the number line and instead plot the frog's journey on a "modulo one clock"—a circle with circumference 1, with 0 at the 12 o'clock position, 0.5 at 6 o'clock, 0.99 at a few minutes before 12 o'clock, etc. (See Figure 1.) This clock only shows the decimal part of a "time." Thus any integer would correspond to 0, and in general, the real number $x$ is represented by $\langle x \rangle$ on the circle. For example, suppose that $\alpha = \pi$. On the number line, the frog's first few positions are 0, $\pi$, and $2\pi$. But on the circle, the positions are (approximately) 0, 0.14, 0.28. For clarity, we will prove Theorem 1 for a specific value of $\epsilon$, say $\epsilon = 0.01$. Our argument can be generalized to any $\epsilon$. Let us divide the circle into $\epsilon$-sized arcs, starting at 0, going clockwise. Thus the first arc is $[0, 0.01)$, the second arc is $[0.01, 0.02)$, etc. Unlike the number line, the circle is finite, and this is the key, for there will be finitely many—namely 100—such arcs, the last one being $[0.99, 1)$, where it is understood that 1 and 0 are the same point—"noon"—on the circle.

Now consider the first 101 frog jumps (not including the starting point). First of all, no jump can land on the boundary of an arc, since these are rational values, and $\alpha$ and all its multiples are irrational. Secondly, no two jumps can land on the same point of the circle. To see why, suppose that, say, $62\alpha$ and $89\alpha$ corresponded to the same point on the circle. In other words, $\langle 62\alpha \rangle = \langle 89\alpha \rangle$, which implies that $89\alpha - 62\alpha = 27\alpha$ is an integer. Of course, this implies that $\alpha$ is rational, a contradiction.

Therefore, the 101 frog jumps land on 101 distinct circle points. Since there are 100 arcs, the pigeonhole principle tells us that one of the arcs contains two distinct points. Again, for clarity, let's use a specific example, and suppose that $\langle 78\alpha \rangle$ and $\langle 15\alpha \rangle$ lie in the same arc. There are two possibilities: either $\langle 78\alpha \rangle > \langle 15\alpha \rangle$, or $\langle 78\alpha \rangle < \langle 15\alpha \rangle$.

[1] On more than one occasion, I've mentioned Theorem 1 to colleagues who then declare it to be "trivial," and who, when challenged to provide a rigorous proof, retreat in embarrassment. A cruel practical joke, true, but a good way to wean people from overusing the T-word.

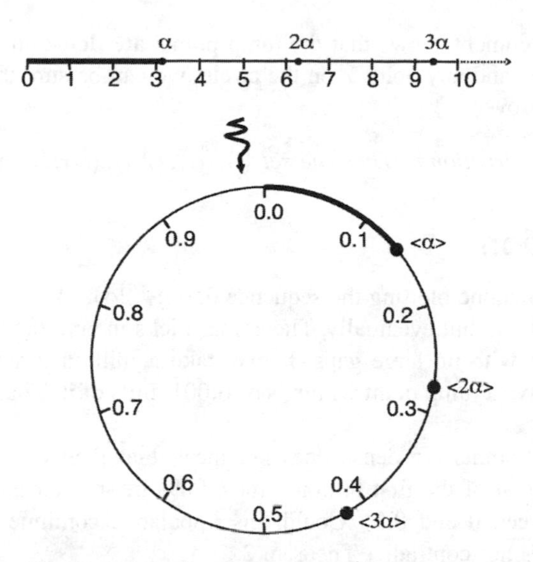


**Figure 5.1.** Consider a frog jumping along the number line with jump length $\alpha = \pi = 3.14159\ldots$. Rather than view this motion on the line, we restrict our attention to the decimal parts of the jumps on the "modulo one clock."

Now let's look at the integer and the decimal parts of the jumps. In the first case, we can write $78\alpha = N + \theta$, and $15\alpha = M + \gamma$, where $N \geq M$ are nonnegative integers, and $1 > \theta > \gamma > 0$ with $\theta - \gamma < 0.01$, since the points lie in the same arc. Subtracting, we have

$$(78 - 15)\alpha = (N - M) + (\theta - \gamma).$$

In other words, $\langle 63\alpha \rangle = \theta - \gamma < 0.01$.

In the second case, everything is the same except that $\gamma > \theta$. Then we have

$$63\alpha = (N - M) + (\theta - \gamma).$$

Since $\theta - \gamma$ is negative (but with absolute value less than 0.01), we write (noting that in this case $N$ must be strictly greater than $M$) $63\alpha = (N - M - 1) + (1 + \theta - \gamma)$, and thus $\langle 63\alpha \rangle = 1 + \theta - \gamma$, so

$$1 - \langle 63\alpha \rangle = \gamma - \theta < 0.01.$$

In other words, the 63rd jump came within 0.01 of a crack (in the first case, to the right of the crack; in the second case, to the left).

It should be clear that this argument can be modified for any $\epsilon$. Just divide the circle into $A$ arcs where $1/A < \epsilon$. Two of the first $A + 1$ jumps will lie in the same arc. After subtracting, we will get a value of $n$ between 1 and $A$ such that the $n$th jump is within $\epsilon$ of a crack. Theorem 1 is proved.

This theorem says that jumps will get arbitrarily close to 0 on the circle. In the example above, jump 63 landed within 0.01 of 0, either to the right (clockwise) or the left (counterclockwise). For the sake of argument, suppose it was to the left (in other words, $63\alpha$ is a number whose decimal part is greater than 0.99). Thus no matter where you start from, after 63 jumps you will end up less than 0.01 counterclockwise units away from your initial position. In other words, when viewed on the modulo 1 circle, the sequence $0, 63\alpha, 2 \cdot 63\alpha, 3 \cdot 63\alpha, 4 \cdot 63\alpha, \ldots$ moves counterclockwise from 0, each point within 0.01 units of its predecessor. Therefore, eventually, jumps will land in all 100 of the length-0.01 arcs.

More generally, this argument shows that the jump points are **dense** in the circle: for any fixed $\epsilon > 0$, no matter how tiny, and any point $x$ on the circle, we can be sure that some jump will land within $\epsilon$ of $x$. We have proved

**Theorem 2** *Let $\alpha > 0$ be irrational. Then the set $\{0, \langle\alpha\rangle, \langle 2\alpha\rangle, \langle 3\alpha\rangle, \ldots\}$ is dense in $[0, 1)$.*

## 3 Equidistribution

Fix an irrational $\alpha$, and imagine plotting the sequence $0, \langle\alpha\rangle, \langle 2\alpha\rangle, \langle 3\alpha\rangle, \ldots$ on the circle. At first, the points are scattered about, but eventually, Theorem 2 kicks in, and the plotted points will cover the circumference nicely, with no large gaps. It may take a million jumps, but eventually every point on the circle will have a jump point within, say, $0.001$ units of it. That's what density implies, after all.

Just because the jump points are dense does not mean that their distribution is uniform. For example, it may happen that of the first billion jumps, 90% lie in the right half of the circle (i.e., have decimal parts between 0 and 0.5). Could this imbalance continue forever? This behavior seems strange, but it does not contradict Theorem 2.

Once again, your intuition suggests that this imbalance cannot be permanent. We would expect that eventually, 50% of the jumps would land in the right half of the circle. Likewise, eventually, 13% of the jumps would land in the interval $[0.70, 0.83)$, etc. In other words, the sequence of jumps is **equidistributed**. Here is the formal statement. We are using the notation $|A|$ for the number of elements in the finite set $A$.

**Theorem 3** *Let $\alpha > 0$ be irrational, and let $I$ be any interval in $[0, 1)$ with length $\ell$. Let $A_n = \{\langle\alpha\rangle, \langle 2\alpha\rangle, \langle 3\alpha\rangle, \ldots, \langle n\alpha\rangle\}$. Then*

$$\lim_{n\to\infty} \frac{|I \cap A_n|}{n} = \ell.$$

Let us study the notation of this theorem. The set $A_n$ is just the set of the first $n$ jumps. The quantity $|I \cap A_n|$ is the number of elements in $A_n$ which also lie in $I$, in other words, the number of jumps which land in $I$. When we divide this by $n$, we are computing the fraction of the first $n$ jumps which landed in $I$. The theorem states that as the number of jumps increases, the fraction approaches the length of $I$, which makes sense, since the entire circle has circumference 1, and the length of an interval is the same as its fraction of the full circumference. For example, the interval $[0.70, 0.83)$ is 13% of the circle, and also has length $0.13$.

Theorem 3 is known as the **Weyl Equidistribution Theorem**, named after its discoverer, H. Weyl, who proved it in 1916. Weyl's original proof used calculus. However, there is a completely "elementary" proof that employs the pigeonhole principle in an ingenious and intricate way. This argument is rather long and would take us too far afield here. See *Concrete Mathematics*, by Graham, Knuth, and Patashnik (Addison-Wesley, 1994) for more details. Meanwhile, we will apply equidistribution to analyze a phenomenon that seems rather far removed from jumping frogs.

## 4 Initial Digits

Mathematicians are adaptable and opportunistic people. When stumped by a problem, they don't give up, but instead look at related problems that they *can* answer. For example, suppose you roll two fair dice and you record the sum. If you repeat this, you will generate a sequence of numbers ranging between 2 and 12. One sequence may look like this: $3, 7, 9, 2, 11, 11, 5, 6, 7, \ldots$. There is no way at all to predict the exact value of any given number in this sequence, but you *can* predict the statistical behavior: in the long run, one-sixth of the numbers will equal 7. Denote the $n$th number in our dice-sum sequence by $s_n$. Then the fraction of sequence values that equal 7 approaches $1/6$ as $n$ gets larger and larger. More formally,

$$\lim_{n\to\infty} \frac{\text{the number of } s_1, s_2, \ldots, s_n \text{ that equal } 7}{n} = \frac{1}{6}.$$

This is what we mean when we say that "There is a $1/6$ probability that a dice sum equals 7."

The Equidistribution Theorem is another example of this: rather than compute the exact location of the points in the sequence $\langle\alpha\rangle, \langle 2\alpha\rangle, \ldots$, we settle for statistical information about the distribution of the locations. You may find equidistribution boring, or at least not too exciting. After all, this theorem confirms something that your intuition most likely took for granted. So let's use equidistribution to discover some unexpected statistics.

Consider the powers of two, the sequence $2^n$, as $n$ ranges through the nonnegative integers:

$$1, 2, 4, 8, 16, 32, 64, 128, 256, 512, 1024, 2048, \ldots.$$

The final—rightmost—digits of the powers of two are completely predictable, forever repeating (after $2^0$) the period-4 sequence $2, 4, 8, 6$. For example, one could determine, in a matter of seconds, that the final digit of $2^{10^{100}}$ is 6. The initial digits are another story. Unlike the dice-sum sequence, the initial digits are not random—they are completely determined by computing the successive powers of two. But are there shortcuts to determining the initial digits? It is not at all obvious, say, how to determine the leftmost digit of $2^{10^{100}}$ without calculating the entire number first.

Suppose you wanted to guess the initial digit of $2^{10^{100}}$. Is there a "best" choice? If you had to guess a dice sum, 7 is the best choice, since that is the most frequent sum. Is there a most frequent initial digit for the powers of two? And if so, what is its frequency? At first you may guess that the initial digits are equidistributed; i.e., each digit has frequency $1/9$. But a quick glance at the first few powers of two suggests that 1 appears more frequently than any other digit.

It's time to get our hands dirty with some empirical work. While $2^{10^{100}}$ may take a lot of work, we can easily compute the first 1000 initial digits. See the table below.

**Initial digits of $2^n$, $n = 0, \ldots, 999$.**

1 2 4 8 1 3 6 1 2 5 1 2 4 8 1 3 6 1 2 5 1 2 4 8 1 3 6 1 2 5 1 2 4 8 1 3 6 1 2 5 1 2 4 8 1 3 7 1 2 5
1 2 4 9 1 3 7 1 2 5 1 2 4 9 1 3 7 1 2 5 1 2 4 9 1 3 7 1 3 6 1 2 4 9 1 3 7 1 3 6 1 2 4 9 1 3 7 1 3 6
1 2 5 1 2 4 8 1 3 6 1 2 5 1 2 4 8 1 3 6 1 2 5 1 2 4 8 1 3 6 1 2 5 1 2 4 8 1 3 6 1 2 5 1 2 4 8 1 3 7
1 2 5 1 2 4 9 1 3 7 1 2 5 1 2 4 9 1 3 7 1 2 5 1 2 4 9 1 3 7 1 3 6 1 2 4 9 1 3 7 1 3 6 1 2 5 1 2 4 8
1 3 6 1 2 5 1 2 4 8 1 3 6 1 2 5 1 2 4 8 1 3 6 1 2 5 1 2 4 8 1 3 6 1 2 5 1 2 4 8 1 3 7 1 2 5 1 2 4 9
1 3 7 1 2 5 1 2 4 9 1 3 7 1 2 5 1 2 4 9 1 3 7 1 3 6 1 2 4 9 1 3 7 1 3 6 1 2 4 9 1 3 7 1 3 6 1 2 5 1
2 4 8 1 3 6 1 2 5 1 2 4 8 1 3 6 1 2 5 1 2 4 8 1 3 6 1 2 5 1 2 4 8 1 3 6 1 2 5 1 2 4 8 1 3 7 1 2 5 1
2 4 9 1 3 7 1 2 5 1 2 4 9 1 3 7 1 3 6 1 2 4 9 1 3 7 1 3 6 1 2 4 9 1 3 7 1 3 6 1 2 5 1 2 4 8 1 3 6 1
2 5 1 2 4 8 1 3 6 1 2 5 1 2 4 8 1 3 6 1 2 5 1 2 4 8 1 3 6 1 2 5 1 2 4 8 1 3 7 1 2 5 1 2 4 9 1 3 7 1
2 5 1 2 4 9 1 3 7 1 2 5 1 2 4 9 1 3 7 1 3 6 1 2 4 9 1 3 7 1 3 6 1 2 4 9 1 3 7 1 3 6 1 2 5 1 2 4 8 1
3 6 1 2 5 1 2 4 8 1 3 6 1 2 5 1 2 4 8 1 3 6 1 2 5 1 2 4 8 1 3 7 1 2 5 1 2 4 8 1 3 7 1 2 5 1 2 4 9 1
3 7 1 2 5 1 2 4 9 1 3 7 1 3 6 1 2 4 9 1 3 7 1 3 6 1 2 4 9 1 3 7 1 3 6 1 2 5 1 2 4 8 1 3 6 1 2 5 1 2
4 8 1 3 6 1 2 5 1 2 4 8 1 3 6 1 2 5 1 2 4 8 1 3 6 1 2 5 1 2 4 8 1 3 7 1 2 5 1 2 4 9 1 3 7 1 2 5 1 2
4 9 1 3 7 1 2 5 1 2 4 9 1 3 7 1 3 6 1 2 4 9 1 3 7 1 3 6 1 2 5 1 2 4 8 1 3 6 1 2 5 1 2 4 8 1 3 6 1 2
5 1 2 4 8 1 3 6 1 2 5 1 2 4 8 1 3 6 1 2 5 1 2 4 8 1 3 7 1 2 5 1 2 4 9 1 3 7 1 2 5 1 2 4 9 1 3 7 1 2
5 1 2 4 9 1 3 7 1 3 6 1 2 4 9 1 3 7 1 3 6 1 2 4 9 1 3 7 1 3 6 1 2 5 1 2 4 8 1 3 6 1 2 5 1 2 4 8 1 3
6 1 2 5 1 2 4 8 1 3 6 1 2 5 1 2 4 8 1 3 6 1 2 5 1 2 4 8 1 3 7 1 2 5 1 2 4 9 1 3 7 1 2 5 1 2 4 9 1 3
7 1 3 6 1 2 4 9 1 3 7 1 3 6 1 2 4 9 1 3 7 1 3 6 1 2 5 1 2 4 8 1 3 6 1 2 5 1 2 4 8 1 3 6 1 2 5 1 2 4
8 1 3 6 1 2 5 1 2 4 8 1 3 6 1 2 5 1 2 4 8 1 3 7 1 2 5 1 2 4 9 1 3 7 1 2 5 1 2 4 9 1 3 7 1 2 5 1 2 4
9 1 3 7 1 3 6 1 2 4 9 1 3 7 1 3 6 1 2 4 9 1 3 7 1 3 6 1 2 5 1 2 4 8 1 3 6 1 2 5 1 2 4 8 1 3 6 1 2 5

This table is rather striking; the initial digits seem almost, but not quite periodic. It is clear that 1 is the most frequent digit. The precise digit frequencies are counted in the following table.

| digit | 1 | 2 | 3 | 4 | 5 | 6 | 7 | 8 | 9 |
|---|---|---|---|---|---|---|---|---|---|
| occurrence | 301 | 176 | 125 | 97 | 79 | 69 | 56 | 52 | 45 |

Will these frequencies hold up if we computed more powers of two? It turns out that they do, but let's try a different experiment. Let's look at powers of seven. Here are the frequencies for the initial digits of $7^n$, $n = 0, \ldots, 999$.

| digit | 1 | 2 | 3 | 4 | 5 | 6 | 7 | 8 | 9 |
|---|---|---|---|---|---|---|---|---|---|
| occurrence | 302 | 176 | 126 | 96 | 78 | 67 | 59 | 50 | 46 |

The two tables are practically identical! This cannot be a coincidence. What is controlling the frequency of the initial digits?

Let's look at the powers of 7, and attempt to compute the frequency of the initial digit 1. What determines whether $7^n$ begins with a 1? If we write a number in scientific notation, that allows us to focus on the initial digit. For example,

$$7^9 = 40353607 = 4.0353607 \times 10^7.$$

If $7^n$ began with a 1, in scientific notation we would have

$$7^n = u \times 10^r,$$

where $1 \leq u < 2$, and $r$ is a nonnegative integer (that we don't care about). The above equation is practically begging you to take logarithms (base 10) of both sides. This yields

$$n \log_{10} 7 = \log_{10} u + r.$$

Since $r$ is a nonnegative integer, the decimal part of $n \log_{10} 7$ is $\log_{10} u$. In other words,

$$\langle n \log_{10} 7 \rangle = \log_{10} u.$$

Since $1 \leq u < 2$, we have $0 \leq \log_{10} u < \log_{10} 2 = 0.30103$.[2] Thus we have determined that

$7^n$ *begins with the digit* 1 *if and only if* $\langle n \log_{10} 7 \rangle$ *lies in the interval* $[0, 0.30103)$.

We have suddenly returned to the world of jumping frogs. Let $\beta = \log_{10} 7$ and let $I$ be the interval $[0, 0.30103)$. If $\beta$ is irrational, then the Equidistribution Theorem says that

$$\lim_{n \to \infty} \frac{|I \cap \{\langle \beta \rangle, \langle 2\beta \rangle, \langle 3\beta \rangle, \ldots, \langle n\beta \rangle\}|}{n} = 0.30103.$$

In other words, the limiting (long-term) frequency of exponents $n$ such that $7^n$ begins with the digit 1 is the same as the limiting frequency of those $n$ such that $\langle \beta \rangle$ lies in the interval $I$, namely 0.30103.

It is easy to prove that $\log_{10} 7$ is irrational. If, on the contrary, $\log_{10} 7 = a/b$ where $a, b$ are integers, we would have $10^{a/b} = 7$ or $10^a = 7^b$, an impossibility, since $10^a$ is even and $7^b$ is odd!

This explains why the initial digit 1 had the same frequency of approximately 30% for both powers of 2 and powers of 7. Let $\gamma = \log_{10} 2$. By the same reasoning as above, $2^n$ begins with the digit 1 if and only if $\langle n\gamma \rangle$ lies in the interval $[0, 0.30103)$. Since $\gamma$ is irrational (exercise!), this will happen with limiting frequency 0.30103.

What about other bases and other initial digits? The same scientific-notation argument yields the following general statement about initial digits of powers.

[2] For simplicity here we're using a five-digit decimal approximation for $\log_{10} 2$. This number is irrational and thus not a terminating decimal.

**Theorem 4** *Let $b$ be a positive real number with $\log_{10} b$ irrational. Let $d \in \{1, 2, 3, \ldots, 9\}$. Then the limiting frequency of exponents $n$ such that $b^n$ begins with the digit $d$ is equal to*

$$\log_{10} \frac{d+1}{d}.$$

For example, this theorem predicts that if you look at enough powers of 2, the fraction that begin with the digit 5 will be $\log_{10} \frac{6}{5} = 0.0792$. And indeed, 7.9% of the first 1000 powers of 2 begin with 5.

# 5 Conclusion: The Strategic Picture

We showed how an intuitive concept, equidistribution, led to a truly surprising discovery—the robust and unequal distribution of initial digits of powers. The mathematics behind this was pretty simple manipulation of logarithms. But *why* did it work? Mathematics involves both tactics (manipulating logarithms, say), and strategies (for example, deciding which questions are worth asking).

There are two strategies operating behind the scenes. One is an adaptable, opportunistic attitude: when you can't compute something precisely, look at statistical distributions instead. The second is an important problem-solving strategy: vary your point of view. If we kept looking at frogs on a number line, we'd have made no progress. Instead, we transformed the problem from lines to circles. We had to give up the integer part of our numbers, but this was worth it, because the circle was finite, allowing us to use the pigeonhole principle. With powers of two, taking logarithms changed our point of view in the same way: we gave up the magnitude (power of 10) part of the scientific notation, and focused on the logarithm of the coefficient instead. Since this number was between 0 and 1, we were able to move powers of two onto a finite, easy-to-analyze circle.

Try your hand at tactics and strategy with the following problems. Some are pretty routine, and others are quite challenging.

1. What happens to the equidistribution theorem if $\alpha$ is rational?

2. Prove Theorem 4.

3. What can you say about the distribution of initial digits of $ab^n$, where $a$, $b$ are constants, and $n = 0, 1, 2, \ldots$?

4. There is a phenomenon called **Benford's Law** which says that approximately 30% of "naturally occurring" numbers begin with digit 1. Think of naturally occurring numbers as constants of nature, like the mass of a whale in kilograms, or the width of the sun in meters. Benford's Law seems empirically true, but is not a rigorously stated mathematical theorem, so it cannot be "proved." However, you can convince yourself of its "truth" by making two assumptions: first, that the initial digits of "naturally occurring" numbers do have a fixed distribution, and second, that this distribution is scale-independent. For example, if we suddenly switched all measurements from metric to the British system, the distribution of digits would not change. Show that from these two assumptions, one can conclude that the initial digits of "naturally occurring" numbers have the same distribution as the initial digits of powers of 2 (or powers of 7, etc.)

5. Let $b$ be a fixed, finite block of digits. For example, $b = 70099876322221$ is one possibility. Let $C(b, n)$ be the number of powers of 2 between $2^0$ and $2^n$ which contain the block $b$ somewhere in their base-10 representation. For example, if $b = 04$, then $C(b, 5) = 0$, since none of the numbers $1, 2, 4, 8, 16, 32$ contain the block 04. In fact, $C(b, n) = 0$ for

$n = 0, 1, 2, \ldots 10$. On the other hand, $C(b, 11) = 1$, since $2^{11} = 2048$ contains 04. The values of $C(b, n)$ do not increase until $n = 20$; then $C(b, n) = 2$, because $2^{20} = 1048576$. Prove the following amazing theorem.

*Let $b$ be any fixed finite block of digits. Then*

$$\lim_{n\to\infty} \frac{C(b,n)}{n} = 1.$$

In other words, no matter what block you pick, eventually, the long term frequency of powers of 2 that contain this block will approach 100%! For example, if you looked at the first $10^{10^{10^{100}}}$ powers of 2, almost all of them would contain the block 31415926.

6. The near-periodicity of the initial digits of the powers of 2 (or any other power) is an intriguing phenomenon to investigate. The period-10 sequence $1, 2, 4, 8, 1, 3, 6, 1, 2, 5$ works perfectly at first, but then it gets "noisy." We can quantify this concept. Let $b$ be a block of digits, such as 1248136125. Imagine tiling the first 1000 digits in the sequence with copies of $b$. If the digits agree with $b$, there are no "errors." Every time a digit disagrees, we count one error. Thus we can tile the first 40 initial digits with copies of $b$ with no errors. But the next 10 digits disagree with $b$ at one place, giving a total of 1 error. The next 10 digits have 2 errors, etc. It may turn out that our copies of $b$ will agree better if we don't strictly tile, but instead skip a few digits here or there. For example, it appears that the 1248136125 stops for a while, but then repeats very nicely. So, imagine superimposing copies of $b$ on the table of initial digits, skipping here and there, and count as errors the "gaps" (the digits not covered by copies of $b$) plus the number of digits that disagree with the $b$-values. You can then, at least in theory, compute the error-frequency of covering the table with copies of $b$. Here's the theorem that you can try to prove:

   *For any $\epsilon > 0$, there exists a block $b$ such that the initial digits of the powers of 2 can be covered by copies of $b$, with error frequency less than $\epsilon$.*

   In other words, there exist better and better patterns (which get longer and longer), and the table of initial digits can thus be thought to be "as near to periodic as can be."

## 6 Acknowledgements and References

None of the ideas in this article are original. They are mostly inspired by a course I took in Ergodic Theory in 1988 at UC Berkeley that was taught by Jonathan King. Indeed, King has written about similar subjects (at a higher level; see, for example, his award-winning article "Three Problems in Search of a Measure," which appeared in the *American Mathematical Monthly*, **101**, (7) (1994), 609–28.

If you would like to see a calculus-based proof of the Equidistribution Theorem, consult *Problems and Theorems in Analysis I* by Pólya and Szegö (Springer-Verlag, 1972).

Problem 5 arose from conversations that I had with Jack Feldman and David Gale (UC Berkeley); Gale wrote about this in the *Mathematical Intelligencer*, **14**, (3) (1992), 62–64.

To learn more about numbers, in particular, irrational numbers, look at two books by Ivan Niven: *Numbers: Rational and Irrational* (Anneli Lax New Mathematical Library, No. 1, MAA, 1961) and *Irrational Numbers* (Carus Mathematical Monographs, No. 11, MAA, 1956).

If you are interested in thinking about mathematical "strategy and tactics," you may want to look at my book, *The Art and Craft of Problem Solving* (Wiley, 1999). This book also has a wealth of material about the pigeonhole principle.

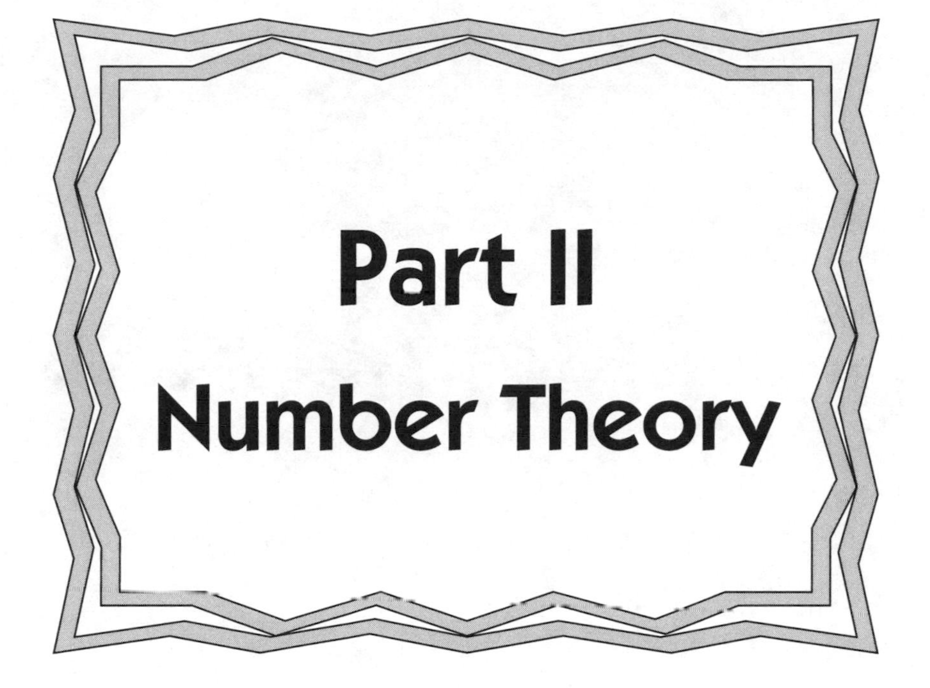

# Part II
# Number Theory

# 6

# Triangles, Squares, Oranges and Cuboids

Peter Stevenhagen
*Universiteit Leiden*

We discuss here three problems in number theory that are of a "Pythagorean" nature. The first is a more or less classical one, and we present various approaches to its solution. The second one was only settled relatively recently using 20th century mathematics. The third problem is still widely open at the moment of writing, and probably requires 21th century mathematics for its solution.

## Pythagoras

Most people know the name of Pythagoras of Samos from his theorem on the lengths of sides of right angled triangles. The modern statement of this theorem, which is $a^2 + b^2 = c^2$, is so compact that it rivals Einstein's $E = mc^2$ in popularity. In fact, there are many people who know both formulas, but have no clue as to what the $c$ stands for in either expression!

Pythagoras, who lived from around 570 B.C. to about 475 B.C., is often regarded as the first pure mathematician, and certainly as the first number theorist. As none of his writings have survived, it is however hard to get a clear picture of his mathematical achievements. From ancient Greek biographies, we know that in his time, he was primarily viewed as a philosopher, a prophet who led a "society" of a scientific-religious nature in Southern Italy. In his younger years, Pythagoras had traveled to Egypt and Babylon, and his society was known for "oriental" customs such as vegetarianism and a refusal to eat beans or to wear clothes made from animal skins.

The society had a code of secrecy and practiced communalism, so it is hard to distinguish between the discoveries of the master and of the inner circle of followers known as "mathematikoi." What certainly goes back to Pythagoras is the philosophical principle that harmony in nature consists of numerical ratios:

*Everything is ordered by number.*

The inspiration for this belief appears to be based on the discovery that strings produce harmonious tones when the ratios of the lengths of the strings are whole numbers. When a string is shortened to half its length, the tone it produces is an octave higher. Similarly, the ratios 3: 2 and 4: 3 correspond to the intervals of the fifth and the fourth. The numbers 1, 2, 3, 4 themselves, from which the basic harmonic intervals are formed, constitute the *tetractys,* the "scale of the sirens" that can be

represented by the *perfect triangle*

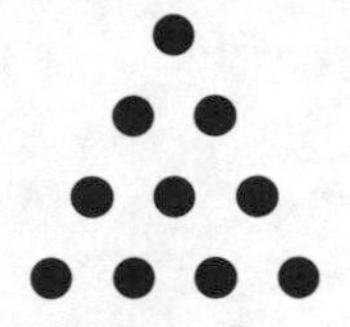

or by the *triangular number*

$$10 = 1 + 2 + 3 + 4.$$

The Pythagoreans viewed 10 as a perfect number since it represented the tetractys—that it is also the number of fingers of the human species may be a coincidence. The left picture below, which is taken from a book of Boethius, shows Pythagoras analyzing the harmony of tones. It is rather different from the bust of Pythagoras on the right, which is remarkable for the "oriental" head covering.

## Number mysticism

Pythagoras' approach to numbers was rather different from the approach that is common in modern number theory courses. Not only did the Pythagoreans regard the number 10 above as a perfect number, most other small numbers had an emotional value as well. Even numbers were regarded as feminine, and odd numbers as masculine. As the "unit" 1 did not count as a number for the ancient Greeks, the sum $5 = 2 + 3$ of the first feminine and the first masculine number was an obvious symbol of marriage.

In a similar way, 17 was a horrible number, being right between the particularly beautiful numbers 16 and 18.

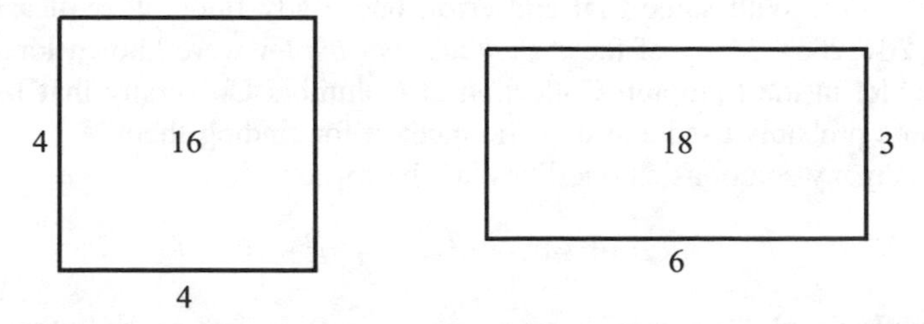


In case you did not realize this right away: the rectangles above having area 16 and 18 also have perimeter 16 and 18, respectively.

**Exercise.** Show that no other rectangles with integral sides have this property.

Already in Pythagoras' day some people made fun of the mystical beliefs of the Pythagoreans, and nowadays number mysticism is no longer part of mathematics. Superstition however remains: multi-story apartment buildings in the United States often have their 14th floor on top of the 12th!

## Theorem of Pythagoras

Already the Babylonians knew around 1700 B.C. about the theorem that is now ascribed to Pythagoras. In fact, we do not know if Pythagoras actually discovered the theorem himself, and if so, how he did this. The classical way to present the theorem (and its proof!) is the picture below, in which

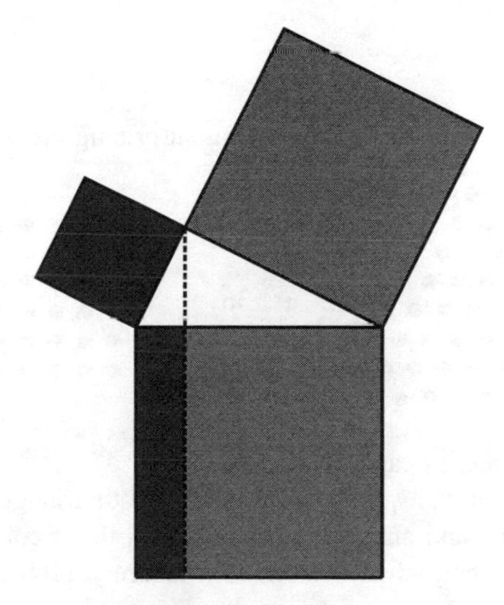

the area of the lower square is the sum of the areas of the other two squares. In our efficient modern notation, which mostly goes back to the 18th century mathematician Leonhard Euler (1707–1783), one simply writes

$$a^2 + b^2 = c^2$$

for the relation between the sides. The idea of using letters ("variables") to denote various quantities is in fact a rather recent one in the history of mathematics! It has enabled us to solve geometric questions "by algebraic means." For this reason, geometry and algebra are inextricably linked in modern mathematics.

Long before Pythagoras people already knew that a triangle having sides of length 3, 4 and 5 was right angled since $(a, b, c) = (3, 4, 5)$ is, as we would say nowadays, a "solution" to the "equation" $a^2 + b^2 = c^2$. With some trial and error, one easily finds other pleasing solutions as $(5, 12, 13)$ or $(23, 264, 265)$. Many of these *Pythagorean triples* were known long ago: there is a Babylonian clay tablet in the Plimpton Collection at Columbia University that lists a number of them, and its authors probably used a systematic method for finding them.

One can make as many solutions as one likes, as the triple

$$(2mn, m^2 - n^2, m^2 + n^2)$$

is a Pythagorean triple for all choices of integers $m > n > 0$. In fact every Pythagorean triple is a multiple of a triple of this form, as one can easily deduce by viewing the triple $(a, b, c)$ as a point $(\frac{a}{c}, \frac{b}{c})$ on the unit circle in the plane with equation $x^2 + y^2 = 1$ having positive coordinates.

**Exercise.** Show that Pythagorean triples correspond to rational points $P$ with positive coordinates on the unit circle as indicated above, provided that we identify a triple $(a, b, c)$ with its scalar multiples $(ka, kb, kc)$. Show that such $P$ are obtained as intersection points of the lines $y = \frac{m}{n}x - 1$ through $(0, -1)$ with rational slope $\frac{m}{n} > 1$ with the unit circle. Deduce that we have

$$P = \left( \frac{2mn}{m^2 + n^2}, \frac{m^2 - n^2}{m^2 + n^2} \right).$$

Finding integral solutions to various equations still lies very much at the heart of modern number theory. Whenever we have solved a problem, a new one naturally suggests itself.

## Figural harmony

Our first Pythagorean problem concerns the following surprising property of the number 36.

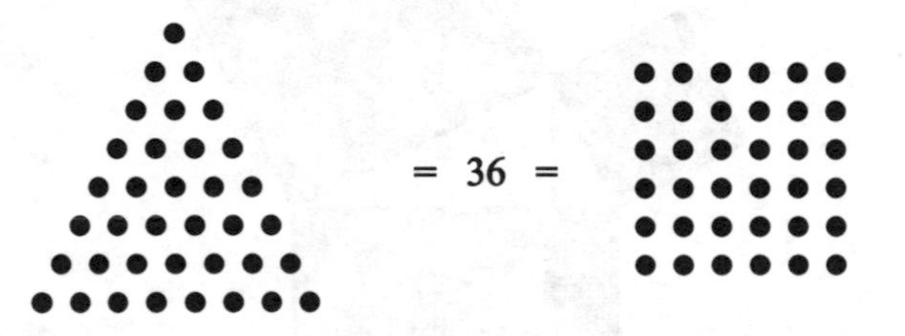


It shows that in arithmetic, triangles can be squares!

From a Pythagorean point of view, a number that is both triangular and a square certainly deserves to be called *harmonic*, and anyone playing with a pile of coins will find that it is rather exceptional for a number to be harmonic. Our first Pythagorean problem will now be obvious.

**Question.** *Which integers are harmonic?*

More specifically, we would like to know whether there are harmonic integers at all besides 36, and come up with a method to produce a complete list of them. For all we know, there can be finitely many or infinitely many harmonic numbers, and in the second case the best we can hope for is an efficient *algorithm* to produce all harmonic numbers in ascending order or, even better, a simple *formula* for the $n$th harmonic number.

As for the existence of harmonic integers besides 36, the mathematically inclined reader may already have noticed that the number 1 is actually both a triangular number and a square! One might be inclined to call this a *trivial* example of a harmonic number, and actually a true Pythagorean would contend that 1 is not to be considered a number at all—it is the unit by which the true numbers are measured. The modern reader, who knows about relatively recent inventions such as the number zero and negative numbers as well, may feel this Pythagorean distinction of 1 to be merely awkward.

## Using your computer

When confronted with the question of existence of small harmonic numbers, the modern mathematician has a tool that often makes life easy: the computer. It is not too difficult to ask a computer to look for harmonic numbers, provided that we can tell it how to decide whether a number is triangular.

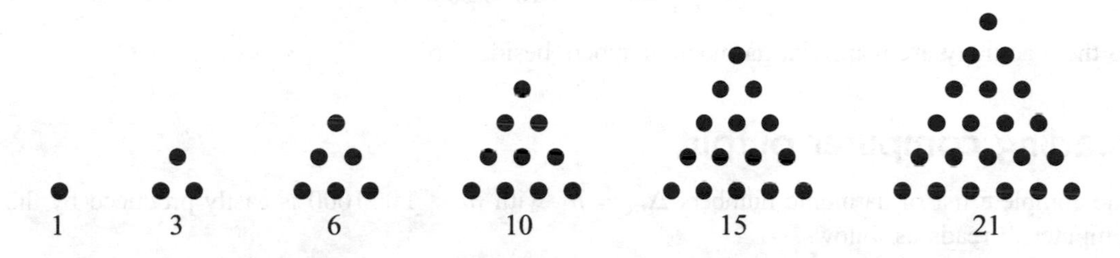


Looking at the first few triangular numbers, one can try to come up with an easy formula for the $m$-th number in the sequence

$$1, 3, 6, 10, 15, 21, 28, 36, \ldots.$$

You may have heard the famous story in this context about the German mathematician Carl Friedrich Gauss (1777–1855). When he was still a small boy in elementary school, the teacher tried to keep his class busy for a while by ordering them to compute the sum $1 + 2 + 3 + \ldots + 99 + 100$ of all positive integers up to 100. In other words, the children had to compute the 100th triangular number, which takes a while by hand if one actually performs the 99 required additions. Gauss thought for a few seconds before writing down the correct answer. In geometric terms, he had realized that doubling the $m$th triangular number yields the "rectangular number" $m(m+1)$. The resulting simple formula

$$\Delta_m = \frac{m(m+1)}{2}$$

for the $m$th triangular number $\Delta_m$ is obvious if one draws the right picture.

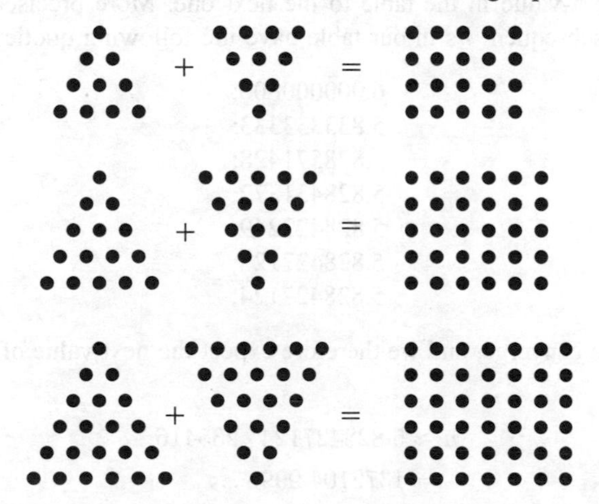

Using Gauss's formula, one may write a little computer program that, depending on the language one uses, will more or less resemble the following program in pseudo-code.

```
for m = 1 to 1000:
  do Δ_m = m(m+1)/2;
  if issquare Δ_m print Δ_m;
endif; enddo; endfor;
```

You will find:

$$\frac{49 \cdot 50}{2} = 1225 = 35^2$$

and

$$\frac{288 \cdot 289}{2} = 41616 = 204^2.$$

So there actually are nontrivial harmonic numbers besides 36!

## Reading computer output

The complete list of harmonic numbers $\Delta_m = n^2$ with $m < 1\,000\,000$ is easily produced by the computer. It reads as follows.

| $\Delta_m = n^2$ | $m$ | $n$ |
|---|---|---|
| 1 | 1 | 1 |
| 36 | 8 | 6 |
| 1225 | 49 | 35 |
| 41616 | 288 | 204 |
| 1413721 | 1681 | 1189 |
| 48024900 | 9800 | 6930 |
| 1631432881 | 57121 | 40391 |
| 55420693056 | 332928 | 235416 |

The question is: what can we deduce from such a table? It seems that there are many harmonic numbers, but that they grow rapidly when written down in ascending order. Can we predict from the table above where the first harmonic number $\Delta_m$ with $m > 1\,000\,000$ is to be expected?

It appears that the growth in the table is quite regular, as $n$ gets about 6 times as large if we move from a harmonic $n$-value in the table to the next one. More precisely, a pocket calculator suffices to tell us that subsequent $n$s in our table have the following quotients:

6.000000000;
5.833333333;
5.828571428;
5.828431372;
5.828427249;
5.828427128;
5.828427124.

These quotients seem to converge, and we therefore expect the next value of $n$ to have the approximate value

$$\begin{aligned} n &\approx 5.828427124 \cdot 235416 \\ &= 1372104.9998\ldots. \end{aligned}$$

And indeed, for $n = 1372105$ we find

$$n^2 = 1372105^2 = 1882672131025$$
$$= \frac{1940449 \cdot 1940450}{2} = \Delta_{1940449}.$$

The corresponding quotient $\frac{1372105}{235416} = 5.8284271247$ yields, correct up to 5 decimals:

$$\frac{1372105}{235416} \cdot 1372105 \approx 7997214$$

as the next harmonic value of $n$. Is it a surprise that our expectations are met again?

$$7997214^2 = 63955431761796$$
$$= \frac{11309768 \cdot 11309769}{2} = \Delta_{11309768}.$$

By what one might call "empirical math," we have arrived at a method to predict the next harmonic number from the previous ones in the list. Those with some faith in nature will feel the method is bound to work in the infinitely many cases further down the list, so that we have found a method to generate all harmonic numbers in ascending order. But the cynical mathematician will observe that there is no guarantee that our method will always work, and that we can only hope that no harmonic numbers will be missed by our method.

## Looking for structure

From a mathematical point of view, we have only scratched the surface, as our *understanding* of why our "method" works is still non-existent. Let us try to gain a deeper understanding of what is going on by looking in a more *structured* way at the integers occurring in our table. What are their striking properties from an arithmetic point of view?

If one looks at the parity of $m$ and $n$, one sees that $m$ and $n$ have the same parity, which alternates between consecutive values. The odd values of $m$ are easily recognized as squares: 1, $49 = 7^2$ and $1681 = 41^2$. A calculator gives us $57121 = 239^2$. The even values are not squares, but one less than a square: $8 = 3^2 - 1$, $288 = 17^2 - 1$, $9800 = 99^2 - 1$. The calculator confirms this: $332928 = 577^2 - 1$. One may also recognize the even values of $m$ as being twice a square: $8 = 2 \cdot 2^2$, $288 = 2 \cdot 12^2$, $9800 = 2 \cdot 70^2$ and finally $332928 = 2 \cdot 408^2$.

Having found that the harmonic values of $m$ are alternatingly of the form $x^2$ and $x^2 - 1$, the next observation is that $x$ is a *divisor* of the corresponding $n$-value. This leads us to rewrite the table in the following form.

| $\Delta_m = n^2$ | $m$ | $n$ |
|---|---|---|
| 1 | $1^2$ | $1 \cdot 1$ |
| 36 | $3^2 - 1$ | $3 \cdot 2$ |
| 1225 | $7^2$ | $7 \cdot 5$ |
| 41616 | $17^2 - 1$ | $17 \cdot 12$ |
| 1413721 | $41^2$ | $41 \cdot 29$ |
| 48024900 | $99^2 - 1$ | $99 \cdot 70$ |
| 1631432881 | $239^2$ | $239 \cdot 169$ |
| 55420693056 | $577^2 - 1$ | $577 \cdot 408$ |

Those who chose to write the even values of $m$ as $m = 2y^2$ would also note that this $y$ divides $n$, leading to $n = x \cdot y$ with $x$ as before. Either way, one finds that every harmonic number is of the form

$$x^2 \cdot y^2$$

with $x$ and $y$ taking on the following values:

$$x : 1, 3, 7, 17, 41, 99, 239, \ldots$$
$$y : 1, 2, 5, 12, 29, 70, 169, \ldots$$

We are now down to the level of much smaller integers, and in order to predict the next values of $x$ and $y$ from the previous ones one needs a final discovery: the sequences for $x$ and $y$ are *recurrent sequences*. More precisely, for $k \geq 2$ their $k+1$th term is given by

$$x_{k+1} = 2x_k + x_{k-1}$$
$$y_{k+1} = 2y_k + y_{k-1}.$$

This is satisfactory as we have now gotten rid of non-integers such as the mysterious limit 5.8284271247.... But do we really understand what is going on?

## Solving equations

We get a better understanding if we realize that what we are doing is finding the integer solutions of the equation

$$\frac{m(m+1)}{2} = n^2.$$

No general method for solving equations in integers exists—but in many cases we can do it! In this case, it helps to read the left-hand side as

$$m \cdot \frac{(m+1)}{2} \quad \text{for } m \text{ odd;}$$

$$\frac{m}{2} \cdot (m+1) \quad \text{for } m \text{ even.}$$

The integers $m$ and $m+1$ have no common factors as they differ by 1. It follows that the two integers on the left hand side of our equation are *coprime*. Now anyone familiar with the factorization of integers into primes will have no problem to prove the following fundamental fact.

**Fact.** *If a product $ab$ of two coprime positive integers $a$ and $b$ is a square, then both $a$ and $b$ are squares.*

It follows immediately that if $\Delta_m$ is a square, then we have $\Delta_m = x^2y^2$ with

$$m = x^2 \quad \text{and} \quad m+1 = 2y^2$$

if $m$ is odd, and

$$m+1 = x^2 \quad \text{and} \quad m = 2y^2$$

if $m$ is even. As $x^2$ and $2y^2$ differ in both cases by 1, we see that $x$ and $y$ are solutions of the equation

$$x^2 - 2y^2 = \pm 1.$$

There is an amazing trick due to Euler to produce infinitely many integer solutions to this equation out of the "trivial" solution $x = y = 1$. To do this, one boldly leaves the realm of the integers and writes the equation in terms of "algebraic integers" of the form $a + b\sqrt{2}$ with $a$ and $b$ integers as

$$(x + y\sqrt{2})(x - y\sqrt{2}) = \pm 1.$$

The trivial solution $x = y = 1$ yields the identity

$$(1+\sqrt{2})(1-\sqrt{2}) = -1,$$

which we can raise to the power $k$ to obtain

$$(1+\sqrt{2})^k(1-\sqrt{2})^k = (-1)^k.$$

Now every power of an "integer" of the form $a+b\sqrt{2}$ is again of this form. Writing out the powers of $1+\sqrt{2}$, we encounter the values for $x$ and $y$ that occur in our final table!

$$\begin{aligned}(1+\sqrt{2})^2 &= 3+2\sqrt{2}\\ (1+\sqrt{2})^3 &= 7+5\sqrt{2}\\ (1+\sqrt{2})^4 &= 17+12\sqrt{2}\\ (1+\sqrt{2})^5 &= 41+29\sqrt{2}\\ (1+\sqrt{2})^6 &= 99+70\sqrt{2}\\ (1+\sqrt{2})^7 &= 239+169\sqrt{2}\end{aligned}$$

Why does this work? Take for instance

$$(1+\sqrt{2})^{10} = 3363+2378\sqrt{2}.$$

Then we have

$$(1-\sqrt{2})^{10} = 3363-2378\sqrt{2}.$$

The product of these numbers is

$$(3363+2378\sqrt{2})(3363-2378\sqrt{2}) = 3363^2 - 2\cdot 2378^2.$$

It also equals

$$(1+\sqrt{2})^{10}(1-\sqrt{2})^{10} = (-1)^{10} = 1.$$

Thus $x = 3363$ and $y = 2378$ satisfy

$$x^2 - 2y^2 = 1,$$

and the square $3363^2 \cdot 2378^2$ is harmonic: it is the $(3363^2 - 1)$th triangular number.

More generally, we see that the integer pairs $(x_k, y_k)$ defined by

$$x_k + y_k\sqrt{2} = (1+\sqrt{2})^k$$

satisfy $x_k^2 - 2y_k^2 = (-1)^k$ for all $k \geq 1$, and that the number $x_k^2 \cdot y_k^2$ is harmonic, being the $x_k^2$th or $(x_k^2 - 1)$th triangular number.

With some extra effort, one can show that we actually find *all* harmonic numbers in this way, so that $x_k^2 y_k^2$ is indeed the $k$th harmonic number in the ordered list of all harmonic numbers.

From $(1+\sqrt{2})^2 = 2(1+\sqrt{2})+1$ we obtain $(1+\sqrt{2})^{k+1} = 2(1+\sqrt{2})^k + (1+\sqrt{2})^{k-1}$ for all $k \geq 1$, and therefore

$$x_{k+1} + y_{k+1}\sqrt{2} = (2x_k + x_{k-1}) + (2y_k + y_{k-1})\sqrt{2}.$$

This explains the recurrences we found for the sequences of the numbers $x_k$ and $y_k$.

**Exercise.** Show that $k$th harmonic number is obtained by squaring the integral part $n$ of the real number

$$\frac{(3+2\sqrt{2})^k}{4\sqrt{2}}.$$

Deduce that the quotient of subsequent values of $n$ tends to $3+2\sqrt{2} = 5.828427124746\ldots$.

## The orange problem

Whenever we successfully solve a problem, there are always similar problems one may hope to be able to deal with in an analogous way. In the case of harmonic numbers, where triangular numbers are squares, one can move on to similar questions for pentagonal and hexagonal numbers, or think of an analogue in three-dimensional space. In the latter case we run into the following greengrocer's problem.

**Orange problem.** *Which numbers of oranges can be piled up both as a "triangular" and as a "square" pyramid?*

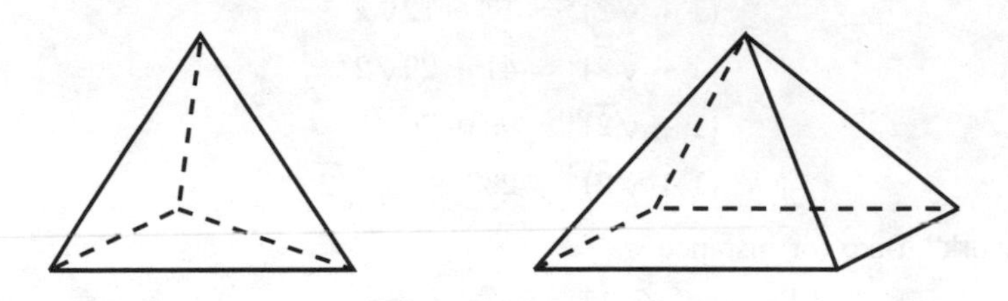

As before, there is the trivial solution 1 corresponding to a single orange. We are of course interested in the larger solutions.

It is an interesting and not too difficult exercise to show that a triangular pyramid with $m$ layers has

$$\frac{m(m+1)(m+2)}{6}$$

oranges. In a similar way, one finds that a square one with $n$ layers has

$$\frac{n(n+1)(2n+1)}{6}$$

oranges. Thus, we need to solve in positive integers $x$ and $y$ the equation

$$m(m+1)(m+2) = n(n+1)(2n+1).$$

As in the case of harmonic numbers, one can write a slightly more complicated computer program to find small solutions, and the reader is invited to do so. However, no solutions $(m, n) \neq (1, 1)$ are found if we check all values of $m$ up to $10^6$ or even $10^9$. Does this mean that there are no such solutions, or have we run into one of these equations for which the smallest nontrivial solution is too large to be found by a computer?

This time, there is no easy Euler-type trick such as the one we employed in the case of harmonic numbers. We suggestively rewrite the equation as

$$x(x+1)(x+2) = y(y+1)(2y+1),$$

so that the modern reader will think of this equation as describing a curve in the $xy$-plane. The given cubic equation describes what is called an *elliptic curve*. Its plot looks as follows.

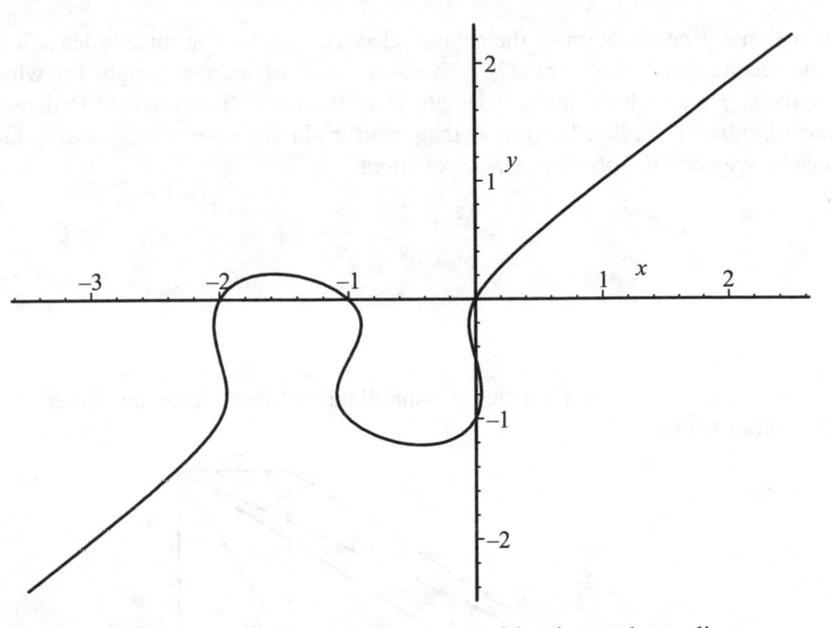


We want to determine the points on this curve having positive integral coordinates.

The arithmetic theory of elliptic curves is very much a 20th century topic. It is an attractive blend of number theory, geometry and analysis. The theory is not yet complete, and some of the striking properties of elliptic curves, such as the famous Birch-Swinnerton-Dyer conjecture, are still unproven. Recently there have been several major breakthroughs. The *modularity* of elliptic curves, from which Fermat's Last Theorem was known to follow, was proved in 1995 by Wiles.

Effectively finding the integral points on an elliptic curve is something that can now be done—at least in principle. In 1988, the Dutch mathematicians Beukers and Top used the existing theory to prove the following theorem.

**Theorem.** *The only integral points on the curve*

$$x(x+1)(x+2) = y(y+1)(2y+1)$$

*are the eight points*

$$\begin{array}{cccc} (-3,-2), & (-2,-1), & (-2,0), & (-1,0) \\ (-1,-1), & (0,-1), & (0,0), & (1,1). \end{array}$$

In order to find a solution to the orange problem, we need integral points $(x, y)$ with *positive* $x$ and $y$. Thus the theorem tells us that it will never work out with more than a single orange. It may come as a surprise that the proof of this simple fact requires a fair amount of modern mathematics.

## Pythagorean cuboids

In the case of Pythagorean triples solving the equation

$$a^2 + b^2 = c^2,$$

we mentioned the complete classification of the integral solutions. They essentially correspond to the set of *rational* points on a specific curve, the unit circle in the plane. When we try to use this complete knowledge to deal with 3-dimensional analogues, we are soon stuck with widely open problems.

A 3-dimensional generalization of the right-angled triangle having three sides $a, b, c$ of integral length is the rectangular box ("cuboid") with sides $a, b, c$ of integral length for which the three face diagonals $x, y, z$ also have integral length. Here we have "interrelated" Pythagorean triples, as each pair of sides of the box forms a Pythagorean triple with the corresponding face diagonal. More precisely, we need to solve the *set* of equations

$$\begin{aligned} a^2 + b^2 &= x^2 \\ b^2 + c^2 &= y^2 \\ a^2 + c^2 &= z^2. \end{aligned}$$

in positive integers $a, b, c, x, y, z$. By a clever manipulation of these formulas, Euler found solutions such as the cuboid below.

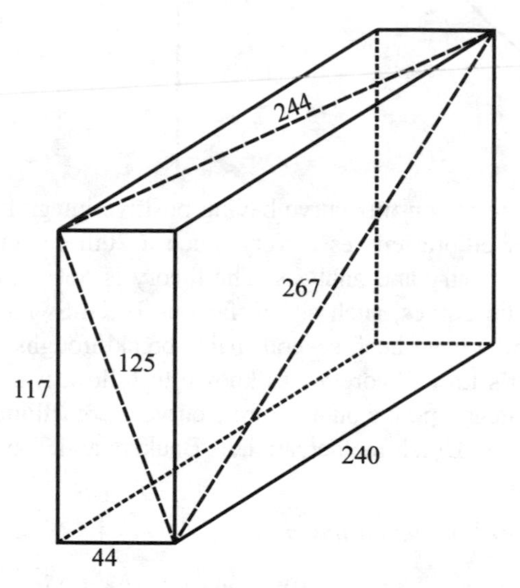


One can even show that there are infinitely many "essentially different" solutions to this set of 3 equations in 6 unknowns.

Suppose now that we require that the full diagonal in such a cuboid has integral length as well. We then obtain a set of 4 equations in 7 unknowns:

$$\begin{aligned} a^2 + b^2 &= x^2 \\ b^2 + c^2 &= y^2 \\ a^2 + c^2 &= z^2 \\ a^2 + b^2 + c^2 &= w^2. \end{aligned}$$

In this more complicated case of *Pythagorean cuboids*, no solutions in positive integers are known to exist. Extensive computer calculations have shown that no "small" solutions exist. But it would just be a wild guess to conjecture that they do not exist at all. In geometric terms, we are here in a situation where we need to find rational points on a *surface*. Finding the rational points on a curve is already a hard and intensively studied problem for which a complete solution is still lacking. For surfaces, the arithmetic theory barely exists, and we may only expect it to be developed in the 21st century. The question as to whether Pythagorean cuboids exist at all is one of the yardsticks by which we can measure the progress of the theory.

## References

For information on the life and work of Pythagoras (and many other mathematicians), one can consult the corresponding webpage at the history site of the University of St Andrews in Scotland:

`http://www-history.mcs.st-andrews.ac.uk/history`.

A very readable book on ancient mathematics that also devotes a dozen pages to Pythagoras and his school is

B. L. van der Waerden, *Science Awakening*, Noordhoff, Groningen, 1954.

Euler's solution of the *Pell equation* $x^2 - 2y^2 = \pm 1$ that we gave appeared in

L. Euler, De solutione problematum Diophantaeorum per numeros integros, *Comm. Acad. Sci. Petropolitanae* **6**, 175–188 (1732/33)

and can be found in Euler's collected works. If you have trouble reading Latin, try one of the more modern texts explaining algebraic number theory.

Finding rational points on elliptic curves is the topic of

J. Silverman, J. Tate, *Rational points on elliptic curves*, Springer, New York, 1992.

For the arithmetic theory of surfaces, we will still have to wait a while before the first textbooks become available.

# 7

# When Is an Integer the Product of Two and of Three Consecutive Integers?

Edward F. Schaefer
*Santa Clara University*

In this chapter we will solve the problem posed in the title, one first solved by Louis Mordell in the 1960s. More interesting than the question itself, perhaps, is the method of solution, which serves to introduce the beautiful subject of elliptic curves. This is a field of lively current research interest and the gateway to techniques used in the recent acclaimed proof of Fermat's Last Theorem and to problems of cryptography.

When is an integer simultaneously the product of two consecutive integers and the product of three consecutive integers? The first example that comes to mind is $6 = 2 \cdot 3 = 1 \cdot 2 \cdot 3$. We could also write $6 = -3 \cdot -2 = 1 \cdot 2 \cdot 3$. The second example that comes to mind is 0. We have

$$\begin{aligned} 0 = \quad 0 \cdot 1 = \quad & 0 \cdot \ \ 1 \cdot 2 \\ -1 \cdot 0 = & -1 \cdot \ \ 0 \cdot 1 \\ = & -2 \cdot -1 \cdot 0. \end{aligned}$$

So we have two representations of 6 as a product of two and of three consecutive integers and six such representations of 0. So far we have 0 and 6. Are there others? Are there infinitely many others or only finitely many? If there are only finitely many, how could we ever show that we have them all?

We are looking for integer solutions to the following equation $y(y+1) = (x-1)x(x+1)$ or $y^2 + y = x^3 - x$. The solution set of such an equation is called an elliptic curve, see Figure 7.1. An elliptic curve is a smooth cubic curve including all of its points at infinity, and one point specified. That definition requires explanation. A cubic curve is a curve described by a polynomial equation in $x$ and $y$ whose highest degree is 3, such as $y = x^3$ or $xy^2 = x + 7$. A cubic curve is smooth if it has no corners and does not cross itself. In Figures 7.2 and 7.3 are the cubic curves $y^2 = x^3$ and $y^2 = x^3 + x^2$; they are not smooth. Ours is smooth.

Points at infinity are limiting points of the curve. A point at infinity is at both ends of all lines of the same slope. In our example ($y^2 + y = x^3 - x$), the slopes on the curve are becoming increasingly vertical. We can draw in vertical lines as in Figure 7.4. Let us force all of those lines to meet at the top and bottom like longitude lines on a globe. We will consider the point at the top

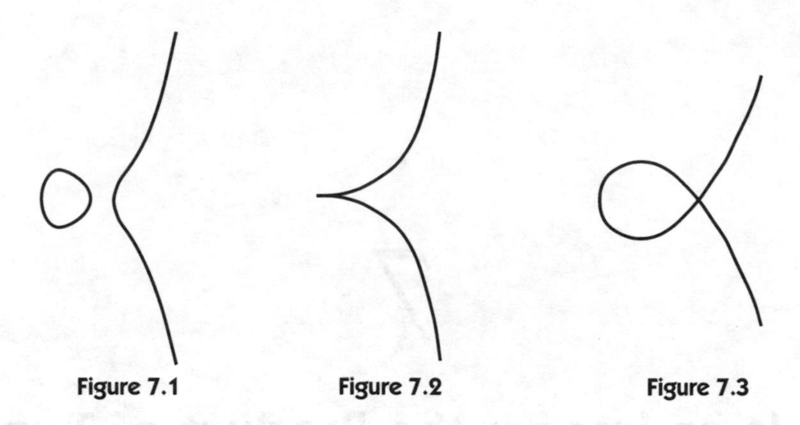

Figure 7.1 Figure 7.2 Figure 7.3

and bottom to be the same point. If we draw in our curve, we see that it will meet that point at infinity; see Figure 7.5. So our curve has that one point at infinity; it is at the top and bottom of every vertical line. (The more advanced reader will realize that we have embedded the curve into projective 2-space and completed it).

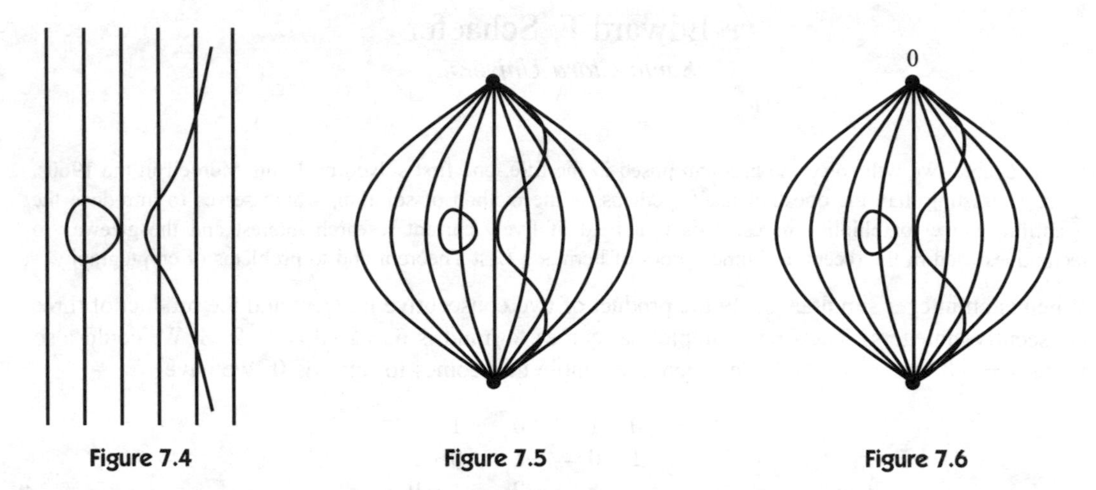


Figure 7.4 Figure 7.5 Figure 7.6

Why are elliptic curves special? Because we can add points on an elliptic curve. First we need to specify which point will act as a 0 with respect to addition. We will call it the 0-point. Any point can be chosen. We will choose the point at infinity as our 0-point and denote it simply as 0. See Figure 7.6. The following rule enables us to add points on an elliptic curve. It works as long as the 0-point is what is called an inflection point. The fact that the point at infinity is an inflection point is not especially easy to see (unless you are familiar with projective geometry), but it is!

Rule: Three points lying on a line sum to 0.

Let us see how that rule enables us to add points. First we intersect our curve with a vertical line; see Figure 7.7. The line meets our curve at $P$, $Q$ and since it is vertical, at 0 also. The rule says $P + Q + 0 = 0$. Therefore $P = -Q$ and $Q = -P$. We see that two points with the same $x$-coordinates are negatives of each other.

Now let us add two points with different $x$-coordinates; see Figure 7.8. We want to add $P + Q$. We draw a line connecting them. That line will meet the curve in a third point $R$. The rule says $P + Q + R = 0$. Therefore $P + Q = -R$. And we know where $-R$ is; it is the other point with the same $x$-coordinate. That is the sum of $P$ and $Q$; see Figure 7.9.

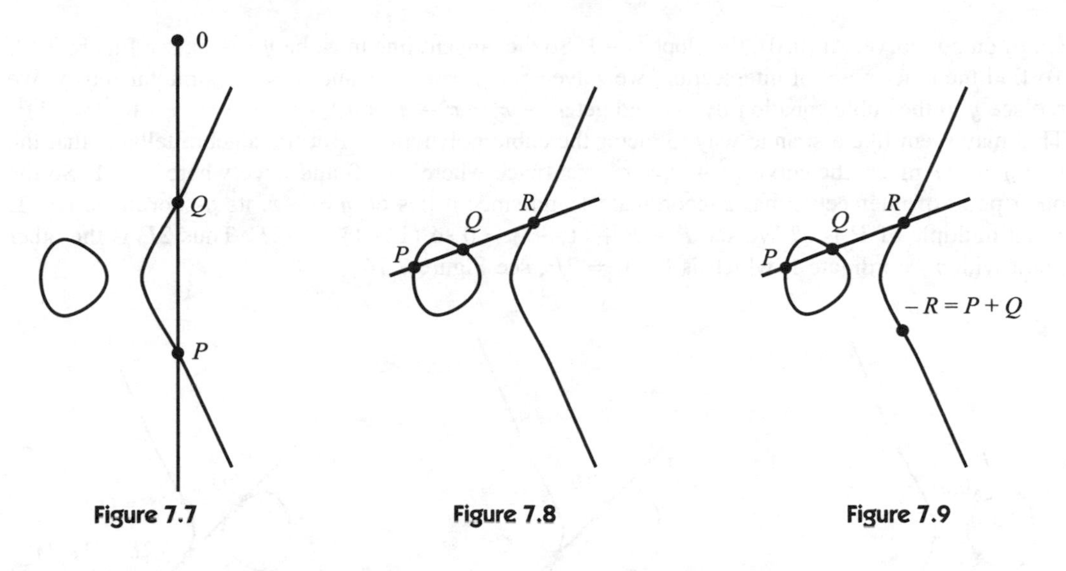


Figure 7.7 Figure 7.8 Figure 7.9

The only thing remaining is to show how to double a point; see Figure 7.10. In order to double $P$, we draw the tangent line at $P$. It meets our curve at one other point $Q$. We will try to convince you in two ways that the tangent line meets the curve twice at $P$. The first argument will be geometric and a later, more convincing argument will be algebraic. If we pin the line down at $Q$ and move it up some on the left then it will meet the egg-shaped piece in two points. Now if we bring it back down, it continues meeting in two points until we get to $P$. Then the line still wants to meet the egg in two points, but they are both $P$. (For the trouble-makers, if we continue to move the line down, it will continue meeting in two more points, but they have imaginary coordinates). So the tangent line meets the curve twice at $P$ and once at $Q$. The rule says $P + P + Q = 0$ or $2P = -Q$. Again, we know where $-Q$ is and that is $2P$; see Figure 7.11.

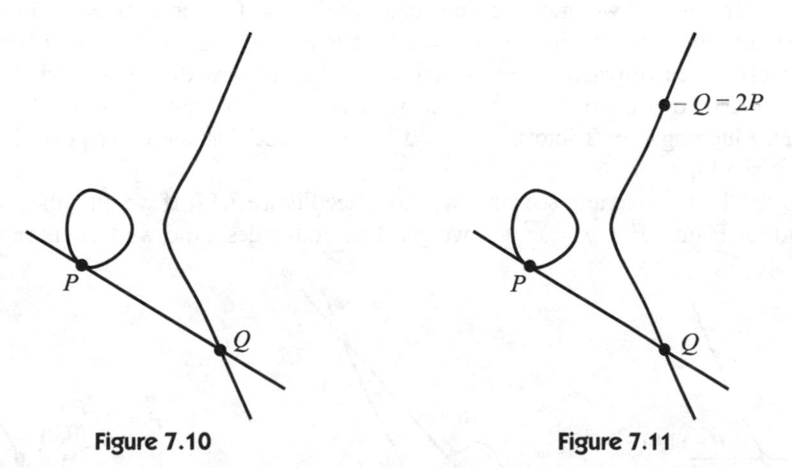


Figure 7.10 Figure 7.11

This addition is well-defined, commutative and associative, though only commutivity is obvious. (For the advanced reader, the points form an abelian group). Now let us try out this addition on our curve. We know the point $P = (0,0)$ is on $y^2 + y = x^3 - x$. The point $-P$ has $x$-coordinate 0 also. By plugging $x = 0$ into $y^2 + y = x^3 - x$, we see that the other possible $y$ is $-1$ so $-P = (0, -1)$; see Figure 7.12. To find $2P$, we need the equation of the tangent line at $P = (0,0)$. For this we turn briefly to calculus. We take the derivative of both sides of our equation with respect to $x$ and get $2yy' + y' = 3x^2 - 1$. So $y' = (3x^2 - 1)/(2y + 1)$ gives the slope of the tangent line at any point

$(x, y)$ on our curve. At $(0,0)$, the slope is $-1$. So the tangent line must be $y = -x$; see Figure 7.13. To find the other point of intersection, we solve $y^2 + y = x^3 - x$ and $y = -x$ simultaneously. We replace $y$ in the cubic equation by $-x$ and get $x^2 - x = x^3 - x$ or $x^3 - x^2 = 0 = (x-0)^2(x-1)^1$. That may seem like a strange way to factor the cubic polynomial. But the algebra tells us that the line $y = -x$ meets the curve $y^2 + y = x^3 - x$ twice where $x = 0$ and once where $x = 1$. So the other point of intersection has $x$-coordinate 1 and since it lies on $y = -x$, its $y$-coordinate is $-1$. What multiple of $P$ is it? We see $P + P + (1, -1) = 0$ so $(1, -1) = -2P$. Thus $2P$ is the other point with $x$-coordinate 1, which is $(1, 0) = 2P$; see Figure 7.14.

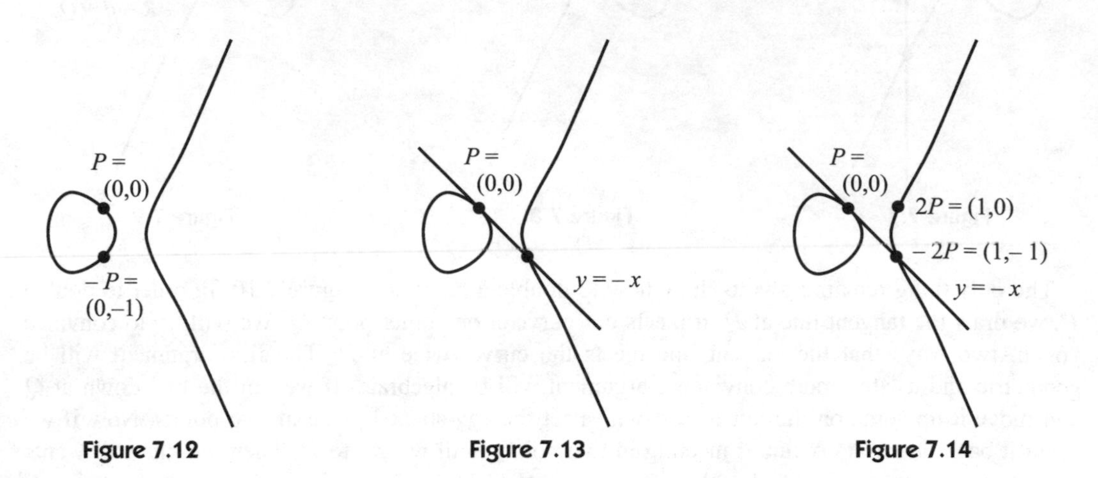


Figure 7.12 Figure 7.13 Figure 7.14

To get $3P$ we draw the line connecting $P = (0,0)$ and $2P = (1,0)$, which is $y = 0$. It meets $y^2+y = x^3-x$ in the third point $(-1,0)$. What multiple of $P$ is this? We see $P+2P+(-1,0) = 0$ so $(-1,0) = -3P$. Thus $3P$ is the other point with $x$-coordinate $-1$, which is $(-1,-1) = 3P$; see Figure 7.15. To find $4P$ we draw the line connecting $P = (0,0)$ and $3P = (-1,-1)$ which is $y = x$. It meets $y^2+y = x^3-x$ where $x^2+x = x^3-x$ or $x^3-x^2-2x = 0 = (x-0)(x+1)(x-2)$. So the line meets the curve where $x = 0$, where $x = -1$ (we knew those), and where $x = 2$. The third point of intersection is on $y = x$ so that point is $(2,2)$. We have $P + 3P + (2,2) = 0$ so $(2,2) = -4P$. Plugging $x = 2$ into $y^2 + y = x^3 - x$ we see that the other possible $y$ is $-3$ so $4P = (2,-3)$; see Figure 7.16.

Let us stop for the moment and see what we have; see Figure 7.17. If we plug the coordinates of $\pm P, \pm 2P$ and $\pm 3P$ into $y^2 + y = x^3 - x$ we get 0 on both sides. And we had six representations

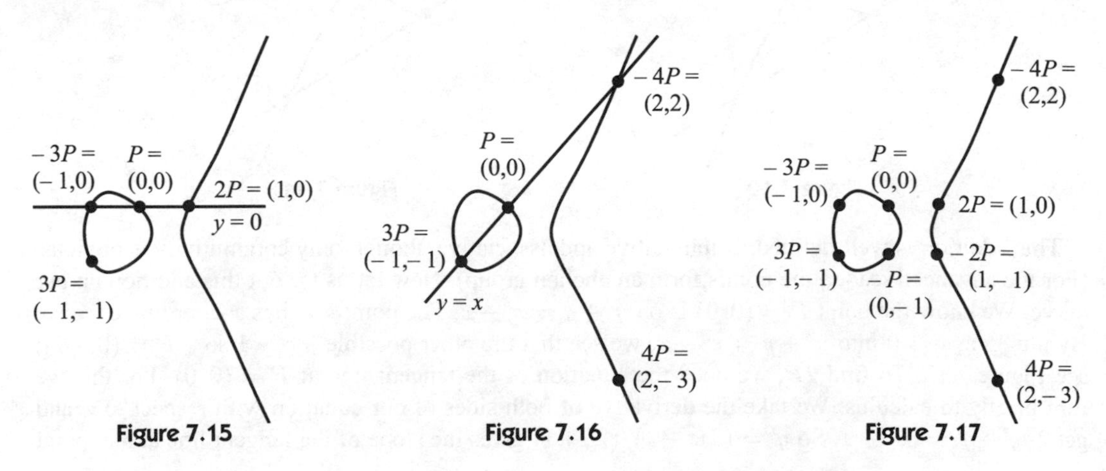


Figure 7.15 Figure 7.16 Figure 7.17

of 0 as the product of two and of three consecutive integers. If we plug the coordinates of $\pm 4P$ into the equation then we get 6 on both sides. And we had two representations of 6 as the product of two and of three consecutive integers. So we have exhausted all of our known solutions, but we can continue adding points.

The reader undoubtedly understands the algebra and geometry by now, so we will simply state the results. We find $5P = (1/4, -5/8)$, a disappointment. Yes, it is a point on our elliptic curve, but the coordinates are not integers so it does not give us a solution to our problem. We find $6P = (6, 14)$ — success! Our equation is $y(y+1) = (x-1)x(x+1)$, so $14 \cdot 15 = 5 \cdot 6 \cdot 7$ and indeed both sides are 210. So we have a new solution to our problem. Let us continue: $7P = (-5/9, 8/27)$, $8P = (21/25, -69/125)$, $9P = (-20/49, -435/343)$, $10P = (161/16, -2065/64)$, $11P = (116/23^2, -3612/23^3)$, $12P = (1357/29^2, 28888/29^3)$. We are starting to get pessimistic since the denominators seem to be getting worse. They do not always get bigger—from $5P$ to $6P$ and from $9P$ to $10P$ they got smaller, but the overall trend seems to be that the denominators are getting worse. On the other hand, maybe there are points on our curve with integer coordinates that are not multiples of $P$.

To resolve this, we need to go back to the theory of elliptic curves. Let $\mathbf{Q}$ denote the rational numbers (those are fractions like $3/5$ and $-7/1$, but not $\sqrt{2}$ or $\pi$). Let $\mathcal{E}$ be an elliptic curve whose defining equation has coefficients that are in $\mathbf{Q}$ (like ours). Let $\mathcal{E}(\mathbf{Q})$ denote the points with coordinates in $\mathbf{Q}$. Mordell proved the following theorem.

**Theorem.** *There is always a finite set of points from which you can get any point in $\mathcal{E}(\mathbf{Q})$ via a finite sequence of line intersections (i.e., point additions).*

(For the more advanced reader, this says that $\mathcal{E}(\mathbf{Q})$ is a finitely-generated abelian group). As an example, let us take the curve $E_1$ given by $y^2 = x^3 + 17$. The points $R = (-2, 3)$ and $S = (2, 5)$ generate $E_1(\mathbf{Q})$. For example, $(4, 9) = R + (-S) = R - S$, $(-1, 4) = S - 2R$, $(5234, 378661) = 3S - 4R$ and $(137/64, -2651/512) = 2S - 4R$.

For our elliptic curve $E$ given by $y^2 + y = x^3 - x$, the points in $E(\mathbf{Q})$ are generated by $P = (0, 0)$. This is unfortunately very difficult to show. We will assume it and move on. This means that every point with rational coordinates is a multiple of $P$. Thus every point with integer coordinates is a multiple of $P$ and so will show up in the list $\pm P, \pm 2P, \pm 3P, \ldots$. In fact, by $6P$ they already have. In order to show that there are no more points with integer coordinates in the above list, we will prove that the denominators are, indeed, getting worse.

That last sentence was a number theoretic statement, so we will need some number theoretic notation. Let $m$ and $n$ be integers. We write $n|m$ if $m$ is a multiple of $n$ and we will say that $n$ divides $m$. Otherwise we write $n \nmid m$ to say $n$ does not divide $m$. So $3|12$ and $3 \nmid 11$. We will need three properties of divides.

i) If $n|m$ then $n|ml$. In other words, if $m$ is a multiple of $n$ then $ml$ is also.

ii) If $n|m$ and $n|l$ then $n|(m \pm l)$. As an example, if you add or subtract two multiples of 3, then the result will also be a multiple of 3.

iii) If $n|m$ but $n \nmid l$ then $n \nmid (m \pm l)$. As an example, if $m$ is a multiple of 3 but $l$ is not, then neither the sum nor the difference of $m$ and $l$ will be a multiple of 3.

We are interested in points with integer coordinates. Let $R = (x, y)$ be in $E(\mathbf{Q})$.

Claim: If one coordinate is an integer, then so is the other.

*Proof.* Let $R = (\frac{a}{b}, \frac{c}{d})$, with $a, b, c$ and $d$ integers, both $b$ and $d$ positive and both fractions in lowest terms. Each rational number can be written uniquely as a fraction in lowest terms with a positive denominator. Since $R$ is a point on our curve, its coordinates satisfy the equation. Plugging them into $y^2 + y = x^3 - x$ and simplifying we get

$$\frac{c^2 + cd}{d^2} = \frac{a^3 - ab^2}{b^3}.$$

Each of these fractions is in lowest terms also. Let us explain why. If the prime number $p$ divides $d^2$, then $p|d$ and so $p|cd$. But since $\frac{c}{d}$ is in lowest terms, $p \nmid c$ and so $p \nmid c^2$. Thus $p|cd$ but $p \nmid c^2$, so $p \nmid c^2 + cd$. So primes in the denominator of the left-hand fraction do not appear in the numerator, hence it is in lowest terms. A very similar argument works for the right-hand fraction also.

Since both fractions are in lowest terms, have positive denominators, and are equal, their denominators must be the same. So $d^2 = b^3$. Now if $d = 1$ then $b = 1$. And if $b = 1$ then (since $d$ is positive) $d = 1$. So if either $b$ or $d$ is 1, then so is the other. In other words, if either $x$ or $y$ is an integer, then so is the other. □

We will call a point with integer coordinates an integer point. Now we want to show that the denominators are getting worse.

Claim: If $R$ is in $E(\mathbf{Q})$ and is not an integer point then $2R$ is not either.

*Proof.* Let $R = (x_1, y_1)$ and $2R = (x_2, y_2)$. Repeating the steps involved in finding the coordinates of the double of a point, it can be shown that

$$x_2 = \frac{x_1^4 + 2x_1^2 - 2x_1 + 1}{4x_1^3 - 4x_1 + 1}$$

Let $x_1 = \frac{a}{b}$ be in lowest terms with $a$ and $b$ integers. Since $R$ is not an integer point, we have $b > 1$. Plugging $x_1 = \frac{a}{b}$ into the above formula we get

$$x_2 = \frac{1}{b}\left(\frac{a^4 + 2a^2b^2 - 2ab^3 + b^4}{4a^3 - 4ab^2 + b^3}\right).$$

Since $b > 1$ there is a prime number $p$ with $p|b$ and since $\frac{a}{b}$ is in lowest terms we have $p \nmid a$. Now $p$ does appear in the denominator of $x_2$ since $p|b$. And since $p|b$ we have $p|2a^2b^2 - 2ab^3 + b^4$. But $p \nmid a$ so $p \nmid a^4$ and thus $p \nmid a^4 + 2a^2b^2 - 2ab^3 + b^4$. So $p$ does not appear in the numerator. Thus $p$ is in the denominator of $x_2$ when written in lowest terms. So we see $2R$ is not an integer point. □

To prove that there are no more integer points, we need only look at the positive multiples of $P$ since the negative multiples have the same $x$-coordinates. We want to show that there are no more integer points among the positive multiples of $P$. For every positive integer, we can factor out 2's until we are left with an odd number. So every positive integer is of the form $2^n \cdot 1$, $2^n \cdot 3$ or $2^n \cdot m$ with $m$ odd and $m > 3$. Thus every positive multiple of $P$ is of the form $2^nP$, $2^n3P$ or $2^nmP$ with $m$ odd and $m > 3$.

Notice in Figure 7.17 that every odd multiple of $P$ is on the egg and every even multiple of $P$ is on the other piece. This will always be true. That is because by the line construction, adding $P$ to a multiple on one piece of the curve will give the next multiple on the other piece. Now the egg is very small and we have found all of the points with integer coordinates on the egg, namely $\pm P$ and $\pm 3P$. So if $m$ is odd and bigger than 3, then $mP$ is not an integer point. So $2mP$ is not an integer point and $4mP$ is not either. In fact $2^nmP$ can not be an integer point so there are no integer points in the list $2^nmP$ with $m$ odd and $m > 3$.

Earlier we found that $8P = (21/25, -69/125)$ is not an integer point. So $16P$ is not and $32P$ is not. So, for $n > 2$, $2^nP$ is not an integer point. Hence, there are no more integer points in the list $2^nP$. Earlier we found that $12P = (1357/29^2, 28888/29^3)$ is not an integer point. So $24P$ is not and $48P$ is not. So, for $n > 1$, $2^n 3P$ is not an integer point. Thus, there are no more integer points in the list $2^n 3P$. We have exhausted all of the possibilities, and so we have found them all. The only integer points are $\pm P$, $\pm 2P$, $\pm 3P$, $\pm 4P$ and $\pm 6P$. Therefore, the only integers that are simultaneously the product of two and of three consecutive integers are 0, 6 and 210.

This was first proved by Mordell in the article

L.J. Mordell, On the integer solutions of $y(y+1) = x(x+1)(x+2)$, *Pacific J. Math.,* **13** (1963), 1347–1351.

The method of solution described here follows a series of exercises in the book

J.H. Silverman, *The Arithmetic of Elliptic Curves*, Springer-Verlag, New York, 1986.

This book also contains the $y^2 = x^3 + 17$ example.

The most elementary book written on elliptic curves is

J.H. Silverman, and J. Tate, *Rational Points on Elliptic Curves*, Springer-Verlag, New York, 1992.

The most elementary books explaining the use of elliptic curves in cryptography are

N. Koblitz, *A Course in Number Theory and Cryptography*, Springer-Verlag, 1987.

and

W. Trappe, and L. Washington, *Introduction to Cryptography with Coding Theory*, Prentice Hall, 2002.

The author is grateful to Nicholas Tran for help on the figures and to Peter Ross for helpful comments on an earlier draft.

# 8

# Right Triangles and Elliptic Curves[1]

Karl Rubin
*Stanford University*

## 1 Introduction

In this lecture we would like to study, and answer as best we can, the following question.

**Question** *Suppose we are given a natural number d. Is there a right triangle with three rational sides and area equal to d?*

We will write $\mathbf{Z}$ for the set of integers $\{\ldots, -3, -2, -1, 0, 1, 2, 3, \ldots\}$, $\mathbf{Z}^+$ for the natural numbers (or positive integers) $\{1, 2, 3, \ldots\}$, and $\mathbf{Q}$ for the set of rational numbers (fractions) $\{a/b : a, b \in \mathbf{Z}, b \neq 0\}$. If $S$ is a set, $\#S$ will denote the number of elements in $S$.

The Pythagorean Theorem says that a triangle with sides $a \leq b \leq c$ is a right triangle if and only if $a^2 + b^2 = c^2$. If it is a right triangle, then its area is $ab/2$. Thus we can rephrase our question as follows.

**Question (restated)** *Suppose $d \in \mathbf{Z}^+$. Are there rational numbers $a, b, c > 0$ such that $a^2 + b^2 = c^2$ and $ab = 2d$?*

If the answer to this question is "yes," then $d$ is called a *congruent number* (not to be confused with the notion of "congruence modulo an integer," which is different).

Table 8.1 gives some examples.

| $d$ | $a$ | $b$ | $c$ |
|---|---|---|---|
| 5 | 3/2 | 20/3 | 41/6 |
| 6 | 3 | 4 | 5 |
| 7 | 35/12 | 24/5 | 337/60 |

**Table 8.1.** Rational right triangles with area $5, 6, 7$

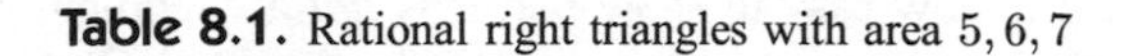

[1]Partially supported by NSF grant DMS-9800881

On the other hand, around 1640 Fermat proved that $d = 1$ is *not* a congruent number. That is, there is no right triangle with rational sides and area equal to 1. (Of course, this is true only because we insist that all three sides of the triangle are rational. The triangle with sides $1, 1, \sqrt{2}$ is a right triangle with area 1, but $\sqrt{2}$ is not a rational number.)

Note that to prove that a number $d$ *is* congruent, one just has to produce the three rational numbers $a$, $b$, $c$. On the other hand, it is not at all obvious how to prove that a given $d$ is *not* a congruent number.

Here is the best answer to our question that current mathematics can provide. It is easy to see that if $d$ and $t$ are natural numbers, than $d$ is a congruent number if and only if $dt^2$ is a congruent number (just scale the right triangle by $t$). Thus it is enough to answer our question for *squarefree* natural numbers $d$, those not divisible by a perfect square other than 1.

**Theorem 1 (Tunnell [8])** *Suppose d is a squarefree natural number, and define*

$$a = \begin{cases} 1 & \text{if } d \text{ is odd} \\ 2 & \text{if } d \text{ is even,} \end{cases}$$

$$n = \#\{(x, y, z) : x, y, x \in \mathbf{Z} \text{ and } x^2 + 2ay^2 + 8z^2 = d/a\},$$

$$m = \#\{(x, y, z) : x, y, z \in \mathbf{Z} \text{ and } x^2 + 2ay^2 + 32z^2 = d/a\}.$$

*If* $n \neq 2m$, *then d is not a congruent number.*

**Conjecture 2** *Suppose d is a squarefree natural number, and let a, n, and m be as in Theorem 1. If* $n = 2m$, *then d is a congruent number.*

(A *conjecture* is a statement that is believed to be true, but that nobody has been able to prove.)

Given $d$ (not too large), it is easy to compute the integers $n$ and $m$. Table 8.2 gives some examples. (Note that $n$ and $m$ are counting solutions $(x, y, x)$ where $x$, $y$, and $z$ are arbitrary integers, positive, negative, or zero). The last three columns of Table 8.2 reflect the fact that

| $d$ | 1 | 2 | 3 | 5 | 6 | 7 | 10 | 11 | 34 | $8k$+5 | $8k$+6 | $8k$+7 |
|---|---|---|---|---|---|---|---|---|---|---|---|---|
| $n$ | 2 | 2 | 4 | 0 | 0 | 0 | 4 | 12 | 8 | 0 | 0 | 0 |
| $m$ | 2 | 2 | 4 | 0 | 0 | 0 | 4 | 2 | 4 | 0 | 0 | 0 |

**Table 8.2.** Some values of $n$ and $m$ in Theorem 1

$x^2 + 2y^2$ can never leave a remainder of 5 or 7 when divided by 8, and $x^2$ can never leave a remainder of 3 when divided by 4. Thus if $d$ is 5, 6, or 7 more than a multiple of 8 then $n = m = 0$.

*Exercise* 1. Check the values of $n$ and $m$ in Table 8.2, and compute $n$ and $m$ for some other $d$.

It follows from Theorem 1 and Table 8.2 that 1, 2, 3, 10, and 11 are not congruent numbers. By Conjecture 2 and Table 8.2, the integer 34 and every integer that is 5, 6, or 7 more than a multiple of 8 *should be* congruent numbers. For $d = 5, 6$ or 7 we know this to be true, thanks to the examples in Table 8.1.

*Exercise* 2. Show that 34 is a congruent number by finding a rational right triangle of area 34.

Thanks to Theorem 1, it is now easy to show that $d$ is *not* a congruent number. But it can be difficult to show that $d$ *is* a congruent number, when Conjecture 2 predicts it to be. For example, Conjecture 2 predicts that $d = 157 = 19 \cdot 8 + 5$ is a congruent number, but the simplest rational right triangle with area 157 has sides

$$\frac{6803298487826435051217540}{411340519227716149383203}, \frac{411340519227716149383203}{21666555693714761309610},$$
$$\frac{224403517704336969924557513090674863160948472041}{8912332268928859588025535178967163570016480830},$$

(see [5], p. 5). Similarly $d = 1063 = 132 \cdot 8 + 7$ is a congruent number, but the simplest rational right triangle with area 1063 has shortest side $a$ where the numerator of $a$ has 104 digits and the denominator has 103 digits ([4][2]).

## 2 Translating the question

In the rest of this lecture we will explain where Theorem 1 and Conjecture 2 come from. The first step is to restate our question in a different form.

*Exercise* 3. Show that if $a, b, c$ are the sides of a rational right triangle with area $d$, then

$$x = \frac{a-c}{b}, \quad y = \frac{2(a-c)}{b^2}$$

are nonzero rational numbers satisfying the equation $dy^2 = x^3 - x$.

*Exercise* 4. Conversely, if $x$ and $y$ are rational numbers satisfying the equation $dy^2 = x^3 - x$ and $y \neq 0$, then

$$\left|\frac{x^2-1}{y}\right|, \quad \left|\frac{2x}{y}\right|, \quad \left|\frac{x^2+1}{y}\right|$$

are the sides of a rational right triangle with area $d$.

Note that the equation

$$dy^2 = x^3 - x \tag{1}$$

has three solutions $(0,0)$, $(1,0)$, and $(-1,0)$ with $y = 0$. We will call these the *trivial* solutions, and we will call a solution of (1) *nontrivial* if $y \neq 0$. Combining the two exercises above gives the following reformulation of the definition of congruent number.

**Proposition 3** *A squarefree natural number d is congruent if and only if the equation* $dy^2 = x^3 - x$ *has a nontrivial rational solution* $(x, y)$.

## 3 Elliptic curves

Fix a squarefree natural number $d$. The equation

$$dy^2 = x^3 - x \tag{2}$$

defines an *elliptic curve*.[3] We will call this elliptic curve $E_d$, and by a rational point on $E_d$ we will mean a pair of rational numbers $(x, y)$ satisfying (2). For more about elliptic curves see [5], [6], [7]. We will continue our investigation of the congruent number problem by applying results from the general theory of elliptic curves. The most important of these is the following process for using two points on $E_d$ to construct a third point.

Suppose that $P$ and $Q$ are two points on $E_d$.

[2]There is an error in [4]; on page 130 the expression before the displayed value of $X$ should read $X^2/1063$.

[3]One usually requires an elliptic curve to be a curve given by an equation $y^2 = x^3 + Ax + B$. Under the change of variables $(x, y) \mapsto (x/d, y/d^2)$ our elliptic curve is equivalent to $y^2 = x^3 - d^2x$.

- Draw the line through $P$ and $Q$.
- In general this line will intersect the curve $E_d$ in three points (because equation (2) has degree 3). Two of these points are $P$ and $Q$; call the third point $R$. Since $P$ and $Q$ are rational points, $R$ will be a rational point as well.

We also want to apply this construction when $P = Q$. In that case by "line through $P$ and $Q$" we mean the tangent line to the curve $E_d$ at $P$ (the limiting line as $Q$ approaches $P$ on $E_d$).

*Exercise 5.* If $P = (x_1, y_1)$ and $Q = (x_2, y_2)$ with $x_1 \neq x_2$, check that

$$R = \left( d\frac{(y_2 - y_1)^2}{(x_2 - x_1)^2} - x_1 - x_2, d\frac{(y_2 - y_1)^3}{(x_2 - x_1)^3} + \frac{x_1y_1 - x_2y_2 + 2(x_2y_1 - x_1y_2)}{x_2 - x_1} \right).$$

If $P = Q = (x, y)$, check that

$$R = \left( \frac{(x^2 + 1)^2}{4x(x^2 - 1)}, -\frac{x^6 - 5x^4 - 5x^2 + 1}{8d^3y^3} \right).$$

For example, if $P = (-1, 0)$ and $Q = (0, 0)$, then $R = (1, 0)$.

It is not obvious, but one can use this construction to prove the following theorem.

**Theorem 4** *If the elliptic curve $E_d$ has a nontrivial rational point (i.e., a rational point with $y \neq 0$), then it has infinitely many rational points.*

Figure 8.1 shows some of the rational points on $E_6$, which can be found by starting with $(-\frac{1}{2}, \frac{1}{4})$ and using the construction above.

Using Exercise 4 and Proposition 3 this leads to the following corollaries.

**Corollary 5** *If there is a rational right triangle with area d, then there are infinitely many rational right triangles with area d.*

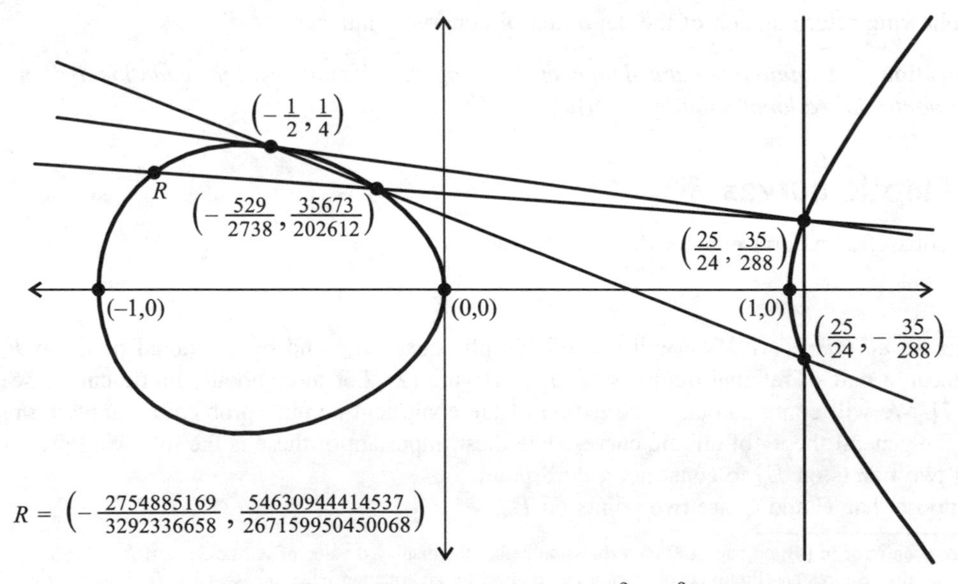


**Figure 8.1.** Some rational points on $6y^2 = x^3 - x$.

**Corollary 6** *A squarefree natural number $d$ is congruent if and only if $E_d$ has infinitely many rational points.*

Table 8.3 lists the shortest side of some rational right triangles with area 6. (If we call the short side $a$, then $b = 12/a$ and $c = \sqrt{a^2 + b^2}$ are the other two sides.) These were computed using the formulas of Exercises 4 and 5. Note the parabolic shape of the table.

$$3$$

$$\frac{7}{10}$$

$$\frac{3404}{1551}$$

$$\frac{2017680}{1437599}$$

$$\frac{3122541453}{2129555051}$$

$$\frac{43690772126393}{20528380655970}$$

$$\frac{353847840904157040 4}{46440507850340968 01}$$

$$\frac{1214980735300888708857264 0}{415611880854896794176960 1}$$

$$\frac{5628773075353652252514840840 03}{9096802581030701081135787921001}$$

$$\frac{98036059631049308485775054091376224 0600}{31849720982909420672712416881546090 0807}$$

$$\frac{1819157495197128710444993870521048471797359 8996}{2850984812127142751980727458173233067600976 0751}$$

$$\frac{219291389196040469380401637407576189535221272585678 18399}{969596010399029433102598494384114956082566977513816 8420}$$

$$\frac{1076784912325042146290273662036091437066100455618812531478882 27347}{803047890581182290757365789767280596270396579819644619336229 42851}$$

$$\frac{638686275366681889789788669730841297975070777160618154198122040800885900 7160}{417650183130159383654288534276869863228771421483222833898076529253870635 8393}$$

Table 8.3. The shortest side of some right triangles with area 6

## 4 Counting points modulo $p$

Thanks to Corollary 6, we can solve the congruent number problem if we can find a way to decide whether a given elliptic curve $E_d$ has infinitely many rational points, or only the three trivial rational points $\{(-1,0),(0,0),(1,0)\}$. There is currently no known way to do this in general, but we will be able to make a good start and decide in many cases.

Instead of asking the difficult question "how often is $dy^2 - (x^3 - x)$ equal to zero?," we will ask, for each prime number $p$, "how often is $dy^2 - (x^3 - x)$ a multiple of $p$?" (of course when it is zero, it is a multiple of $p$ for every $p$).

More precisely, for every prime number $p$ define an integer

$$N_p(d) = \#\{(x, y) : 0 \le x, y < p,\ dy^2 - (x^3 - x) \text{ is a multiple of } p\} + 1.$$

For example, when $d = 1$ we have $N_5(1) = 8$ because the 7 points

$$\{(0,0), (1,0), (2,1), (2,4), (3,2), (3,3), (4,0)\}$$

all have the property that $dy^2 - (x^3 - x)$ is a multiple of 5, and no other pairs in the appropriate range have this property. Table 8.4 lists the values of $N_p(d)$ for some primes $p$, and certain $d$.

| | | $p$ | | | | | | | | |
|---|---|---|---|---|---|---|---|---|---|---|
| | | 5 | 7 | 11 | 13 | 17 | 19 | 1000003 | 1000033 | 1000037 |
| | 1 | 8 | 8 | 12 | 8 | 16 | 20 | 1000004 | 998208 | 998056 |
| $d$ | 2 | 4 | 8 | 12 | 20 | 16 | 20 | 1000004 | 998208 | 1002020 |
| | 3 | 4 | 8 | 12 | 8 | 20 | 20 | 1000004 | 998208 | 1002020 |

**Table 8.4.** Some values of $N_p(d)$

*Exercise* 6. Check some of the values in Table 8.4.

**Idea (Birch and Swinnerton-Dyer [1])** *Suppose $d$ is a squarefree positive integer. The more rational points $E_d$ has, the larger the $N_p(d)$ will be "on average", as $p$ varies.*

To make sense of this,[4] we need to measure the "average" size of the $N_p(d)$, for fixed $d$, as $p$ varies. Birch and Swinnerton-Dyer computed, for large values of $X$, the product over all primes $p < X$ of $N_p(d)/p$:

$$\pi_d(X) = \prod_{p<X} \frac{N_p(d)}{p}.$$

Figure 8.2 shows the function $\pi_d(X)$ for $d = 1, 2, 3, 5, 6, 7$, with $X$ plotted on a logarithmic scale. As the idea of Birch and Swinnerton-Dyer suggests, $\pi_d(X)$ is larger for the three congruent numbers $d = 5, 6, 7$ than for the non-congruent numbers $d = 1, 2, 3$.

Birch and Swinnerton-Dyer observed that there is a better way to measure the average size of the $N_p(d)$. Hasse defined a function of a complex variable attached to $E_d$, called the $L$-function of $E_d$, by the following infinite product over all primes $p$ not dividing $2d$:

$$L(E_d, s) = \prod_{p \nmid 2d} \left(1 - (p + 1 - N_p(d))p^{-s} + p^{1-2s}\right)^{-1}. \tag{3}$$

This infinite product converges if the real part $\mathrm{Re}(s)$ of the complex number $s$ is bigger than $3/2$, and there is a natural way to extend the function $L(E_d, s)$ to all complex numbers $s$ (called the analytic continuation: the only function defined by a convergent power series in $s$ that agrees with $L(E_d, s)$ when $\mathrm{Re}(s) > 3/2$).

Note that *formally*, if we put $s = 1$ in the infinite product (3) we get $\prod_{p \nmid 2d} \frac{p}{N_p(d)}$, which is essentially the limit of the values $\pi_d(X)^{-1}$, if that limit exists. This doesn't *prove* a connection between $L(E_d, 1)$ and the $\pi_d(X)$, because the product (3) need not converge at $s = 1$, but it led Birch and Swinnerton-Dyer to the following conjecture.

[4] Because we are getting closer to the frontier of current research, more background and experience may be required to read the rest of this section. Don't be discouraged if you aren't yet familiar with all of these concepts, just try to get a feel for the ideas involved.

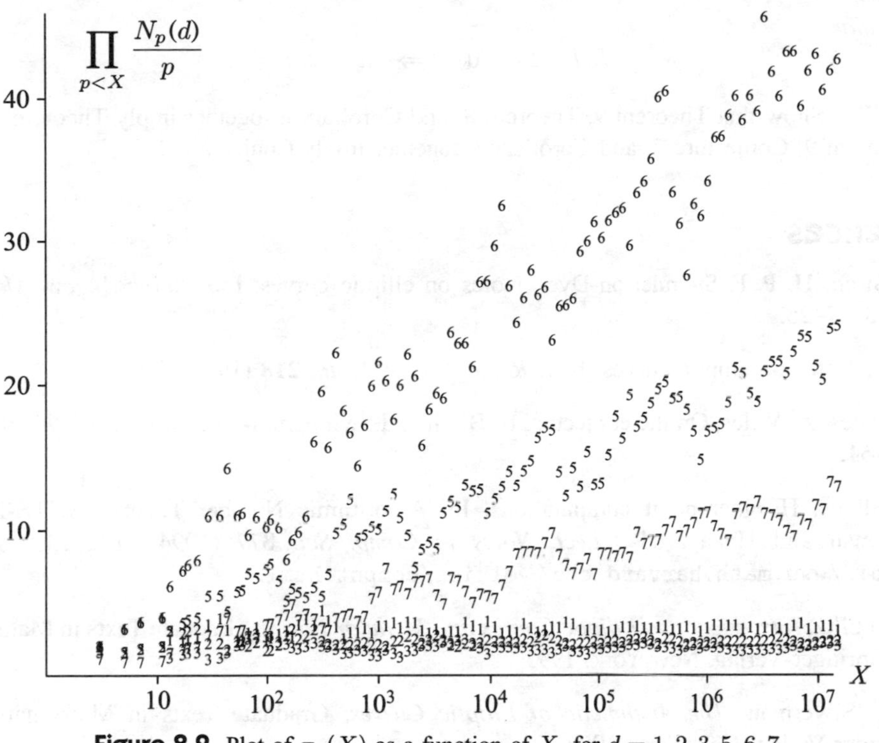


**Figure 8.2.** Plot of $\pi_d(X)$ as a function of $X$, for $d = 1, 2, 3, 5, 6, 7$.

**Conjecture 7 (Birch and Swinnerton-Dyer)** *The elliptic curve $E_d$ has infinitely many rational points if and only if $L(E_d, 1) = 0$.*

Equivalently (by Corollary 6), the Birch and Swinnerton-Dyer conjecture predicts that $d$ is a congruent number if and only if $L(E_d, 1) = 0$.

**Theorem 8 (Coates and Wiles [3])** *If $E_d$ has infinitely many rational points, then $L(E_d, 1) = 0$.*

Unfortunately it is still an open problem to prove the converse (that if $L(E_d, 1) = 0$ then $E_d$ has infinitely many rational points).

We next need a good way to evaluate $L(E_d, 1)$. This is provided by the following theorem.

**Theorem 9 (Tunnell [8])** *If $d$ is a squarefree positive integer, then*

$$L(E_d, 1) = \frac{(n - 2m)^2 a\Omega}{16\sqrt{d}}$$

*where*

$a = 1$ *if $d$ is odd,* $a = 2$ *if $d$ is even,*

$n = \#\{(x, y, z) : x, y, z \in \mathbf{Z}$ *and* $x^2 + 2ay^2 + 8z^2 = d/a\}$,

$m = \#\{(x, y, z) : x, y, z \in \mathbf{Z}$ *and* $x^2 + 2ay^2 + 32z^2 = d/a\}$,

$$\Omega = \int_1^\infty \frac{dx}{\sqrt{x^3 - x}} \approx 2.6220575542921198\ldots$$

*In particular*

$$L(E_d, 1) = 0 \iff n = 2m.$$

*Exercise* 7. Show that Theorem 9, Theorem 8, and Corollary 6 together imply Theorem 1. Show that Theorem 9, Conjecture 7, and Corollary 6 together imply Conjecture 2.

## References

[1] B. Birch, H. P. F. Swinnerton-Dyer, Notes on elliptic curves. I, *J. Reine Angew. Math.* **212** (1963), 7–25.

[2] ——, Notes on elliptic curves. II, *J. Reine Angew. Math.* **218** (1965), 79–108.

[3] J. Coates, A. Wiles, On the conjecture of Birch and Swinnerton-Dyer, *Invent. Math.* **39** (1977), 223–51.

[4] N. Elkies, Heegner point computations. In: Algorithmic Number Theory (ANTS-I) 1994, Adleman and Huang, eds., *Lect. Notes in Comp. Sci.* **877** (1994), 122–133; see also `http://www.math.harvard.edu/~elkies/compnt.html`.

[5] N. Koblitz, *Introduction to Elliptic Curves and Modular Forms,* Graduate Texts in Mathematics **97**, Springer-Verlag, New York, 1993.

[6] J. H. Silverman, *The Arithmetic of Elliptic Curves,* Graduate Texts in Mathematics **106**, Springer-Verlag, New York, 1986.

[7] J. H. Silverman, J. Tate, *Rational Points on Elliptic Curves,* Undergraduate Texts in Mathematics, Springer-Verlag, New York, 1992.

[8] J. B. Tunnell, A classical Diophantine problem and modular forms of weight $3/2$, *Invent. Math.* **72** (2) (1983), 323–334.

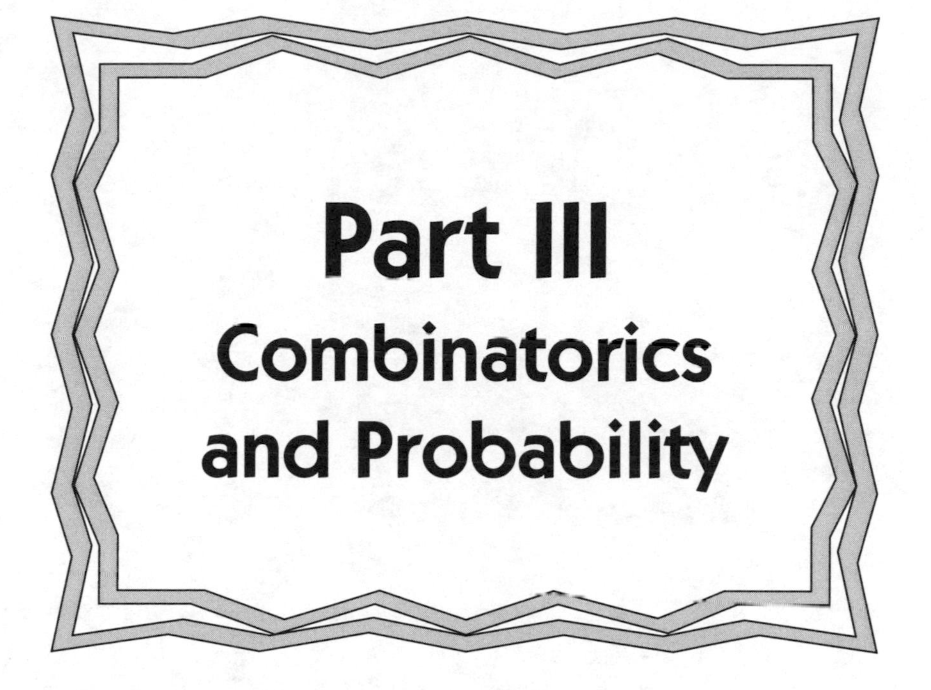

# Part III
## Combinatorics and Probability

# 9

# Proofs that Really Count: The Magic of Fibonacci Numbers and More

Arthur T. Benjamin & Jennifer J. Quinn
*Harvey Mudd College* *Occidental College*

## A Magic Trick

A mathemagician hands a sheet of paper as in Figure 9.1 to a volunteer and says, "Secretly write a positive integer in Row 1 and another positive integer in Row 2. Next, add those numbers together and put the sum in Row 3. Add Row 2 to Row 3 and place the answer in Row 4. Continue in this fashion until numbers are in Rows 1 through 10. Now, using a calculator if you wish, add all the numbers in Rows 1 through 10 together." While the spectator is adding, the mathemagician glances at the sheet of paper for just a second, then instantly reveals the total. "Now using a calculator, divide the number in Row 10 by the number in Row 9, and announce the first three digits of your

| 1 | |
|---|---|
| 2 | |
| 3 | |
| 4 | |
| 5 | |
| 6 | |
| 7 | |
| 8 | |
| 9 | |
| 10 | |
| TOTAL | |

**Figure 9.1.** Enter positive integers in Rows 1 and 2. The number in each successive row is the sum of the numbers in the previous two rows.

| | |
|---|---|
| 1 | $x$ |
| 2 | $y$ |
| 3 | $x+y$ |
| 4 | $x+2y$ |
| 5 | $2x+3y$ |
| 6 | $3x+5y$ |
| 7 | $\mathbf{5x+8y}$ |
| 8 | $8x+13y$ |
| 9 | $13x+21y$ |
| 10 | $21x+34y$ |
| TOTAL | $\mathbf{55x+88y}$ |

**Figure 9.2.** The sum of the 10 numbers is Row 7 times 11.

answer. What's that you say? 1.61? Now turn over the paper and look what I have written." The back of the paper says "I predict the number 1.61".

A direct explanation of this trick involves nothing more than high school algebra. For the first part, observe in Figure 9.2 that if Row 1 contains $x$ and Row 2 contains $y$ then the total of Rows 1 through 10 will sum to $55x + 88y$. As luck (or is it something more?) would have it, the number in Row 7 is $5x + 8y$. Consequently, the grand total is simply 11 times Row 7, and with practice, even large numbers can be mentally multiplied by eleven.

As for the ratio, it's all about adding fractions badly. For any two fractions $\frac{a}{b} < \frac{c}{d}$ with positive numerators and denominators, the quantity $\frac{a+c}{b+d}$ is called the *mediant* (sometimes called the *freshman sum*) and it's easy to show that

$$\frac{a}{b} < \frac{a+c}{b+d} < \frac{c}{d}.$$

Consequently, the ratio of (Row 10)/(Row 9) satisfies

$$1.615\ldots = \frac{21}{13} = \frac{21x}{13x} < \frac{21x+34y}{13x+21y} < \frac{34y}{21y} = \frac{34}{21} = 1.619\ldots$$

This magic trick is an application of some special properties of the Fibonacci numbers 1, 1, 2, 3, 5, 8, 13, 21, 34, 55, 89,..., where each number is the sum of the previous two. Their many beautiful patterns are a constant source of amazement. For instance, the magic trick above is assisted by the fact that the sum of the first $n$ Fibonacci numbers is one less than the $(n+2)$nd Fibonacci number. Here, we reveal interpretations of Fibonacci numbers and related sequences to demystify their secrets—requiring nothing more than the ability to count.

## Fibonacci Numbers

How many sequences of 1's and 2's sum to $n$? Let's call the answer to this counting question $f_n$. For example, $f_4 = 5$ since 4 can be created in the following 5 ways:

$$1+1+1+1, \quad 1+1+2, \quad 1+2+1, \quad 2+1+1, \quad 2+2.$$

| 1 | 2 | 3 | 4 | 5 | 6 |
|---|---|---|---|---|---|
| 1 | 11 | 111 | 1111 | 11111 | 111111 |
| | 2 | 12 | 112 | 1112 | 11112 |
| | | 21 | 121 | 1121 | 11121 |
| | | | 211 | 1211 | 11211 |
| | | | 22 | 122 | 1122 |
| | | | | 2111 | 12111 |
| | | | | 212 | 1212 |
| | | | | 221 | 1221 |
| | | | | | 21111 |
| | | | | | 2112 |
| | | | | | 2121 |
| | | | | | 2211 |
| | | | | | 222 |
| $f_1 = 1$ | $f_2 = 2$ | $f_3 = 3$ | $f_4 = 5$ | $f_5 = 8$ | $f_6 = 13$ |

**Table 9.1.** $f_n$ and the sequence of 1's and 2's summing to $n$ for $n = 1, 2, \ldots, 6$.

Table 9.1 illustrates the values of $f_n$ for small $n$. The pattern is unmistakable; $f_n$ begins like the Fibonacci numbers. In fact, it will continue to grow like Fibonacci numbers, that is for $n > 2$, $f_n$ satisfies

$$f_n = f_{n-1} + f_{n-2}.$$

To see this, we consider the first number in our sequence. If the first number is 1, the rest of the sequence sums to $n - 1$, so there are $f_{n-1}$ ways to complete the sequence. If the first number is 2, there are $f_{n-2}$ ways to complete the sequence. Hence, $f_n = f_{n-1} + f_{n-2}$.

For our purposes, we prefer a more visual representation of $f_n$. By considering the 1's as representing *squares* and the 2's as representing *dominoes*, $f_n$ counts the number of ways to *tile* a board of length $n$ with squares and dominoes. For simplicity, we call a length $n$ board an *n-board.* Thus $f_4 = 5$ enumerates the tilings given in Figure 9.3.

**Figure 9.3.** All five square-domino tilings of the 4-board.

We let $f_0 = 1$ count the empty tiling of the 0-board. Thus for $n \geq 0$, we have a combinatorial interpretation of the $n$th Fibonacci number:

*$f_n$ counts the number of ways to tile a length $n$ board with squares and dominoes.*

This interpretation allows many Fibonacci identities to be proved by asking a counting question and answering it in two different ways. Since both expressions are answers to the same question, they must be equal. For example, the sum of consecutive Fibonacci numbers can be explained as follows:

**Identity 1** $f_0 + f_1 + f_2 + \cdots + f_n = f_{n+2} - 1$

**Question:** How many tilings of an $(n + 2)$-board use at least one domino?

**Answer 1:** There are $f_{n+2}$ tilings of an $(n+2)$-board. Excluding the "all square" tiling gives $f_{n+2} - 1$ tilings with at least one domino.

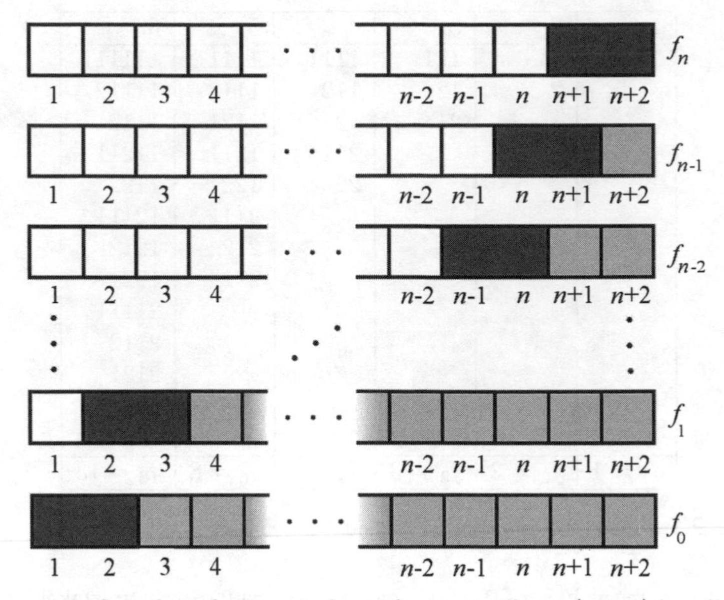


**Figure 9.4.** To see that $f_0 + f_1 + f_2 + \cdots + f_n = f_{n+2} - 1$, tile an $(n+2)$-board with squares and dominoes and consider the location of the last domino.

**Answer 2:** Consider the location of the last domino. There are $f_k$ tilings where the last domino covers cells $k+1$ and $k+2$. This is because cells 1 through $k$ can be tiled in $f_k$ ways, cells $k+1$ and $k+2$ must be covered by a domino, and cells $k+3$ through $n+2$ must be covered by squares. Hence the total number of tilings with at least one domino is $f_0 + f_1 + f_2 + \cdots + f_n$. See Figure 9.4.

Since our logic is impeccable in both answers, they must be equal and the identity follows.

Many Fibonacci identities depend on the notion of breakability at a given cell. We say that a tiling of an $n$-board is *breakable* at cell $k$, if the tiling can be broken into two tilings, one covering cells 1 through $k$ and the other covering cells $k+1$ through $n$. On the other hand, we call a tiling *unbreakable* at cell $k$ if a domino occupies cells $k$ and $k+1$. For example, the tiling of the 10-board in Figure 9.5 is breakable at cells $1, 2, 3, 5, 7, 8, 10$, and unbreakable at cells $4, 6, 9$. Notice that the tiling of an $n$-board (henceforth abbreviated an *$n$-tiling*) is always breakable at cell $n$. We apply these ideas to the next identity.

**Identity 2** $f_{m+n} = f_m f_n + f_{m-1} f_{n-1}$.

**Question:** How many tilings of an $(m+n)$-board exist?

**Answer 1:** There are $f_{m+n}$ $(m+n)$-tilings.

**Answer 2:** Consider breakability at cell $m$.

An $(m+n)$-tiling that is breakable at cell $m$ is created from an $m$-tiling followed by an $n$-tiling. There are $f_m f_n$ of these.

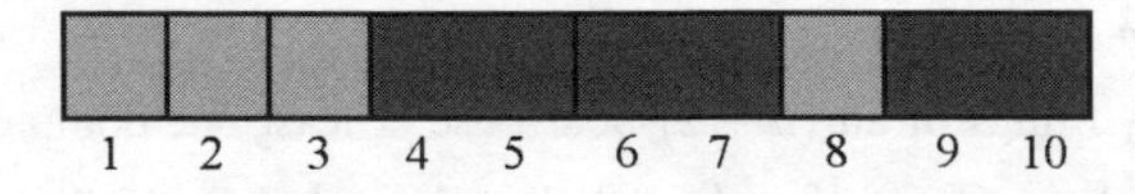


**Figure 9.5.** A 10-tiling that is breakable at cells 1, 2, 3, 5, 7, 8, 10 and unbreakable at cells 4, 6, 9.

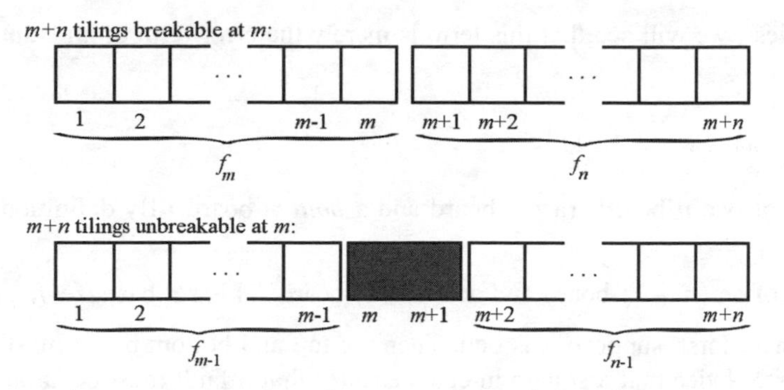


**Figure 9.6.** To prove $f_{m+n} = f_m f_n + f_{m-1} f_{n-1}$ count $(m+n)$-tilings based on whether or not they are breakable or unbreakable at $m$.

An $(m+n)$-tiling that is unbreakable at cell $m$ must contain a domino covering cells $m$ and $m+1$. So the tiling is created from an $(m-1)$-tiling followed by a domino followed by an $(n-1)$-tiling. There are $f_{m-1}f_{n-1}$ of these.

Since a tiling is either breakable or unbreakable at cell $m$, there are $f_m f_n + f_{m-1} f_{n-1}$ tilings altogether. See Figure 9.6.

Another combinatorial proof technique is to interpret both sides of an identity as sizes of two different sets and then find a one-to-one correspondence between them. We apply this idea to the next identity and we introduce the useful technique of *tail swapping*.

Consider the two 10-tilings offset as in Figure 9.7. The first one tiles cells 1 through 10; the second one tiles cells 2 through 11. We say that there is a *fault* at cell $i$, $2 \leq i \leq 10$, if both tilings are breakable at cell $i$. We say there is a fault at cell 1 if the first tiling is breakable at cell 1. Put another way, the pair of tilings has a fault at cell $i$, $1 \leq i \leq 10$, if neither tiling has a domino covering cells $i$ and $i+1$. The pair of tilings in Figure 9.7 has faults at cells 1, 2, 5, and 7. We define the *tail* of a tiling to be the tiles that occur after its last fault. Observe that if we swap the tails of Figure 9.7 we obtain the 11-tiling and the 9-tiling in Figure 9.8, and it has the same faults.

Looking at Identity 3, it may appear that the $(-1)^n$ term prevents us from proving it combina-

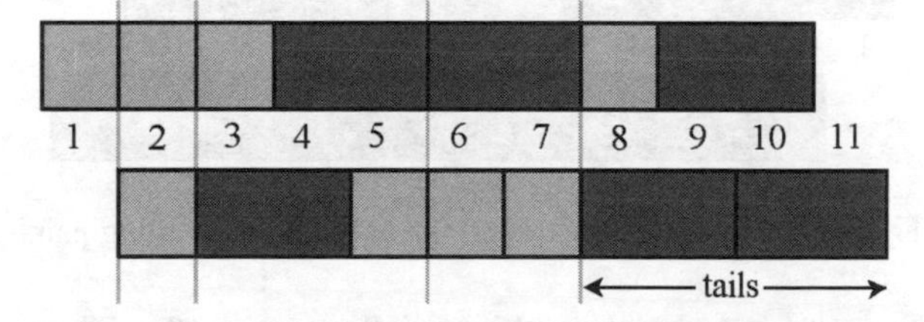


**Figure 9.7.** Two 10-tilings with their faults (indicated with gray lines) and tails.

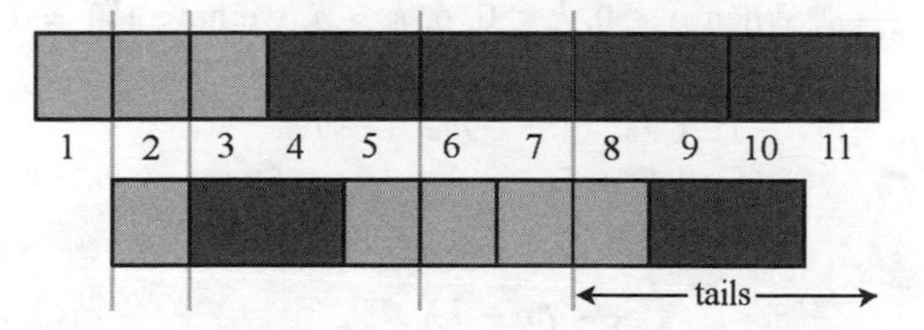


**Figure 9.8.** After tail-swapping, we have an 11-tiling and a 9-tiling with exactly the same faults.

torially. Nonetheless, we will see that this term is merely the error term of an "almost" one-to-one correspondence.

**Identity 3** $f_n^2 = f_{n+1}f_{n-1} + (-1)^n$

**Set 1:** Tilings of two $n$-boards (a *top* board and a *bottom* board.) By definition, this set has size $f_n^2$.

**Set 2:** Tilings of an $(n+1)$-board and an $(n-1)$-board. This set has size $f_{n+1}f_{n-1}$.

**Correspondence:** First, suppose $n$ is odd. Then the top and bottom board must each have at least one square. Notice that a square in cell $i$ ensures that a fault must occur at cell $i$ or cell $i-1$. Swapping the tails of the two $n$-tilings produces an $(n+1)$-tiling and an $(n-1)$-tiling with the same tails. This produces a 1-to-1 correspondence between all pairs of $n$-tilings and all tiling pairs of sizes $n+1$ and $n-1$ that have faults. Is it possible for a tiling pair of sizes $n+1$ and $n-1$ to be "fault free"? Yes, when all the dominoes are in "staggered formation" as in Figure 9.9. Thus, when $n$ is odd, $f_n^2 = f_{n+1}f_{n-1} - 1$.

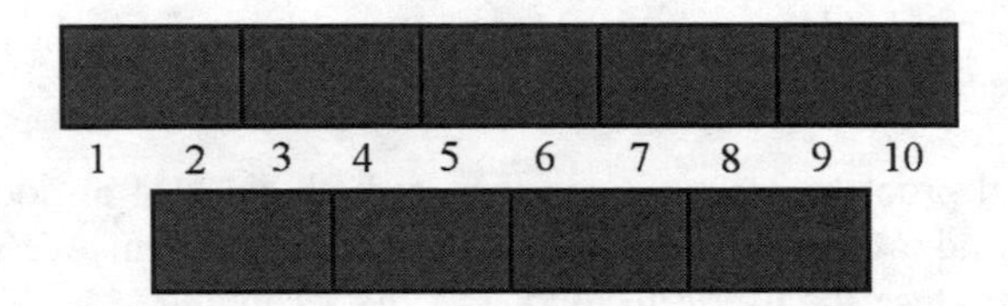


**Figure 9.9.** When $n$ is odd, the only fault-free tiling pair.

Similarly, when $n$ is even, tail swapping creates a 1-to-1 correspondence between faulty tiling pairs. The only fault-free tiling pair is the all domino tiling of Figure 9.10. Hence, $f_n^2 = f_{n+1}f_{n-1} + 1$. Considering the odd and even cases together produces our identity.

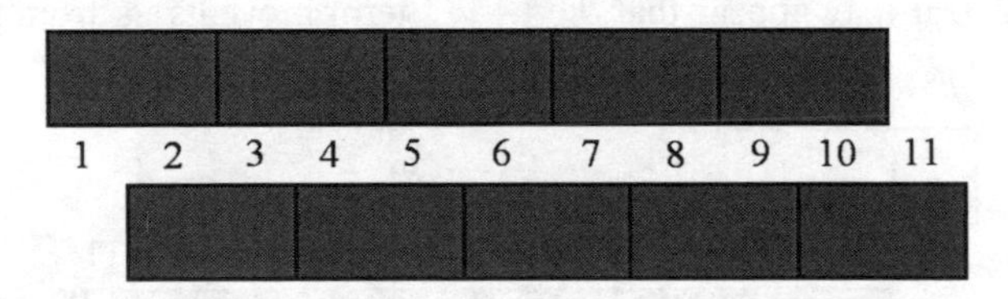


**Figure 9.10.** When $n$ is even, the only fault-free tiling pair.

We invite readers to try their hand at combinatorially proving the Fibonacci identities below. Recall that $\binom{n}{k}$ counts the number of ways to select $k$ elements from an $n$ element set. For $0 \le k \le n$, $\binom{n}{k} = \frac{n!}{k!(n-k)!}$. When $n < 0$, $k < 0$, or $k > n$, we have $\binom{n}{k} = 0$.

$$f_1 + f_3 + \cdots + f_{2n-1} = f_{2n} - 1.$$

$$f_0 + f_2 + f_4 + \cdots + f_{2n} = f_{2n+1}.$$

$$f_{n+2} + f_{n-2} = 3f_n.$$

$$\sum_{k=0}^{n} \binom{n-k}{k} = f_n.$$

$$\sum_{i=0}^{n}\sum_{j=0}^{n}\binom{n-i}{j}\binom{n-j}{i} = f_{2n+1}.$$

$$f_n + f_{n-1} + f_{n-2} + 2f_{n-3} + 4f_{n-4} + 8f_{n-5} + \cdots + 2^{n-2}f_0 = 2^n.$$

## Lucas Numbers

Close companions to the Fibonacci numbers, are the Lucas numbers 2, 1, 3, 4, 7, 11, 18, 29, 47, 76, 123,..., where each term is the sum of the previous two terms but the initial conditions are different. As we shall see Lucas numbers operate like Fibonacci numbers running in circles.

Let us combinatorially define $L_n$ to be the number of ways to tile a circular board of length $n$ with (slightly curved) squares and dominoes. For example $L_4 = 7$ as illustrated in Figure 9.11. Clearly there are more ways to tile a *circular n-board* than a straight $n$-board since it is now possible for a single domino to cover cells $n$ and 1. We define an *n-bracelet* to be a tiling of a circular $n$-board. A bracelet is *out-of-phase* when a single domino covers cells $n$ and 1 and *in-phase* otherwise. In Figure 9.11, we see that there are 5 in-phase 4-bracelets and 2 out-of-phase 4-bracelets. Figure 9.12 illustrates that $L_1 = 1$, $L_2 = 3$, and $L_3 = 4$. Notice that there are two ways to create a 2-bracelet with a single domino – either in-phase or out-of-phase.

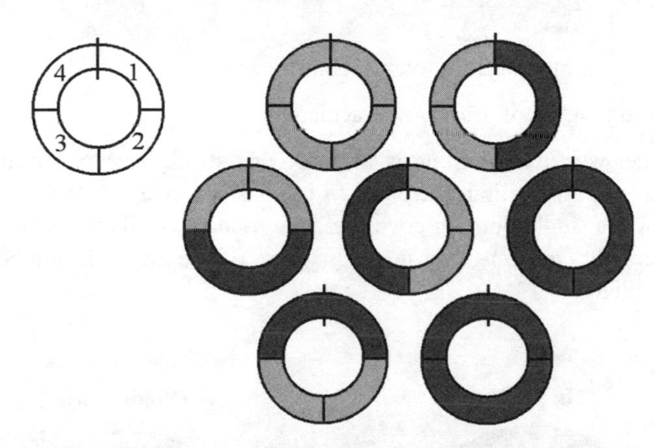


**Figure 9.11.** A circular 4-board and its 7 bracelets. The first 5 bracelets are in-phase and the last 2 are out-of-phase.

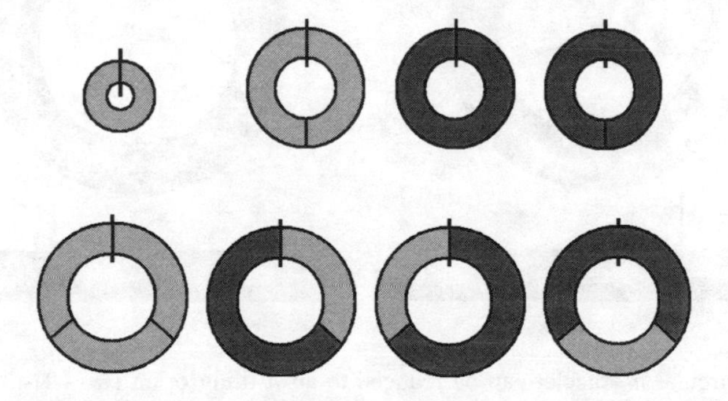

**Figure 9.12.** There are one 1-bracelets, three 2-bracelets, and four 3-bracelets.

From our initial data, the number of $n$-bracelets looks like the Lucas sequence. To prove that they continue to grow like the Lucas sequence, we must argue that for $n \geq 3$,

$$L_n = L_{n-1} + L_{n-2}.$$

To see this we simply consider the *last tile* of the bracelet. We define the *first* tile to be the tile that covers cell 1, which could either be a square, a domino covering cells 1 and 2, or a domino covering cells $n$ and 1. The second tile is the next tile in the clockwise direction, and so on. The last tile is the one that precedes the first tile. Since it is the first tile, not the last, that determines the phase of the tiling, there are $L_{n-1}$ $n$-bracelets that end with a square and $L_{n-2}$ $n$-bracelets that end with a domino. By removing the last tile, we produce smaller bracelets.

To make the recurrence valid for $n = 2$, we define $L_0 = 2$, and interpret this to mean that there are two empty tilings of the circular 0-board, an in-phase 0-bracelet and an out-of-phase 0-bracelet. Thus for $n \geq 0$, we have a combinatorial interpretation of the $n$th Lucas number:

*$L_n$ counts the number of ways to tile a circular board of length $n$ with squares and dominoes.*

As one might expect, there are many identities with Lucas numbers that resemble Fibonacci identities. In addition, there are many beautiful identities where Lucas and Fibonacci numbers interact.

**Identity 4** $L_n = f_n + f_{n-2}$.

**Question:** How many tilings of a circular $n$-board exist?

**Answer 1:** By definition, there are $L_n$ $n$-bracelets.

**Answer 2:** Consider whether the tiling is in-phase or out-of-phase. Since an in-phase tiling can be straightened into an $n$-tiling, there are $f_n$ in-phase bracelets. Likewise, an out-of-phase $n$-bracelet must have a single domino covering cells $n$ and 1. Cells 2 through $n-1$ can then be covered as a straight $(n-2)$-tiling in $f_{n-2}$ ways. Hence the total number of $n$-bracelets is $f_n + f_{n-2}$. See Figure 9.13.

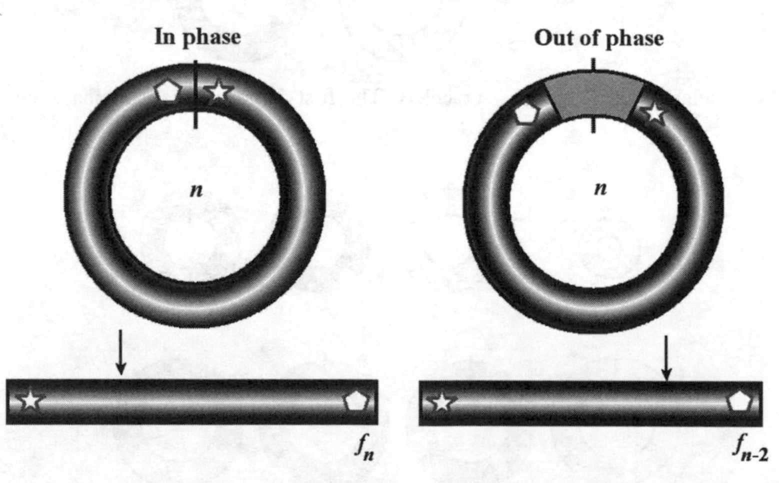


**Figure 9.13.** Every circular $n$-bracelet can be reduced to an $n$-tiling or an $(n-2)$-tiling, depending on its phase.

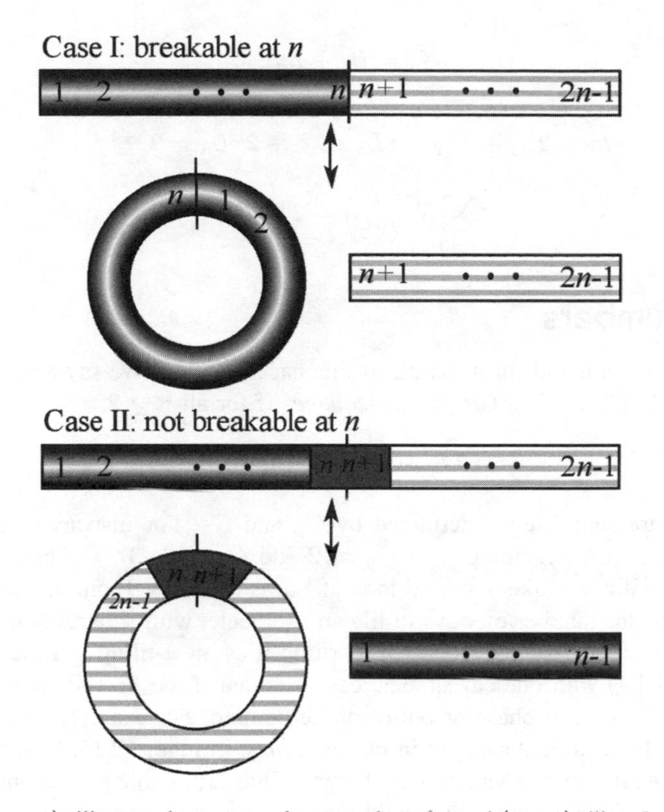


**Figure 9.14.** A $(2n-1)$-tiling can be converted to an $n$-bracelet and $(n-1)$-tiling. In our correspondence, the $n$-bracelet is in-phase if and only if the $(2n-1)$-tiling is breakable at cell $n$.

**Identity 5** $f_{2n-1} = L_n f_{n-1}$.

**Set 1:** Tilings of a $(2n-1)$-board. This set has size $f_{2n-1}$.

**Set 2:** Bracelet-tiling pairs $(B, T)$ where the bracelet has length $n$ and the tiling has length $n-1$. This set has size $L_n f_{n-1}$.

**Correspondence:** Given a $(2n-1)$-board $T^*$, there are 2 cases to consider, as illustrated in Figure 9.14.

Case I: If $T^*$ is breakable at cell $n$, then glue the right side of cell $n$ to the left side of cell 1 to create an in-phase $n$-bracelet $B$, and cells $n+1$ through $2n-1$ form an $(n-1)$-tiling $T$.

Case II: If $T^*$ is unbreakable at cell $n$, then cells $n$ and $n+1$ are covered by a domino which we denote by $d$. Cells 1 through $n-1$ become an $(n-1)$-tiling $T$ and cells $n$ through $2n-1$ are used to create an out-of-phase $n$-bracelet with $d$ as its first tile.

This correspondence is easily reversed since the phase of the $n$-bracelet indicates whether Case I or Case II is invoked. So the correspondence is a bijection and Set 1 and Set 2 have the same size.

The reader may wish to prove these Lucas identities combinatorially.

$$L_{m+n} = f_m L_n + f_{m-1} L_{n-1}.$$

$$5f_n = L_n + L_{n+2}.$$
$$L_n^2 = L_{n+1}L_{n-1} + (-1)^n \cdot 5.$$
$$L_0 + 2L_1 + 4L_2 + 8L_3 + \cdots + 2^n L_n = 2^{n+1} f_n.$$
$$\sum_{k=0}^{n} 5^k \binom{n}{2k} = 2^{n-1} L_n.$$

## Gibonacci Numbers

Gibonacci number is shorthand for generalized Fibonacci number. We say a sequence of nonnegative integers $G_0, G_1, G_2, \ldots$ is a *Gibonacci sequence* if for all $n \geq 2$,

$$G_n = G_{n-1} + G_{n-2}.$$

Such sequences are completely determined by $G_0$ and $G_1$. For instance, the Lucas sequence is the Gibonacci sequence beginning with $G_0 = 2$ and $G_1 = 1$. To see how to interpret these numbers combinatorially, we take a second look at Lucas numbers. From the previous section we know that $L_n$ counts the number of ways to tile an $n$-bracelet with squares and dominoes. Notice that we can "straighten out" an $n$-bracelet, by writing it as an $n$-tiling starting with the first tile (the tile covering cell 1) with one caveat. The caveat is that if the first tile is a domino, we need to indicate whether it is an in-phase or out-of-phase domino. For example, the seven 4-bracelets of Figure 9.11 have been straightened out in *phased tilings* in Figure 9.15. In summary, $L_n$ counts the number of *phased* $n$-tilings where an initial domino has **2** possible phases and an initial square has **1** possible phase. The next theorem should then come as no surprise.

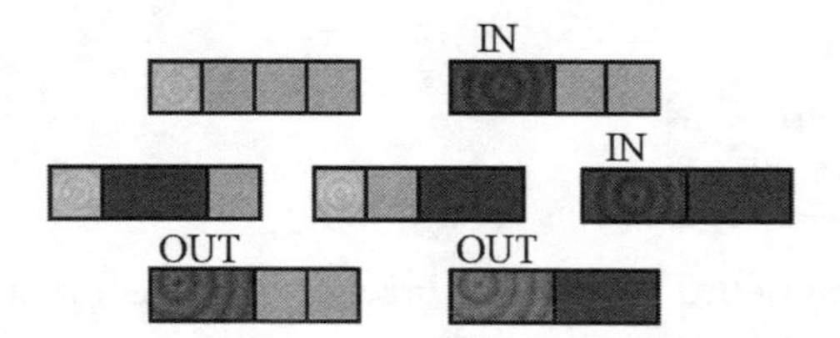


**Figure 9.15.** The seven 4-bracelets can be straightened out to become "phased" 4-tilings.

**Theorem** Let $G_0, G_1, G_2, \ldots$ be a Gibonacci sequence with nonnegative integer terms. For $n \geq 1$, $G_n$ counts the number of $n$-tilings, where the initial tile is assigned a phase. There are $G_0$ choices for a domino phase, and $G_1$ choices for a square phase.

*Proof.* Let $a_n$ denote the number of phased $n$-tilings with $G_0$ and $G_1$ phases for initial dominoes and squares, respectively. Clearly, $a_1 = G_1$. A phased 2-tiling consists of either a phased domino ($G_0$ choices) or a phased square followed by an unphased square ($G_1$ choices). Hence $a_2 = G_0 + G_1 = G_2$. To see that $a_n$ grows like Gibonacci numbers we consider the last tile, which immediately gives us $a_n = a_{n-1} + a_{n-2}$. □

In order for our theorem to be valid when $n = 0$, we combinatorially define the number of phased 0-tilings to be $G_0$, the number of domino phases. Using this combinatorial interpretation of $G_n$, we observe that many identities become transparent. For instance, from the shape of the first tile of a phased tiling (see Figure 9.16), it immediately follows that

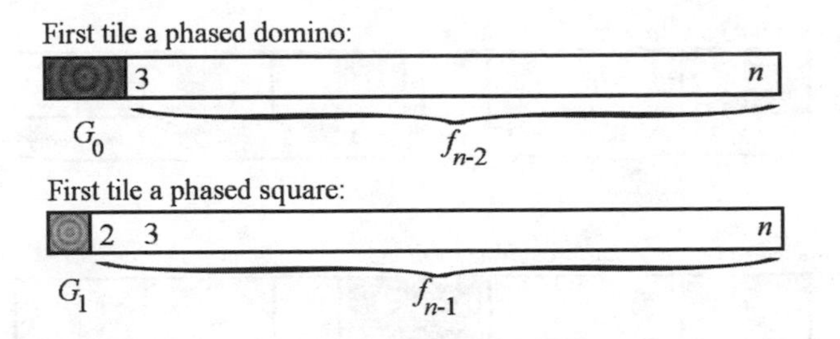


**Figure 9.16.** A phased $n$-tiling either begins with a phased domino or a phased square.

**Identity 6** $G_n = G_0 f_{n-2} + G_1 f_{n-1}$.

The next two identities are generalizations of Identities 1 and 2 respectively.

**Identity 7** $\sum_{k=0}^{n} G_k = G_{n+2} - G_1$.

**Question:** How many phased $(n+2)$-tilings contain at least one domino?

**Answer 1:** There are $G_{n+2}$ phased $(n+2)$-tilings including the $G_1$ tilings consisting of only squares. So there are $G_{n+2} - G_1$ tilings with at least one domino.

**Answer 2:** Consider the location of the last domino. For $0 \leq k \leq n$, there are $G_k$ tilings where the last domino covers cells $k+1$ and $k+2$ as illustrated in Figure 9.17. Notice that when the last domino covers cells 1 and 2, it must have one of $G_0$ phases. So the argument is still valid.

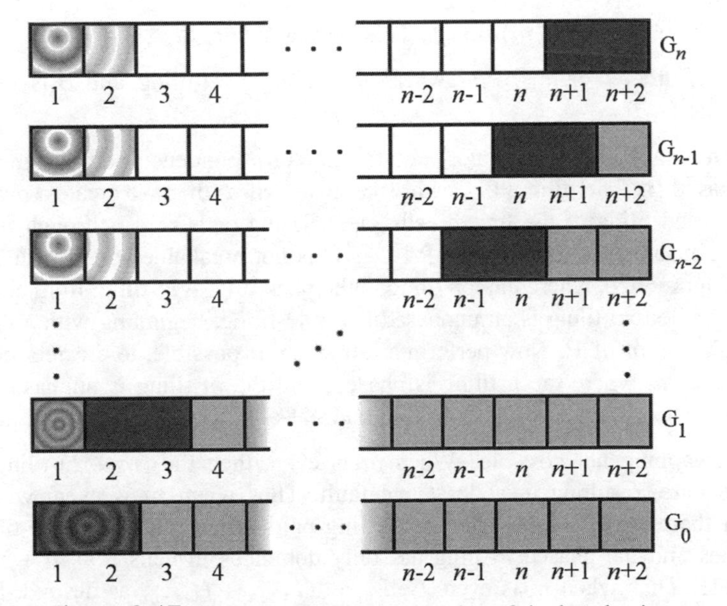


**Figure 9.17.** Here we consider the location of the last domino.

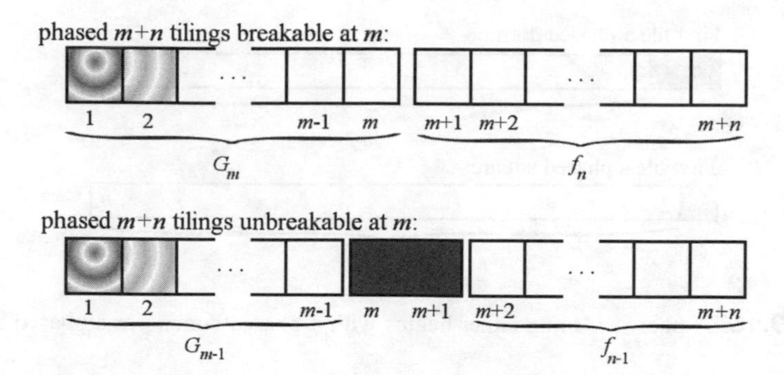


**Figure 9.18.** A phased $(m+n)$-tiling is either breakable or unbreakable at cell $m$.

**Identity 8** $G_{m+n} = G_m f_n + G_{m-1} f_{n-1}$.

**Question:** How many phased $m+n$ tilings exist?

**Answer 1:** By definition, there are $G_{m+n}$ such tilings.

**Answer 2:** Consider whether or not the phased $(m+n)$-tiling is breakable at cell $m$. See Figure 9.18. The number of breakable tilings is $G_m f_n$ since such a tiling consists of a phased $m$-tiling followed by a standard $n$-tiling. The number of unbreakable tilings is $G_{m-1} f_{n-1}$ since such tilings contain a phased $(m-1)$-tiling, followed by a domino covering cells $m$ and $m+1$, followed by a standard $(n-1)$-tiling. Altogether, there are $G_m f_n + G_{m-1} f_{n-1}$ $(m+n)$-tilings.

The next identity uses tail-swapping on phased tilings to create an almost one-to-one correspondence with a nontrivial error term.

**Identity 9** *For* $0 \le m \le n$, $G_{n+m} = G_n L_m + (-1)^{m-1} G_{n-m}$.

**Set 1:** The set of phased $(n+m)$-tilings. This set has size $G_{n+m}$.

**Set 2:** The set of ordered pairs $(A, B)$, where $A$ is a phased $n$-tiling, and $B$ is an $m$-bracelet. This set has size $G_n L_m$.

**Correspondence:** We create an almost one-to-one correspondence between these two sets. Let $P$ be a phased $(n+m)$-tiling. If $P$ is breakable at cell $n$, then we create a phased $n$-tiling $A$ from the phased tiling of the first $n$ cells of $P$. Using cells $n+1$ through $n+m$ create $B$, an in-phase $m$-bracelet, as in Figure 9.19. If $P$ is not breakable at cell $n$, then create the tiling pair of Figure 9.20, where the top tiling is the phased $(n-1)$-tiling from cells 1 through $n-1$ of $P$. The bottom tiling is an unphased $(m+1)$-tiling, beginning with a domino, from cells $n$ through $n+m$ of $P$. Now perform a tail swap, if possible, to create a pair of tilings with sizes $n$ and $m$, where the $n$-tiling is phased, and the $m$-tiling is unphased, but begins with a domino. These become a phased tiling and out-of-phase bracelet in the natural way.

When is tail-swapping not possible? When $m$ is even, then the $(m+1)$-tiling must have at least one square, resulting in at least one fault. Thus when $m$ is even, we can always tail-swap, but there are $G_{n-m}$ unachievable tiling pairs where the bottom $m$-tiling consists of all dominoes and the phased $n$-tiling has only dominoes in cells $n-m+1$ through $n$. See Figure 9.21. Thus when $m$ is even, $G_n L_m = G_{n+m} + G_{n-m}$ as desired. By a similar argument, when $m$ is odd, $G_{n+m} = G_n L_m + G_{n-m}$.

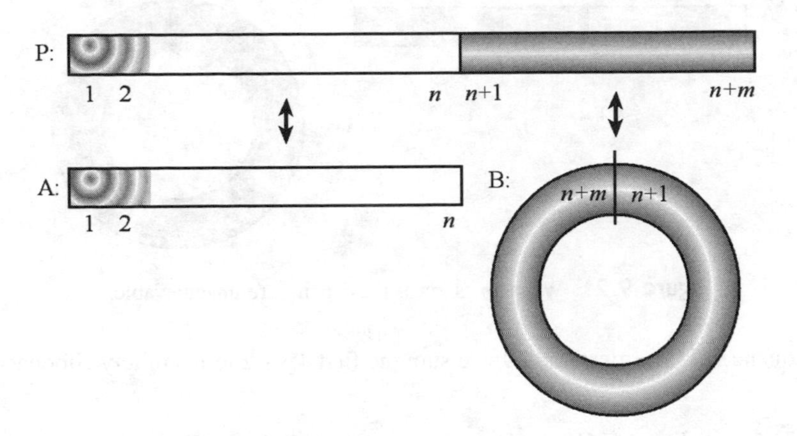


**Figure 9.19.** A breakable phased $(n+m)$-tiling naturally becomes a phased $n$-tiling with an in-phase $m$-bracelet.

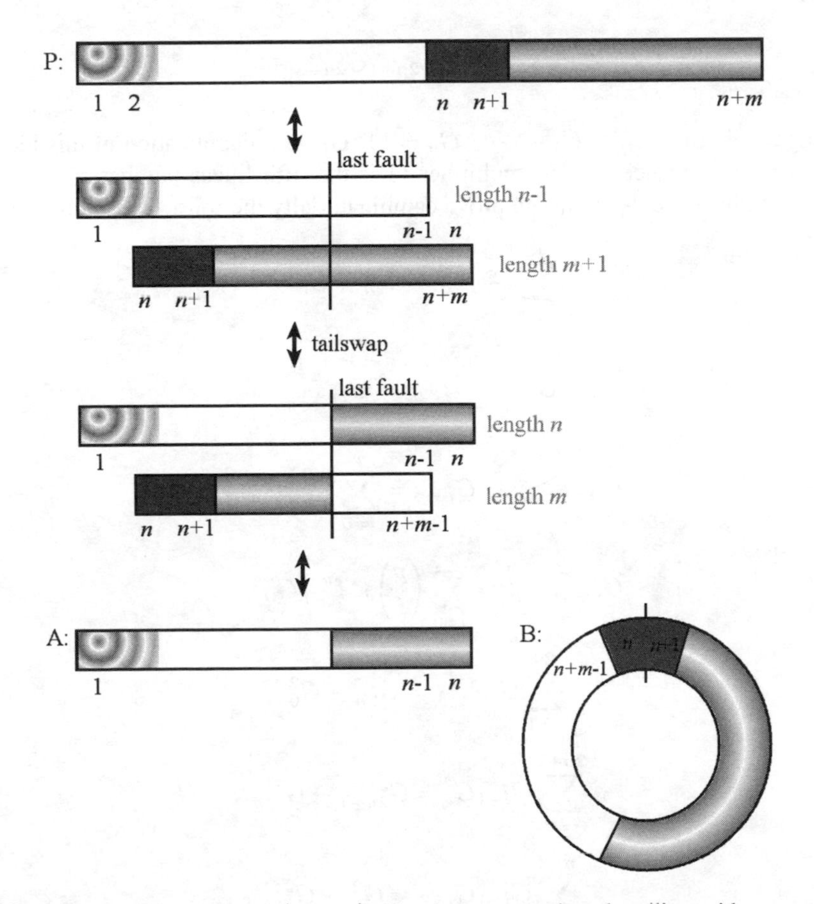


**Figure 9.20.** An unbreakable phased $(n+m)$-tiling becomes a phased $n$-tiling with an out-of-phase $m$-bracelet.

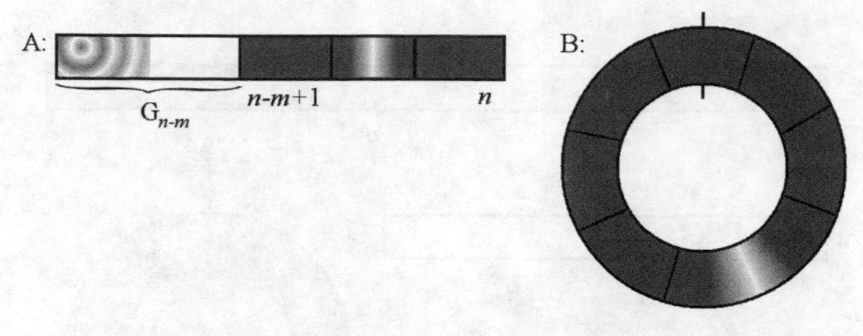


**Figure 9.21.** When $m$ is even, these pairs are unachievable.

As a consequence of Identities 7 and 9, we sum the first $4n+2$ terms of any Gibonacci sequence as

$$\sum_{i=0}^{4n+1} G_i = G_{4n+3} - G_1 = G_{2n+2}L_{2n+1} + (-1)^{2n}G_1 - G_1 = G_{2n+2}L_{2n+1}$$

leading to the following identity.

**Identity 10**

$$\sum_{i=0}^{4n+1} G_i = L_{2n+1}G_{2n+2}.$$

Our opening magic trick, $G_0 + G_1 + \cdots + G_9 = 11 \cdot G_6$, is an application of this identity when $n = 2$. So it is no coincidence that the multiplier 11 is the fifth Lucas number.

Once more, we challenge the reader to prove combinatorially the following Gibonacci identities.

$$\sum_{k=1}^{n} G_{2k-1} = G_{2n} - G_0.$$

$$G_1 + \sum_{k=0}^{n} G_{2k} = G_{2n+1}.$$

$$\text{For } n \geq p, \quad G_{n+p} = \sum_{i=0}^{p} \binom{p}{i} G_{n-i}.$$

$$G_{m+(t+1)p} = \sum_{i=0}^{p} \binom{p}{i} f_t^i f_{t-1}^{p-i} G_{m+i}.$$

$$\sum_{i=1}^{2n} G_i G_{i-1} = G_{2n}^2 - G_0^2$$

$$\sum_{i=2}^{2n+1} G_{i-1}G_i = G_{2n+1}^2 - G_1^2.$$

$$\sum_{i=1}^{n-1} G_{i-1}G_{i+2} = G_n^2 - G_1^2.$$

$$G_{n+1}G_{n-1} - G_n^2 = (-1)^n (G_1^2 - G_0 G_2).$$

$$G_{n+1} + G_n + G_{n-1} + 2G_{n-2} + 4G_{n-3} + 8G_{n-4} + \cdots 2^{n-1}G_0 = 2^n(G_0 + G_1).$$

Let $G_0, G_1, G_2, \ldots$ and $H_0, H_1, H_2, \ldots$ be Gibonacci sequences. Then for $n, h, k \geq 0$,

$$G_m H_n - G_{m-1}H_{n+1} = (-1)^m [G_0 H_{n-m+2} - G_1 H_{n-m+1}].$$

Combinatorial proofs of all the previously listed identities (and more) can be found in the references at the end of this paper or in our book [5].

## Open Problems

The techniques presented here are simple but powerful. Counting tilings enables us to visualize relationships between Fibonacci numbers and their generalizations. This approach facilitates a clearer understanding of existing identities and can be extended in a number of ways. By introducing colored tiles of various lengths, we can interpret sequences generated by linear recurrences with constant coefficients [1]. By allowing some of the squares of our tiling to be stacked up to a certain height, we can combinatorially interpret simple continued fractions [6]. By introducing an element of randomness, even the irrationally looking Binet formulas

$$f_n = \frac{1}{\sqrt{5}}\left[\left(\frac{1+\sqrt{5}}{2}\right)^{n+1} - \left(\frac{1-\sqrt{5}}{2}\right)^{n+1}\right]$$

and

$$L_n = \left[\left(\frac{1+\sqrt{5}}{2}\right)^{n} + \left(\frac{1-\sqrt{5}}{2}\right)^{n}\right],$$

can be rationalized [1, 2].

To indicate the power of our approach, the classic book *Fibonacci & Lucas Numbers and the Golden Section* by Steven Vajda [10] contains 118 identities involving Fibonacci, Lucas, and Gibonacci numbers. These identities are proved by a myriad of algebraic methods—induction, generating functions, hyperbolic functions, to name a few. Although *none* are proved combinatorially in the book, we have used tiling to explain 91 of these identities—and counting!

We leave the reader with some of the more tantalizing identities which have thus far resisted combinatorial explanations.

$$\sum_{i=0}^{2n} \binom{2n}{i} f_{2i-1} = 5^n f_{2n-1}$$

$$\sum_{i=0}^{2n} \binom{2n}{i} f_{i-1}^2 = 5^{n-1} L_{2n}$$

$$\sum_{i=0}^{2n} \binom{2n}{i} L_{2i} = 5^n L_{2n}$$

We have every confidence that these too will be combinatorially explained someday. You can count on it.

## References

[1] Benjamin, A.T., C.R.H. Hanusa, and F.E. Su, Linear recurrences through tilings and Markov chains, *Utilitas Mathematica*, 64 (2003), 3–17.

[2] Benjamin, A.T., G.M. Levin, K. Mahlburg, and J.J. Quinn, Random Approaches to Fibonacci Identities, *Amer. Math. Monthly*, 107 (2000), no. 6, 511–516.

[3] Benjamin, A.T., and J.J. Quinn, Recounting Fibonacci and Lucas Identities, *College Math. J.*, 30 (1999), no. 5, 359–366.

[4] ——, Fibonacci and Lucas Identities through Colored Tilings, *Utilitas Mathematica*, 56 (1999), 137–142.

[5] ——, *Proofs that Really Count: The Art of Combinatorial Proof*, Mathematical Association of America, Dolciani Series, Washington, DC, 2003.

[6] Benjamin, A.T., J.J. Quinn, and F.E. Su, Counting on Continued Fractions, *Math. Mag.*, 73 (2000), no. 2, 98–104.

[7] ——, Generalized Fibonacci Identities through Phased Tilings, *Fibonacci Quarterly*, 38 (2000), no. 3, 282–288.

[8] Benjamin, A.T., and J. Rouse, Recounting Binomial Fibonacci Identities, in *Application of Fibonacci Numbers*, Vol. 9, Kluwer Academic Publishers, 2003.

[9] Brigham, R.C., R.M. Caron, P.Z. Chinn, and R.P. Grimaldi, A Tiling Scheme for the Fibonacci Numbers, *J. Recreational Math.*, 28 (1996–97), no. 1, 10–16.

[10] Vajda, S., *Fibonacci & Lucas Numbers, and the Golden Section: Theory and Applications*, John Wiley and Sons, New York, 1989.

# 10

# Juggling Patterns, Passing, and Posets

Joe Buhler & Ron Graham
*Reed College* *UC San Diego*

During the middle of the 1980's a system for describing periodic juggling patterns appeared independently in at least three circles of jugglers. Although given different names by these groups, these patterns have become generally known as "site swaps" and we will use that terminology here.

Site swaps can be described abstractly in a fashion independent of their connection to juggling; they are in fact succinct descriptions of certain kinds of permutations on (one might even say dynamical systems on) the set **Z** of all integers. This gives rise to some pretty mathematics, which we would like to explain in this article.

We'll start with the basic results, which have their roots in permutations of infinite sets and elementary combinatorics. Although these could be described in completely mathematical terms, we both like to juggle, and will use this excuse to persist with the juggling terminology throughout.

Several years ago we had the idea of applying these ideas to patterns with two or more people. Although we never became truly proficient at nontrivial multi-person site swaps (the single person ones are hard enough!) we did discover a broad generalization of the basic counting theorem of site swaps. The result considers permutations of, and colorings of, partially ordered sets (posets), and is contained in the final theorem of this paper. Curiously, we do not yet know a juggling interpretation of this general result.

We warn the reader that mathematics shares one feature with juggling: for maximum enjoyment, you have to try it yourself. In particular, combinatorial arguments tend to be a bit nerve-wracking on first viewing, and real comprehension requires active participation. We encourage the reader in this direction by including exercises, and hope that you tackle them as you read; some of them are answered later in the article, so that if you don't try them right away, you may lose the chance to play along.

Site swaps were discovered independently by Paul Klimek in Santa Cruz, California (in 1982), by Bruce Tiemann and Bengt Magnusson at Caltech (in 1985) and by Colin Wright and other mathematics graduate students in the Cambridge University juggling club (in 1985). Tiemann and Magnusson investigated their theory and (especially) practice and were instrumental in popularizing them within the juggling community. Juggling has many fascinating aspects; however, the discussion here will focus almost exclusively on mathematics, and especially on the ideas necessary to extend site swap theorems to posets. Readers interested in juggling, or its history, or other connections between mathematics and juggling, might want to consult [1], [3], [6], [8], and references included therein. In addition, the web site `www.juggling.org` is a treasure chest of all sorts of information about juggling.

# 1 Juggling sequences

Imagine the throw times of our juggling balls to be discrete and equally spaced points in time. If a ball that is thrown at time $t$ is next thrown at time $t'$, we draw an arrow from time $t$ to time $t'$. This is a very idealized view of juggling—it doesn't really matter what the objects are, how they are thrown, or when they are caught, etc. However, as often happens in mathematical models of the real world, sparseness can lead to surprising and acute analyses of underlying patterns.

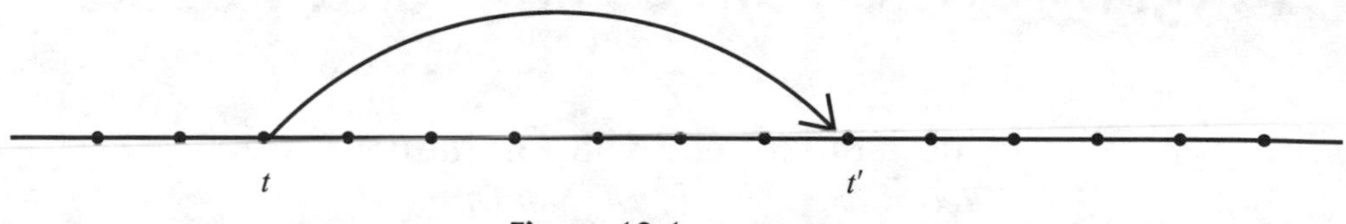


**Figure 10.1.** One throw

In addition to the arrows themselves, it is useful to annotate each throw with its "height" $t - t'$; roughly, this is how long the ball stays in the air. If no ball is thrown at time $t$ we include a loop from $t$ to $t$, with height $t - t = 0$. Several examples follow.

In all of these cases, the sequence of throws is periodic, and it is customary to name the trick, as indicated in the diagrams below, by giving the throw heights in a period. The first pattern is the 3-ball cascade (perhaps the most basic juggling pattern), and the second pattern is the 4-ball fountain (waterfall); there is an obvious generalization to the most basic or "canonical" $n$-ball pattern.

In the standard interpretation of the diagrams, the throws take place alternately from the right and left hands. Although the 3-ball cascade diagram and the 4-ball fountain diagram are similar in appearance, they actually are quite different in the real world. In the cascade a ball is thrown from

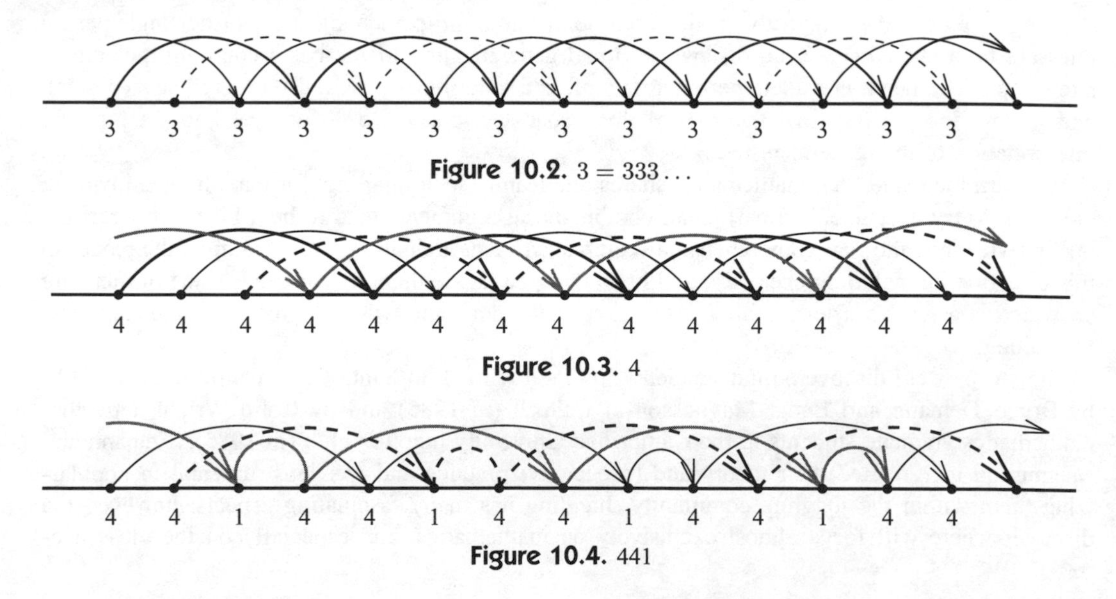


**Figure 10.2.** $3 = 333\ldots$

**Figure 10.3.** 4

**Figure 10.4.** 441

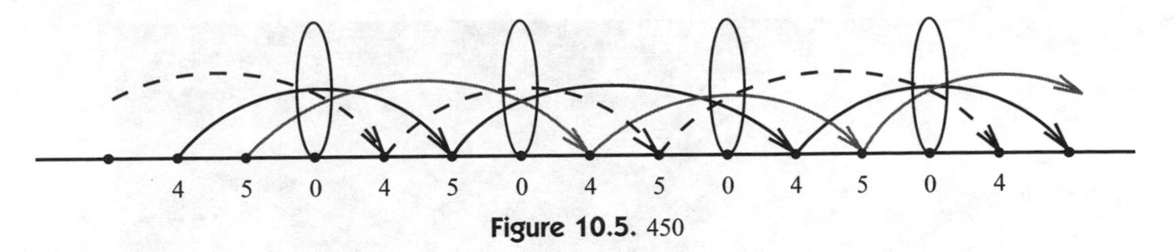


**Figure 10.5.** 450

one hand to another (since a ball is next thrown 3 time ticks later by the opposite hand), whereas in the fountain balls are thrown from each hand to itself (since a ball is next thrown 4 ticks later by the same hand).

These diagrams can be interpreted mathematically in several ways; we choose to think of them as rightward-moving permutations $f$ of the set

$$\mathbf{Z} = \{\ldots, -3, -2, -1, 0, 1, 2, 3, \ldots\}$$

of all integers. Thus, a juggling pattern determines a function $f: \mathbf{Z} \to \mathbf{Z}$, where $f(t)$ is the next throw-time of the ball thrown at the time $t \in \mathbf{Z}$. From Figure 10.2, we see that the 3-ball cascade "3" is the permutation $f(t) = t + 3$, and the most basic $n$-ball pattern is $f(t) = t + n$.

We remind the reader that a permutation $f: \mathbf{Z} \to \mathbf{Z}$ is a bijection, which means that it is one-to-one (distinct points are mapped to distinct points), and onto (every $u$ in $\mathbf{Z}$ is in the image). In symbols: if $f(t) = f(t')$ then $t = t'$, and for every $u$ there is a $t$ such that $f(t) = u$.

The permutations in the figures above have the special property that they are rightward-moving, or increasing. This means that $f(t) \geq t$ for all times $t$.

In the context of juggling, the fact that $f$ is a permutation means that two balls never land at the same time, and that at each instant in time either a ball is caught and thrown (or the hand is empty). If the hand is empty at time $t$, i.e., no ball arrives or is thrown at time $t$ then $f(t) = t$. It is sometimes mathematically slightly more convenient to disallow these "empty" 0-throws by restricting attention to patterns that satisfy $f(t) > t$. However, the theory of these "positive" or "no empty hands" patterns is essentially the same as the theory that we develop here, since it is easy to check that the relation $g(t) = f(t) + 1$ establishes a one-to-one correspondence between juggling patterns $f$, and positive juggling patterns $g$. Since 0-throws do occur in juggling practice we will stick with the convention $f(t) \geq t$.

**Exercise.** Give an example of a permutation of $\mathbf{Z}$ that is not rightward-moving.
**Exercise.** How many balls are being juggled in each of the juggling patterns above?
**Exercise.** Find a juggling pattern that only contains throws with throw heights 7 and 3.

As mentioned above, in the usual two-hand juggling patterns, the odd and even throw times usually correspond to throws with the right and left hands. However, there are many other juggling realizations of a pattern $f$; for instance, one might imagine a one-handed juggler making all of the throws.

Jugglers really need to know only the throw heights

$$j(t) = j_f(t) = f(t) - t$$

at each time $t$. We will assume that our patterns are periodic in the sense that $j$ is periodic, i.e., that there is an $n$ such that $j(t+n) = j(t)$ for all $t$. In particular, the values $j(0), j(1), \ldots, j(n-1)$ uniquely determine $j$ and also determine $f(t) = t + j(t)$.

If $t$ is an integer then let $[t]_n = t \bmod n$ denote the remainder when $t$ is divided by $n$, i.e., the unique integer $a$ such that $0 \leq a < n$ and $t - a$ is divisible by $n$. If $n$ is understood, as it

Buhler and Graham check out the theory

often will be, we will denote this by $[t]$. Finally, we let $\overline{n}$ be shorthand for the set of possible remainders $\{0, 1, 2, \ldots, n-1\}$.

**Definition 1** A juggling sequence, or site swap, is a finite sequence $j$ of nonnegative integers, usually written as a string

$$j(0)\, j(1)\, \ldots\, j(n-1)$$

such that the mapping $f\colon \mathbf{Z} \to \mathbf{Z}$ defined by

$$f(t) = t + j\big([t]_n\big)$$

is a permutation of the integers.

Formally, a site swap is a function $j: \overline{n} \rightarrow \mathbf{Z}$ with nonnegative values, but informally we usually just write a site swap as a string of symbols if the meaning is clear. Thus the site swap "534" is shorthand for the sequence $j$ with $j(0) = 5$, $j(1) = 3$, and $j(2) = 4$.

To actually juggle a juggling sequence $j$ of length $n$, a juggler just makes throws of height $j([t]_n)$ at time $t$.

**Exercise.** Find a finite sequence of nonnegative integers that is not a juggling sequence.
**Exercise.** Which sequences of length 2 are juggling sequences?
**Exercise.** Show that $[[a]_n + [b]_n]_n = [a+b]_n$, and $[[a]_n [b]_n]_n = [ab]_n$.

**Remark.** Tiemann and Magnusson introduced the term "site swap" since they visualized the balls interchanging "sites" in the cyclic ordering of balls of the canonical pattern.

## 2 When is a sequence a site swap?

For a sequence of nonnegative integers to be a site swap it is clearly necessary that the arrival times $t + j(t)$ be distinct for $0 \le t < n$. Somewhat surprisingly, this isn't sufficient. For instance, the sequence 346 has distinct arrival times for $0 \le t < 3$:

$$0 + j(0) = 3, \quad 1 + j(1) = 5, \quad 2 + j(2) = 8.$$

However, if we look at the corresponding permutation $f$ we see that

$$f(2) = 2 + j(2) = 8 = 4 + j([4]_3) = 4 + j(1) = f(4).$$

Thus there is a "collision" and $f$ isn't a permutation, and 346 isn't a juggling sequence.

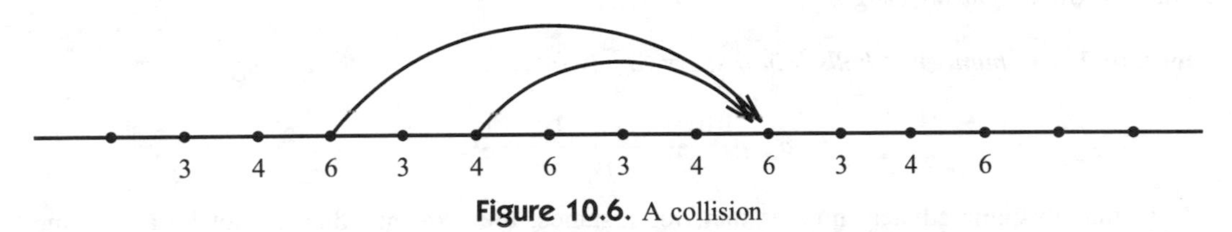


**Figure 10.6.** A collision

Fix a positive integer $n$. We show that if a sequence $j(t)$ gives a function $f(t) = t + j([t]_n)$ that has a collision, then already there is a collision modulo $n$ in the first $n$ values. Indeed, suppose that there is a collision, so that $x + j([x]) = x' + j([x'])$ for some $x$ and $x'$. Writing $x = t + ny$, $x' = t' + ny'$, for $0 \le t, t' < n$, we see that $t + j(t) = t' + j(t') + n(y' - y)$. This implies that if $f$ has a collision, then $f$ modulo $n$ has a collision in its first $n$ values. Therefore, if the $t + j(t)$ are distinct modulo $n$ for $0 \le t < n$ then $j$ is a juggling sequence.

**Exercise.** Show that the converse is true, i.e., that if there are integers $t$ and $t'$ such that

$$t + j(t) \equiv t' + j(t') \bmod n, \qquad 0 \le t, t' < n,$$

then $f(t) := t + j(t)$ isn't a bijection, thereby proving the following theorem.

**Theorem 2** *A finite sequence $j$ of nonnegative integers $j(0) \ldots j(n-1)$ is a juggling sequence if and only if the integers $[t + j(t)]_n$, $0 \le t < n$, are distinct.*

**Exercise.** Show that any cyclic permutation of a juggling sequence is also a juggling sequence.

**Exercise.** Find a site swap whose reversal, obtained by reading the numbers in reverse order, is not a site swap.

By the theorem, if $j$ is a juggling sequence then the function from $\overline{n}$ to itself defined by

$$\pi_j(t) = [t + j(t)] = f(t) \bmod n$$

is a permutation.

To check whether a sequence of integers is in fact a site swap it is convenient to write $t$ and $j(t)$ in rows, add modulo $n$, and check whether there are duplications. For instance, one finds that 345 is a juggling sequence, but that 543 is not:

$$\begin{array}{r} 3\ 4\ 5 \\ +\ \underline{0\ 1\ 2} \\ 0\ 2\ 1 \end{array} \qquad \begin{array}{r} 5\ 4\ 3 \\ +\ \underline{0\ 1\ 2} \\ 2\ 2\ 2 \end{array}$$

## 3 How many balls are there?

How many balls are there in the juggling pattern $f$ corresponding to a site swap $j$?

From looking at the original examples it should be reasonably clear that the number is the number of distinct infinite paths in the diagram. For instance, 3, 450, and 441 are 3-ball tricks, whereas 4 and 345 are 4-ball tricks.

Experiments should convince you that the number of balls is the average throw height. We will prove this assertion by reducing an arbitrary juggling pattern to the basic pattern $f(t) = t + b$, where $b$ is the number of balls, by simple transformations that leave the number of balls and the average throw height unchanged.

**Theorem 3** *The number of balls $b$ in a site swap $j(0)\dots j(n-1)$ is the average*

$$b = \frac{j(0) + \dots + j(n-1)}{n}.$$

Note that this immediately implies that, for instance, the sequence 344 cannot be a juggling sequence since its average isn't an integer!

To prove the theorem, we start by noting that this formula is obvious if the juggling pattern is $f(t) = t + b$, i.e., the constant site swap

$$b = j(0) = j(1) = \dots = j(n-1).$$

Now suppose that the sequence is not constant. Then we can find two adjacent terms that are "out of order" in the sense that $j(t) > j(t+1)$. For instance, in the sequence 5744 the 7 and 4 are out of order since the first ball thrown lands after the second ball thrown.

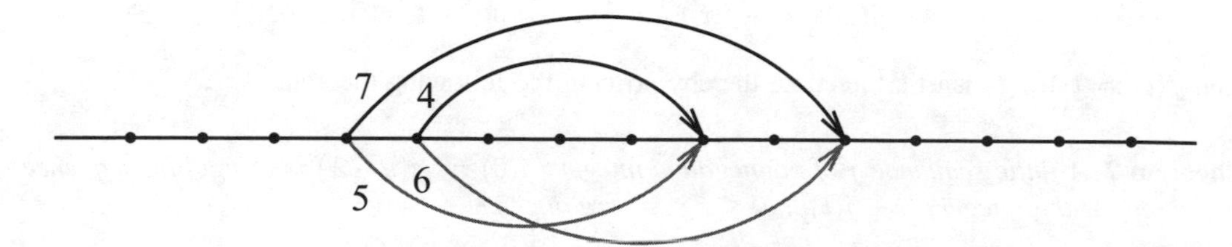


**Figure 10.7.** Switching two throws

Then we fix the out-of-order pair by interchanging their arrival times. From the above diagram it is clear that 5744 becomes 5564. More generally, if $a$, $b$ is a consecutive pair with $a > b+1$ and $a$ is (one of) the largest term(s) in the site swap, then we replace the pair $a$, $b$ with the pair $b+1$, $a-1$. This transformation leaves the number of balls unchanged, leaves the average of the sequence values unchanged, and decreases the largest term or the number of terms equal to the largest term. By continuing to apply this transformation, sooner or later we will arrive at a constant sequence, which has no out-of-order pairs.

In order to carry this out, it may be necessary to remember that a sequence is really a cycle, and count the first and last terms as a consecutive pair. For instance, in the sequence 4457, the first and last terms are out of order, and we apply the above transformation to those two numbers.

In any event, these reductions finally reduce to the basic trick, thereby proving the theorem.

**Exercise.** How many such steps are needed to convert 41479 to 66666 by this procedure?
**Exercise.** List all 3-ball juggling sequences of period 3.
**Exercise.** Find a sequence $j(0)\,j(1)\,j(2)\,j(3)$ of length 4 that is not a juggling sequence, but nonetheless has distinct $t+j(t)$, and whose average is an integer.
**Exercise.** Show that any sequence of integers whose average is an integer can be rearranged to be a juggling sequence. (Warning: this problem seems to be very hard; the only proof that we know is based on a paper by the famous group theorist Marshall Hall from 1952; his paper was on abelian groups and predates site-swaps by 30 years.)

## 4 How many site swaps are there with $b$ balls and period $n$?

This question is much more involved than the earlier ones, and will take several sections to answer.

If $j$ is a juggling sequence of length $n$ then $\pi(t) = [t + j(t)]$ is a permutation of $\overline{n}$.

It is well known that there are $n!$ permutations of an $n$-element set. (There are $n$ choices for the image of the first element, $n-1$ for the second elements, and so on, giving a total of

$$n \cdot (n-1) \cdots 2 \cdot 1 = n!$$

permutations.) The collection of all permutations of $\overline{n}$ is usually denoted $S_n$ and is called the symmetric group on $n$ objects.

**Exercise.** Find the permutation $\pi$ of $\overline{5}$ associated with the juggling sequence 41479.

For permutations $\pi$, let $\mathbf{JS}(\pi, b)$ denote the number of juggling sequences with $b$ balls whose associated permutation is $\pi$, and let $\mathbf{JS}(n, b)$ denote the number of all juggling sequences of period $n$ with $b$ balls. Therefore

$$\mathbf{JS}(n,b) = \sum_{\pi} \mathbf{JS}(\pi, b)$$

where the sum is over all $\pi$ in $S_n$.

To construct a juggling sequence whose associated permutation is $\pi$ we can more or less reverse the procedure used to find $\pi$ from $j$. That is, we first write $\pi(t)$ and $t$ in rows and subtract to get a row $\pi(t) - t$. The new row sums to 0 (and could loosely be interpreted as a 0-ball juggling pattern in which one is allowed to throw balls backwards in time!).

**Exercise.** Prove that the numbers $\pi(t) - t$, $0 \le t < n$, sum to 0.

For example, for the permutation $\pi$ of $\overline{3}$ that interchanges 1 and 2 we get

$$\begin{array}{rrrrl} & 0 & 2 & 1 & \quad \pi(t) \\ - & 0 & 1 & 2 & \quad t \\ \hline & 0 & 1 & -1 & \quad \pi(t) - t \end{array}$$

This is clearly not a site swap since it has a negative term. However, if we add $n$'s (where here $n = 3$) to make all entries nonnegative, then we will have a site swap. If we want a $b$-ball site swap then we must add $b$ different $n$'s (since the original sum is 0).

To summarize: all possible $b$-ball juggling sequences with permutation $\pi$ are obtained by adding $n$ to entries of $\pi(t) - t$ subject to two conditions: every entry must ultimately be nonnegative, and $n$ is added $b$ times.

In the example above for $n = 3$, we have to add 3 to the entries. Clearly we have to add at least one 3 since the last entry is negative. There is precisely one way to do this to get a 1-ball juggling sequence: add 3 to the last entry to get 012.

There are three ways to add two 3's to get 2-ball juggling sequences, since we add one 3 to the last entry and can add the other to any of the three entries; the three 2-ball juggling sequences corresponding to $\pi$ are 312, 042, and 015.

**Exercise.** How many 3 and 4-ball sequences are associated with this specific $\pi$?

**Exercise.** How many 3 and 4-ball sequences are associated with the permutation $\pi$ that has $\pi(0) = 2, \pi(1) = 0, \pi(2) = 1$?

Clearly the problem of calculating $\mathbf{JS}(\pi, b)$ depends on how many negative integers there are in the row $\pi(t) - t$. Specifically, for each $t$ where $\pi(t) - t$ is negative we are required to add at least one $n$.

**Definition 4** Let $\pi \in S_n$ be a permutation of $\overline{n}$. Then an element $t$, $0 \le t < n$, is a **drop** of $\pi$ if $\pi(t) < t$.

**Exercise.** How many permutations in $S_3$ have 0 drops? 1 drop? 2 drops? 3 drops?

Suppose that $\pi \in S_n$ has $k$ drops. To make a site swap of period $n$ and $b$ balls that is associated with $\pi$ we start with the row $\pi(t) - t$, and add an $n$ at each of the $k$ locations of a drop; after this, every entry is nonnegative. At this point, we choose $b - k$ arbitrary locations (repetitions allowed) and add an $n$ at those locations.

**Theorem 5** *Suppose that $\pi \in S_n$ has $k$ drops. Then:*

$$\mathbf{JS}(\pi, b) = \binom{b - k + n - 1}{n - 1}.$$

The binomial coefficient in the theorem seems a bit unlikely. Recall that the binomial coefficient

$$\binom{r}{s} = \frac{r!}{s!(r-s)!} = \frac{r(r-1)\cdots(r-s+1)}{s!}, \qquad 0 \le s \le r$$

(read "$r$ choose $s$") is the number of ways of choosing $s$ things from a set of $r$ things, with repeated choice not allowed and the order of choice ignored; this is the same as the number of $r$-element subsets of an $s$-element set. (*Proof.* There are $r$ choices for the first element, $r - 1$ for the second, and $r - s + 1$ for the $s$th element; each subset can be ordered in $s!$ ways, so if $S$ is the number of subsets then $s!S = r(r-1)\cdots(r-s+1)$.)

As you may already know, binomial coefficients can be constructed recursively "row by row," and written in an especially pleasing format called Pascal's Triangle.

**Exercise.** What fraction of 5-card poker hands are "flushes" (contain cards from only one suit)? What fraction of 5-card poker hands are "straights" (five consecutive cards)? Which hand is ranked more highly in poker, and why?

**Exercise.** Show that the binomial coefficients are "symmetric" in the sense that

$$\binom{r}{s} = \binom{r}{r-s}$$

in two ways: algebraically, using the formula above, and combinatorially, by arguing that the number of subsets of an $r$-element set with $s$ elements is the same as the number of subsets with $r-s$ elements.

In any event, the combinatorial interpretation of the binomial coefficient on right-hand side of the theorem doesn't seem to have much to do with juggling sequences. First, note that if the permutation has 0 drops (i.e., is the identity permutation $\pi(t) = t$), then the theorem asserts that the number of ways to choose $b$ things out of $n$, with repetition allowed (and order irrelevant), is

$$\binom{b+n-1}{n-1}.$$

More generally, by our earlier remarks, we have to count the number of ways that we can choose $b-k$ objects out of $n$, with repetition allowed (and order does not matter).

Somewhat surprisingly, the binomial coefficient can be interpreted in this fashion; this combinatorial interpretation is standard, but not nearly as well-known as the usual interpretation. In order to model the case in which repetition is allowed but the order does not matter, we consider nondecreasing sequences of not-necessarily-distinct integers, such as

$$112444$$

in which six objects—two 1's, one 2, and three 4's—have been chosen from $\{1, 2, 3, 4\}$.

**Lemma 1** *The number of nondecreasing sequences of length $r$ chosen from $\{1, \ldots, s\}$ is*

$$\binom{s+r-1}{r-1} = \binom{s+r-1}{s}.$$

*Proof.* The equality between the two formulas is just the symmetry of the binomial coefficients in a previous exercise.

To show that the number of sequences is given by the binomial coefficient, encode each sequence by replacing numbers by stars, separating each transition from $i$ to $i+1$ with a bar. For instance, the selection of $1, 1, 2, 4, 4, 4$ from $\{1, 2, 3, 4\}$ is encoded

$$* * \mid * \mid\mid * * *$$

A little thought reveals that there is a one-to-one correspondence between the stated sequences and these encodings. The encodings contain $r+s-1$ terms ($s$ stars, and $r-1$ bars separating them), and the number of such encodings is therefore the binomial coefficient stated in the theorem: the number of ways to choose the bars (or stars) in $s+r-1$ positions. □

Let $\delta_n(k)$ denote the number of elements of $S_n$ that have $k$ drops. The number $\mathbf{JS}(n, b)$ of juggling sequences of length $n$ and $b$ balls is the sum of $\mathbf{JS}(\pi, b)$ over all permutations $\pi$, so we immediately obtain the following corollary of the theorem.

**Corollary 6**

$$\mathbf{JS}(n, b) = \sum_{k=0}^{n-1} \delta_n(k) \binom{b-k+n-1}{n-1}.$$

For instance, for $n = 3$ one finds by examination of all six permutations that $\delta_3(0) = 1$, $\delta_3(1) = 4$, $\delta_3(2) = 1$. Thus

$$\mathbf{JS}(3, b) = 1 \cdot \binom{b+2}{2} + 4 \cdot \binom{b+1}{2} + 1 \cdot \binom{b}{2} = 3b^2 + 3b + 1.$$

**Exercise.** Verify that $\mathbf{JS}(4, b) = (b+1)^4 - b^4$.

## 5 Drops versus Descents

To go further we need to be able to say something about $\delta_n(k)$. To do this, we have to consider a superficially unrelated concept.

**Definition 7** An integer $t$, $0 \leq t < n - 1$, is a **descent** of a permutation

$$\pi : \quad \pi(0)\,\pi(1) \ldots \pi(n-1)$$

if $\pi(t) > \pi(t+1)$. The number of permutations in $S_n$ with $k$ descents is denoted

$$\left\langle {n \atop k} \right\rangle .$$

These numbers are called Eulerian numbers [4] and arise in many different contexts.
Here is a table of the drops and descents of 3-object permutations.

| $\pi$ | # drops | # descents |
|---|---|---|
| 1 2 3 | 0 | 0 |
| 1 3 2 | 1 | 1 |
| 2 1 3 | 1 | 1 |
| 2 3 1 | 1 | 1 |
| 3 1 2 | 2 | 1 |
| 3 2 1 | 1 | 2 |

**Exercise.** Make an analogous table for $n = 4$.

Although the number of drops of a permutation is not the same as the number of descents, the total number of permutations with $k$ drops is the same as the total number of permutations with $k$ descents.

**Theorem 8** *For all positive integers $n$ and $k$, with $0 \leq k < n$,*

$$\delta_n(k) = \left\langle {n \atop k} \right\rangle .$$

The proof of this assertion involves a curious bijection between permutations $\pi$ with $k$ drops and permutations $\sigma$ with $k$ descents.

First, we note that a permutation $\pi$ can be written as a product of "cycles." For example a permutation $\pi \in S_6$ could be given in "two-row" notation as

| $i$ | 0 | 1 | 2 | 3 | 4 | 5 |
|---|---|---|---|---|---|---|
| $\pi(t)$ | 4 | 5 | 0 | 3 | 2 | 1 |

The theory works!

It could also be succinctly described in row-form as 450321. Note that $\pi$ takes 0 to 4, 4 to 2, and 2 back to 0, and takes 1 to 5 and 5 back to 1. The corresponding cycle notation is $(024)(15)(3)$. The one-cycle $(3)$ corresponds to a fixed point, and is often omitted, but for our purposes below we need to explicitly include 1-cycles.

We define a map $\pi \mapsto \sigma$ from $S_n$ to itself as follows: Write $\pi$ as a product of cycles, with each cycle written with its largest element at the beginning, and the cycles written in the order in which their largest elements are increasing. Then erase the parentheses to get the row-form of a permutation $\sigma$.

For instance from the permutation $\pi = 450321$ above we get the "canonical" cycle form $(3)(402)(51)$ and hence $\sigma$ has row-form 340251.

**Exercise.** Let $\pi$ be the permutation that reverses the row $0 \ldots n-1$, i.e., $\pi(t) = n-1-i$. Describe the corresponding $\sigma$.

Note that $\pi$ can be reconstructed uniquely from $\sigma$. Indeed, from the row-form of $\sigma$ we start with a left parenthesis, and then close the cycle, and begin another, exactly when we see a new record high-point in the string of integers. Thus 340251 gets mapped to $(3)(402)(51)$ as desired.

A little thought confirms that this peculiar procedure exactly undoes the original mapping; since we have mutually inverse mappings from $S_n$ to itself we see that this correspondence is bijective.

Further thought shows that the drops of $\pi$ correspond to the descents of $\sigma$; in particular if $\pi$ has $k$ drops then $\sigma$ has $k$ descents. Indeed, any drop $\pi(t) < t$ shows up as consecutive "out-of-order" numbers in the cycle form of $\pi$ and hence is a descent in $\sigma$. On the other hand, any descent in $\sigma$ must occur within a cycle of $\pi$ (since the transitions from one cycle to another are increases), and hence corresponds to a drop of $\pi$.

**Exercise.** Prove that

$$\left\langle {n \atop k} \right\rangle = \left\langle {n \atop n-k-1} \right\rangle .$$

(Hint: consider the relationship between the descents of a permutation and the permutation obtained by reversing its row-form.)

**Exercise.** Give a combinatorial proof of the basic rule of formation of Pascal's Triangle:

$$\binom{n+1}{k} = \binom{n}{k} + \binom{n}{k-1}.$$

(Hint: Let $a$ be a specific element of an $n+1$-element set, and consider $k$-subsets that, respectively, do and do not contain $a$.)

**Exercise.** Give a combinatorial proof of the following identity:

$$\left\langle {n \atop k} \right\rangle = (k+1) \left\langle {n-1 \atop k} \right\rangle + (n-k) \left\langle {n-1 \atop k-1} \right\rangle .$$

## 6 Worpitzky's identity

An important formula, sometimes called Worpitzky's identity ([4]), gives a relationship between the Eulerian numbers and perfect powers, and has a close relationship to the formula given earlier for the number $\mathbf{JS}(n, b)$ of site swaps with $b$ balls.

**Theorem 9** *If $n$ and $b$ are nonnegative integers then*

$$b^n = \sum_{k=0}^{n-1} \left\langle {n \atop k} \right\rangle \binom{b+k}{n}.$$

All of the terms in the theorem have combinatorial interpretations, and it is natural to ask for a combinatorial proof; such a proof is given below. We have been unable to locate a combinatorial proof in the literature, but it seems likely that it can be found somewhere.

Before giving the proof, we observe that the identity implies a slick formula for the number of site swaps with fewer than $b$ balls. Indeed, if $\mathbf{JS}_{<}(n, b)$ denotes the number of patterns with period $n$ with fewer than $b$ balls, then Corollary 6, and the relationship between $\delta_n(k)$ and Eulerian numbers, gives

$$\begin{aligned}
\mathbf{JS}_{<}(n, b) &= \sum_{a=0}^{b-1} \mathbf{JS}(n, a) \\
&= \sum_{a=0}^{b-1} \sum_{k=0}^{n-1} \left\langle {n \atop k} \right\rangle \binom{a-k+n-1}{n-1} \\
&= \sum_{k=0}^{n-1} \left\langle {n \atop k} \right\rangle \sum_{a=0}^{b-1} \binom{a-k+n-1}{n-1}.
\end{aligned}$$

By induction using the recursion for binomial coefficients, or purely combinatorially, or otherwise, one can show that

$$\sum_{i=0}^{m} \binom{i+n-1}{n-1} = \binom{m+n}{n}.$$

**Exercise.** Write the first 6 rows of Pascal's triangle, and circle the entries involved on the left-hand side of the identity for $m = n = 3$. Explain how the identity follows from the law of formation of Pascal's triangle. Prove the identity in general.

After a little algebraic juggling (replacing $a - k$ by $i$ and letting $i$ run from 0 to $b - k - 1$) we find that we now have

$$\mathbf{JS}_<(n,b) = \sum_{k=0}^{n-1} \left\langle {n \atop k} \right\rangle \binom{n+b-k-1}{n}.$$

Changing variables by replacing $k$ by $n - k - 1$ gives

$$\mathbf{JS}_<(n,b) = \sum_{k=0}^{n-1} \left\langle {n \atop n-k-1} \right\rangle \binom{b+k}{n}.$$

Using the symmetry

$$\left\langle {n \atop k} \right\rangle = \left\langle {n \atop n-k-1} \right\rangle$$

of the Eulerian numbers and Worpitzky's identity

$$b^n = \sum_{k=0}^{n-1} \left\langle {n \atop k} \right\rangle \binom{b+k}{n}$$

we arrive at our main counting theorem.

**Theorem 10**

$$\mathbf{JS}_<(n,b) = b^n.$$

**Remarks:** The proof of this formula has taken us through a grand tour of parts of combinatorics. However, the elegance of the result suggests that one can hope for a purely combinatorial ("bijective") proof of the result, in which juggling sequences of length $n$ and fewer than $b$ balls are put in one-to-one correspondence with words of length $n$ with symbols chosen from an alphabet with $b$ letters. In fact, several people have constructed such bijections; see the appendix in [3] and also the references cited therein. However, the proof chosen here provides a perfect warm-up for the generalizations to posets. Also, the bijective proof of Worpitzky's identity in the next section could in principle be combined with the earlier arguments to give a bijective proof of the theorem.

From the fact that $\mathbf{JS}(n,b) = \mathbf{JS}_<(n,b+1) - \mathbf{JS}_<(n,b)$ we immediately deduce a formula for $\mathbf{JS}(n,b)$ from the main theorem.

**Corollary 11**

$$\mathbf{JS}(n,b) = (b+1)^n - b^n.$$

We note that our definitions count cyclic permutations of a site swap as being different from each other. A juggler might not care to make this distinction, and it is easy to use "Möbius inversion" to give a formula for the number of patterns of length $n$ with $b$ balls "up to cyclic permutation" (see [3]).

**Exercise.** Look up Möbius inversion in your favorite combinatorics book, and find an explicit formula for the number of patterns of length $n$ with $b$ balls "up to cyclic permutation."

# 7 A combinatorial proof of Worpitzky's identity

Algebraic proofs of Worpitzky's identity

$$b^n = \sum_{k=0}^{n-1} \left\langle {n \atop k} \right\rangle \binom{b+k}{n}$$

are possible, but in keeping with the spirit of this note, we give a combinatorial proof. The term $b^n$ on the left-hand side of the identity counts the number of $n$-tuples where each component comes from a $b$-element alphabet. Thus we must exhibit an explicit one-to-one-correspondence between $n$-tuples of integers $(x_1, \ldots, x_n)$, $0 \le x_i < b$, and objects counted by terms in the sum on the right-hand side of the identity.

Given an $n$-tuple $(x_1, \ldots, x_n)$ of integers between 0 and $b-1$, we determine a unique permutation $\sigma$ of $\{1, 2, \ldots, n\}$ by the conditions that

$$x_{\sigma(1)} \le x_{\sigma(2)} \le \cdots \le x_{\sigma(n)}$$

and equality $x_{\sigma(i)} = x_{\sigma(i+1)}$ occurs if and only if $\sigma(i) > \sigma(i+1)$.

For example, for $n = 3$ there are $3! = 6$ permutations; in the following table each permutation is followed by a "template" that describes the conditions on the $x_i$ that give rise to that permutation according to the above rule. The key facts are that the "$\le$" possibilities occur exactly at the descents of the corresponding permutation, and that any $\sigma$ corresponds to exactly one template.

| $\sigma$ | template |
|---|---|
| 123 | $x_1 < x_2 < x_3$ |
| 132 | $x_1 < x_3 \le x_2$ |
| 213 | $x_2 \le x_1 < x_3$ |
| 231 | $x_2 < x_3 \le x_1$ |
| 312 | $x_3 \le x_1 < x_2$ |
| 321 | $x_3 \le x_2 \le x_1$ |

**Exercise.** Show that there are $\binom{b}{3}$ ways to choose $x_i$ in the first template, i.e., integers $x_i$ between 1 and $b$ such that $x_1 < x_2 < x_3$.

**Exercise.** Show that there are $\binom{b+2}{3}$ ways to choose $x_i$ in the last template in the table.

**Exercise.** Show that there are $\binom{b+1}{3}$ ways to satisfy each of the other four templates, and verify that the total number of ways to fill in the templates is $b^3$.

**Exercise.** How many ways are there to fill in the template

$$x_1 \le x_2 < x_3 \le x_4$$

with integers from 1 to $b$?

Given a permutation $\sigma$, and the associated template, how many ways can one fill in the template with $x_i$? If $\sigma$ has $k$ descents then there are $k$ possible equalities in the template; to fill in the template we must choose $m$ of those $k$ locations where we will actually have equality, and then choose $n - m$ distinct elements of $\{1, \ldots, b\}$ to fill into the template. In other words, we choose a total of $n$ objects out of the set

$$\{1, \ldots, b\} \cup D,$$

where $D$ is the $k$-element set of descent locations. There are $\binom{b+k}{n}$ ways to choose $n$ objects out of $b+k$.

Therefore the total number of all $n$-tuples $(x_1, \ldots, x_n)$ is

$$\sum_{\sigma} \binom{b+k_\sigma}{n},$$

summed over all permutations $\sigma$, where $k_\sigma$ is the number of descents of $\sigma$. Collecting terms with the same number, $k$, of descents shows that the grand total is given by the sum on the right-hand side of Worpitzky's identity, i.e.,

$$b^n = \sum_{\sigma} \binom{b+k_\sigma}{n} = \sum_{k=0}^{n-1} \left\langle {n \atop k} \right\rangle \binom{b+k}{n}.$$

This proves the identity for positive integers $b$. Since both sides of the identity are polynomials in $b$, and the polynomials are equal for all positive integers $b$, it follows that the two polynomials are equal.

## 8 Posets

Suppose that two jugglers are allowed to throw objects at each other. More precisely, suppose that their throw times coincide and that they can throw a ball either to themselves or their partner. Then a juggling pattern (without collisions) could be viewed as a rightward moving permutation of a double-line.

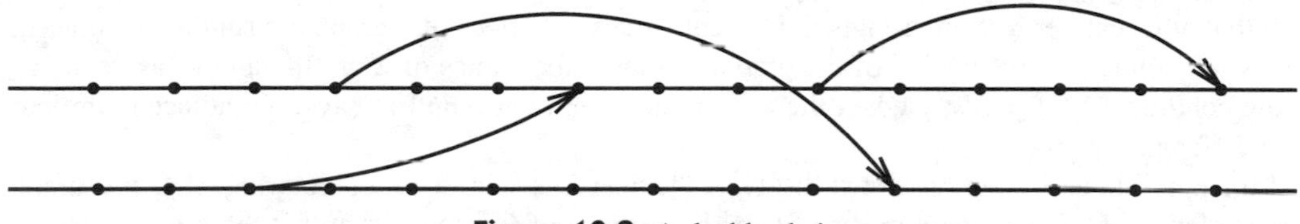

**Figure 10.8.** A double chain

Note that two throws are made at every time, since each juggler makes a throw. If we require that the pattern be periodic, then we can imitate the previous analysis and look at permutations of a finite "double-chain" of length $n$ and with $2n$ positions.

We carried this analysis through and were surprised to find that the total number of $b$-ball patterns of period $n$ again had an elegant description: there are $b^n(b-1)^n$ patterns with $b$ objects of length $n$. The simplicity of this surprised us. What's going on?

After some further experimentation, we discovered that the result extends to a much more general situation, involving partially ordered sets (posets). However, as noted earlier, we don't as yet have a juggling interpretation for permutations on posets.

A **poset** is a set on which there is a relation $<$ that is transitive ($x < y$ and $y < z$ imply that $x < z$), irreflexive ($x < x$ is false), and anti-symmetric (at most one of $x < y$ and $y < x$ is true).

The relation $<$ on $P$ has all of the properties of less-than on numbers except that for given $x$ and $y$ in $P$ we do not require that one of $x < y$ and $y < x$ always be true.

One example is the finite "chain" consisting of the integers $1, \ldots, n$ with the usual $<$ relation; in this case every pair of distinct elements is comparable, i.e., for $x \neq y$, either $x < y$ or $y < x$.

Another example is the set of all subsets of the $n$-element set $\overline{n}$ where the $<$ relation is "is a proper subset of". There are $2^n$ elements of this poset, and many pairs of elements are not comparable.

If $P$ is a finite poset, then its **incomparability graph** $G_P$ has the elements of $P$ as vertices, and there is an edge joining vertices $x$ and $y$ if and only if $x$ and $y$ are incomparable in $P$, i.e., neither $x < y$ or $y < x$ is true.

If $P$ is the $n$-element chain given above, then $G_P$ has $n$ vertices and no edges. If $P$ is the set of subsets of $\overline{n}$, under strict inclusion, then an edge joins two subsets if neither is contained in the other.

Let $P$ be a finite poset. If $\pi$ is a permutation of the elements of $P$, then an element $x$ is a **drop** of $\pi$ if $\pi(x) < x$. We let $\delta_P(k)$ be the number of permutations $\pi$ of $P$ that have $k$ drops.

**Exercise.** Let $P$ be the double-chain of length 3. In other words, $P$ has elements $\{0, 1, 2, 0', 1', 2'\}$ with the order relation induced by the usual relation on the underlying integers. Thus $i$ and $i'$ are incomparable, but every other pair of elements $x, y$ satisfies $x < y$ or $y < x$. How many permutations of $P$ have $k$ drops, for $k = 0, 1, 2, 3, 4, 5$?

By analogy with the case of a finite chain of length $n$, and the two-juggler analysis referred to above, we are led to consider the expression

$$\sum_k \delta_P(k) \binom{b+k}{n}.$$

In order to evaluate this we need a new idea!

# 9 Graph Colorings

A legal coloring of a graph $G$ is a mapping $\lambda: G \to \mathbf{Z}^+$ from the vertices of $G$ to positive integers such that adjacent vertices are assigned different values. (Vertices of a graph are said to be adjacent if they are connected by an edge of the graph.) Usually, the values of $\lambda$ are thought of as "colors", so the condition $\lambda(v) \neq \lambda(v')$ when there is an edge between $v$ and $v'$ says that adjacent vertices have different colors.

If the possible colors are restricted to $\{1, \ldots b\}$, then $\lambda$ is said to be a $b$-coloring. If $G$ is a graph then we let $\chi_G(b)$ denote the number of legal $b$-colorings of $G$. It turns out that the function $\chi_G(b)$ is always a polynomial in $b$, and this first arose in the literature nearly a hundred years ago in connection with attempts to prove the celebrated 4-color map-coloring conjecture (which is now a theorem!).

The goal of this section is to prove that the sum mentioned at the end of the previous section is given by the value, at $b$, of the chromatic polynomial of the incomparability graph $G_P$ of the poset $P$, i.e., that

$$\sum_k \delta_P(k) \binom{b+k}{k} = \chi_{G_P}(b)\ .$$

**Exercise.** Verify that if $P$ is the double-chain of length $n$ referred to above then $\chi_{G_P}(b) = b^n(b-1)^n$.

To prove this we follow the earlier outline: we count legal $b$-colorings of $G_P$.

First we need to make one technical remark in order to clarify later notation. We explicitly number the elements of $P$ in such a way that if the $i$th vertex is less than the $j$th vertex in $P$ then $i < j$ (but not necessarily conversely). This is sometimes called a **linear extension** of the partial order on $P$ ([7, p. 110]).

**Exercise.** Prove that every partial order on a finite set can be extended to a complete order. (An order relation "$<$" on a finite set is **complete** if for all $x$ and $y$ either $x < y$ or $y < x$ is true.)

For notational convenience we will think of $P$ as $\{0,\ldots,n-1\}$, but this means that when we refer to an order on $P$ we have to clarify whether this refers to the original partial order on $P$, or the stronger linear order on the corresponding integers; we will refer to the latter as the "total" order on $P$.

Let $\lambda$ be a coloring of the elements of $P$ that gives a legal $b$-coloring of the incomparability graph $G$ associated with a poset $P$, i.e., for all $x$, $1 \le \lambda(x) \le b$, and for all $x$ and $y$, if $x$ and $y$ are incomparable in $P$ then $\lambda(x) \neq \lambda(y)$.

We will now show that $\lambda$ determines a unique permutation $\sigma = \sigma(\lambda)$ of $\{0,\ldots,n-1\}$ in a manner that exactly generalizes our earlier construction.

Let $\sigma$ be a permutation such that

$$(1) \qquad \lambda(\sigma(0)) \le \lambda(\sigma(1)) \le \cdots \le \lambda(\sigma(n-1)),$$

and if $\sigma(i)$ and $\sigma(i+1)$ have the same color, i.e., $\lambda(\sigma(i)) = \lambda(\sigma(i+1))$, then

$$(2) \qquad \sigma(i) > \sigma(i+1) \text{ in } P.$$

A little thought shows that this ordering of the vertices is uniquely determined: inside any run of vertices with the same color our definitions imply that the vertices are linearly (completely) ordered as elements of $P$, and (2) requires that the order within a run reverses the order in $P$, so that the vertices are ordered in a uniquely specified fashion, and $\sigma$ is well-determined.

Thus to every legal coloring $\lambda$ we have associated a permutation $\sigma$ of $P$. We say that $i$ is a **descent** for $\sigma$ if $\sigma(i) > \sigma(i+1)$ in $P$.

Note that the notion of a descent depends on the choice of a linear extension of the partial order on $P$, whereas the notion of a drop depends only on the partial order itself; in this sense the notion of a drop seems more fundamental. However, the basic fact remains the same.

**Lemma 2** *The number of permutations $\sigma$ of $P$ with $k$ descents is equal to the number of permutations $\pi$ with $k$ drops.*

*Proof.* The earlier argument works. Briefly, given $\pi$ we construct $\sigma$ by writing $\pi$ in cycle form, putting the largest element at the beginning of each cycle, and ordering by cycles so that their first elements increase. (Here we use the total order on $P$ coming from the integers.)

Given $\sigma$ we build $\pi$ by forming cycles, closing a cycle when we see a larger number than any that we've seen before. (Again, using the total order on $P$.)

One then checks as before that these constructions are mutually inverse.

Moreover, if $j = \sigma(i)$, $k = \sigma(i+1)$, and $j > k$ (using the order on $P$), then $jk$ must be internal to a cycle of $\pi$, so that $j > k = \pi(j)$; the converse is easily checked. Thus drops of $\pi$ correspond exactly to descents of $\sigma$. □

Strictly speaking, this result is not needed for the proof of the theorem, but it does show that the number of permutations with a given number of descents is independent of the choice of linear extension, i.e., that it is intrinsic to the original partial order.

As in the earlier case, a permutation $\sigma$ determines an "inequality template," and the number of colorings of the template is $\binom{b+k}{n}$ if $\sigma$ has $k$ descents. All in all, we have proved our main theorem for posets.

**Theorem 12**

$$\sum_k \delta_P(k)\binom{b+k}{n} = \chi_{G_P}(b).$$

Taking $b = -1$, we see that the value $\chi_G(-1)$ of the chromatic polynomial is equal to $(-1)^n\delta_P(0)$. By a classic theorem of Stanley, $(-1)^n\chi_G(-1)$ is the number of "acyclic orientations" of the graph $G$, i.e., the number of ways to choose a direction for the edges of $G$ so that there are no directed cycles. Thus drop-free permutations (which certainly sound like something that jugglers should like) correspond to acyclic orientations. It is natural to ask for a bijective proof of this result. Lemma 3 implies that it suffices to find a bijection between descent-free permutations and acyclic orientations.

**Exercise.** Let $L$ be a descent-free permutation of $P$, written in the form of a list of all $n$ elements of $P$ in which no element is larger than its successor. Construct an orientation $\theta$ of the graph $G$ by orienting an edge $\{x, y\}$ from $x$ to $y$ exactly when $x$ occurs earlier than $y$ in the list $L$. Show that $\theta$ is acyclic. Show that $L$ can be uniquely recovered from $\theta$ as follows: let the first element of $L$ be the smallest source in $G$. (A vertex is a source if it has no incoming arrows; you need to verify that in fact there is a smallest source.) Remove this element from $P$ and $G$, and apply the same procedure to the smaller poset to determine the second element of the list. Etc.

**Exercise.** Let $P$ be the union of two chains of length 2, i.e., the poset on $\{a, b, c, d\}$ whose only relations are $a < b$ and $c < d$. Explicitly identify the drop-free and descent-free permutations of $P$, and the acyclic orientations of $G$, and identify how they correspond under the preceding bijections.

**Acknowledgment:** The authors would like to thank Peter Doyle and Oliver Schirokauer for helpful comments on this paper.

## References

[1] J. Buhler and R. Graham, Fountains, Showers, and Cascades, *The Sciences*, Jan-Feb, 1984, 44–51.

[2] ——, A note on the binomial drop polynomial of a poset, *Journal of Combinatorial Theory, Series A* **66** (1994), 321–326.

[3] J. Buhler, D. Eisenbud, R. Graham, and C. Wright, Juggling Drops and Descents, in *Organic Mathematics*, 133–154, CMS Proc., **20**, AMS, Providence, RI; appeared originally, without the appendix, in *Amer. Math. Monthly* **101** (1994), 507–519.

[4] R. Graham, D. Knuth, and O. Patashnik, *Concrete Mathematics*, Addison Wesley, 1989.

[5] Marshall Hall, Jr., A combinatorial problem on abelian groups, *Proc. Amer. Math. Soc.* **3** (1952), 584–587.

[6] A. Lewbel and P. Beek, The Science of Juggling, *Scientific American*, November 1995, **273**, 92–97.

[7] R. Stanley, *Enumerative Combinatorics*, Volume I, Cambridge University Press, 1997.

[8] M. Truzzi, On keeping things up in the air, *Natural History*, December 1979, 44–55.

# 11

# Platonic Divisions of Space[1]

Jean Pedersen
*Santa Clara Univeristy*

## 1 Introduction

The main purpose of this article is to ask, and answer by elementary methods, the question: *How many bounded and unbounded regions in space result when the planes of the Platonic solids are extended in space?* To set the scene we first discuss, in Section 2, the analogous question in the plane; that is, we ask, and answer, the question "how many bounded and unbounded planar regions result when the sides of regular polygons are extended in the plane."

In order to make the material more self-contained we include, in Section 3, a description of regular polygons and regular polyhedra; by providing the latter, we believe the reader will obtain a clear picture of the Platonic Solids.

In Section 4 we answer the main question, except in the case of the icosahedron, where we refer the reader to [3], [7], or [13] in order to find out how others calculated the number of bounded regions that result when the face planes of the icosahedron are extended in space.

In Section 5 we suggest some questions for the interested reader to ponder.

The author wishes to emphasize that there are no new results in this paper (see [13]), but the approach is more elementary than the classical discussions.

## 2 Divisions of the Plane

The reader might like to try answering the following question before reading further.

*How many bounded and unbounded regions result when the sides of a regular $n$-gon are extended in the plane?*

**Solution** An examination of Figure 11.1 should convince you that the number of unbounded regions created by the extended sides of a regular $n$-gon, $n \geq 3$, is $2n$. You might also observe

[1] This article covers the material in the author's BAMA presentation at San Jose State University on February 6, 2002. However, the author previously published some of the content in [10]. Figures 11.14 and 11.15 are reprinted from [1], while the star part of Figure 11.13 is reprinted from [6]. The rest of the figures were kindly produced by Sylvie Donmoyer for this article.

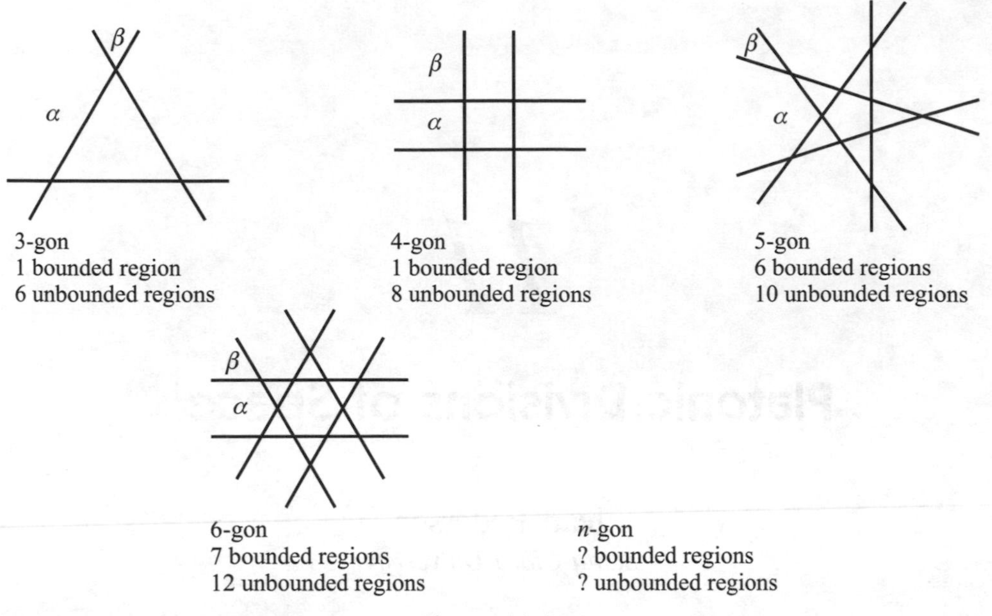


**Figure 11.1.**

that, for any given $n$, the unbounded regions come in precisely two types (labeled as $\alpha$ and $\beta$ in Figure 11.1) which alternate as you scan around the periphery of the figure.

The trickier part of the problem is to find the number of bounded regions. The problem seems to break naturally into the following two cases.

**Case 1** Notice that when $n$ is odd (with $n \geq 3$), each of the $n$ sides, when extended, will intersect each of the other $(n-1)$ sides. Thus, when $n$ is odd, think of drawing the extended sides that produce the final polygon, one at a time, and record the number of new bounded regions created at that stage. We then have the following table where, starting with the third line we see that the number of new finite regions increases one at a time until, finally, there are $(n-2)$ new regions created by the $n$th line.[2]

| The introduction of the extended side number | creates the following *new* bounded regions |
|---|---|
| 1 | 0 |
| 2 | 0 |
| 3 | 1 |
| 4 | 2 |
| 5 | 3 |
| 6 | 4 |
| • | • |
| • | • |
| • | • |
| $n$ | $n-2$ |

[2]It may help you to look at a particular but not special case, like the 7-gon, to see what this table means.

Thus, summing all the entries in the second column, we see that the total number of bounded regions, $B_n$, for a regular $n$-gon, with $n$ odd, is given by

$$B_n = 1 + 2 + 3 + \cdots + (n-2). \tag{1}$$

Using the formula[3] $\sum_{i=1}^{N} i = \frac{N(N+1)}{2}$ with $N = (n-2)$ we see that (1) can be rewritten rather neatly as

$$B_n = \frac{(n-2)(n-1)}{2}. \tag{2}$$

It is interesting to note that formula (2) is, in fact, true for $n \geq 1$ although, in some sense, the argument started producing finite regions only when $n = 3$.

**Case 2** Now we turn to the case where $n$ is even. We first observe that, in this case, each of the $n$ sides when extended will intersect each of the other $(n-2)$ sides that are not parallel to it. For convenience let us assume $n = 2k$, $k \geq 2$, and proceed much the same way as in Case 1, except that now we need to take into account the fact that, when we draw the $(k+1)$st extended side, it will be parallel to the first extended side that was drawn. Thus we have the following table, naturally broken into two parts, where, starting with the third line of Part 1 the number of new finite regions increases one at a time until, in the middle there are $(k-2)$ new regions created by the extended $k$th side. Then in Part 2—because the $(k+1)$st side is parallel to the first side—there are also $(k-2)$ new regions created by the extended $(k+1)$st side at the top of the second column in the table.

| The introduction of the extended side number | creates the following *new* bounded regions |
|---|---|
| 1 | 0 |
| 2 | 0 |
| 3 | 1 |
| 4 | 2 |
| 5 | 3 |
| • | • |
| • | • |
| • | • |
| $k$ | $k-2$ |

**Part 1**

| The introduction of the extended side number | creates the following *new* bounded regions |
|---|---|
| $k+1$ | $k-2$ |
| $k+2$ | $k-1$ |
| $k+3$ | $k$ |
| • | • |
| • | • |
| • | • |
| $2k-2$ | $2k-5$ |
| $2k-1$ | $2k-4$ |
| $2k$ | $2k-3$ |

**Part 2**

Thus, by summing all the entries in the second columns of Part 1 and Part 2, we see that the total number of bounded regions, $B_{2k}$, for a regular $2k$-gon is

$$B_{2k} = (1 + 2 + 3 + \cdots + (2k-3)) + (k-2), \tag{3}$$

where the last term occurs because of the repeat at the top of the table in Part 2.

Again we use the formula $\sum_{i=1}^{N} i = \frac{N(N+1)}{2}$, but this time with $N = 2k-3$, to write (3) as

$$\begin{aligned} B_{2k} &= \frac{(2k-3)(2k-2)}{2} + (k-2) \\ &= (2k-3)(k-1) + (k-2) \\ &= 2(k-1)^2 - 1 \end{aligned}$$

[3] See [12] for a nice discussion of how to derive this and other summation formulas.

Recalling that $n = 2k$, we see that

$$B_n = \frac{1}{2}(n-2)^2 - 1,$$

where $n$ is an even number $\geq 4$.

## 3 Regular Polyhedra

First let us be clear about what we mean by a polyhedron. We begin by recalling that a polygon consists of edges hinged together at vertices and does not include its interior. Similarly, we say that a polyhedron consists of faces hinged together at edges, and it does not include its interior. Thus, strictly speaking, a polyhedron is a *surface* not a *solid*, although it is sometimes loosely referred to as a solid (for example, the Platonic Solids, which we are about to discuss).

A polygon is **connected**, meaning that it is all in one piece; the polyhedra we are about to consider are also connected in this sense. Further, just as we lay particular emphasis on *convex* polygons, we will also confine our attention here to *convex polyhedra.* We call a polyhedron **convex** if given any two points $P$ and $Q$ of the polyhedron, the straight line segment $PQ$ consists of points of the polyhedron or its interior.

It is customary to name polyhedra in a manner similar to the way we name polygons—that is, just as we incorporated the number of sides of a polygon into its name, we incorporate the number of faces a polyhedron possesses into its name. Here are the names of some of the better known polyhedra. You will note that the part preceding *hedron* designates the number of faces (*poly* means "many" in Greek). In the case of convex polyhedra, any of its faces can be used as a base when a model is set on a flat surface. This explains the use of the term *hedron* which is the Greek word meaning "base" or "seat."

| A polyhedron with: | is called a: |
|---|---|
| 4 faces | tetrahedron |
| 5 faces | pentahedron |
| 6 faces | hexahedron |
| 7 faces | heptahedron |
| 8 faces | octahedron |
| 10 faces | decahedron |
| 12 faces | dodecahedron |
| 14 faces | tetracaidecahedron (cai means "and") |
| 15 faces | pentacaidecahedron |
| 16 faces | hexacaidecahedron |
| 20 faces | icosahedron |

It is well known that a *regular* convex polygon (with 3 or more sides) is a polygon with all sides and with all angles equal.[4] But now we want to know: **What would be an appropriate analogous requirement for a *regular* polyhedron?**

We reason that, since the faces of polyhedra are analogous to the sides of a polygon, it should make sense to require that every face on a regular polyhedron should be the same regular polygon. We can readily see, however, that this wouldn't be a strong enough requirement. For example,

[4]The triangle is the only case in which each of these conditions implies the other.

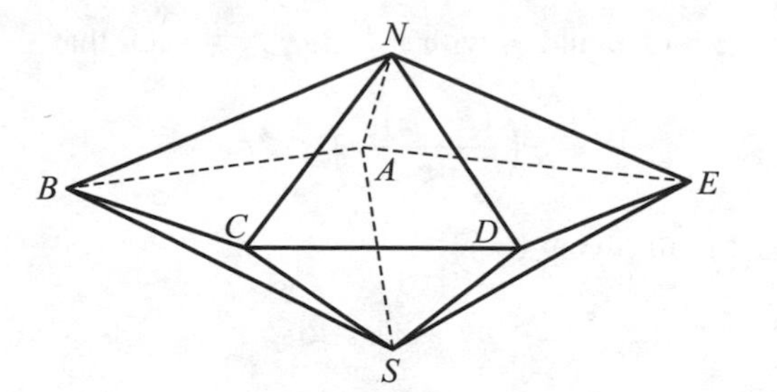


**Figure 11.2.** A pentagonal dipyramid constructed from 10 equilateral triangles

observe the pentagonal dipyramid shown in Figure 11.2. If you view the pentagonal dipyramid at the North Pole, or at the South Pole, you see 5 triangles coming together at $N$ and $S$; but if you view it from a side vertex you see 4 triangles coming together at $A$, $B$, $C$, $D$, and $E$. Surely a regular polyhedron should look the same at every vertex (or the same when you look straight at any edge or face). Let us impose this restriction on vertices and then see, first of all, if any polyhedra exist that satisfy our restriction and, second, if the resulting polyhedra deserve to be called "regular."

Let $p$ stand for the number of sides of each (regular) face. Then it is well known (see page 90 of [4] for an elementary explanation) that the interior angle of the regular $p$-gon, measured in radians,[5] is

$$\frac{(p-2)\pi}{p}.$$

Now if we consider the arrangement of $q$ of these regular $p$-gons about a single vertex, we see that the sum of the angles about each vertex is

$$q\left(\frac{(p-2)\pi}{p}\right), \qquad p \geq 3, q \geq 3.$$

The requirement of convexity now comes into play. It is a fact, first remarked by Euclid (which we illustrate, but do not prove here), that the sum of the face angles about any vertex on a convex polyhedron must be less than $2\pi$. However, we can see from Figure 11.3 that the converse is not true, that is, the sum of the face angles at every vertex may be less than $2\pi$ without the polyhedron being convex.

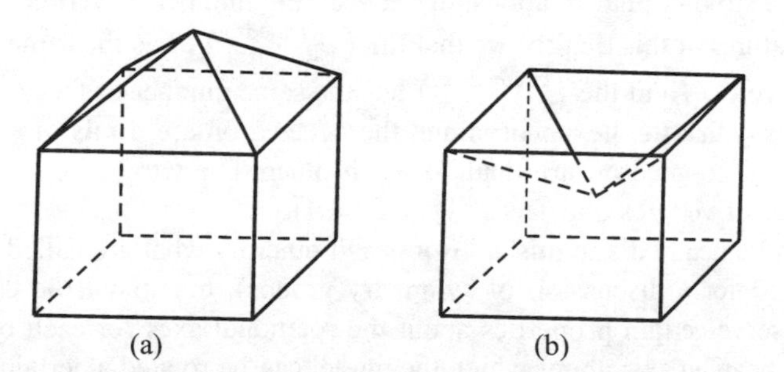


**Figure 11.3.** (a) Convex polyhedron (b) Nonconvex polyhedron with the same face angles at each vertex as its convex cousin on the left

[5]Readers not familiar with radian measure should just remember that $180° = \pi$ radians; henceforth the word "radians" is almost always suppressed.

Let us now look for values of $p$ and $q$, with $p \geq 3$, $q \geq 3$, such that

$$q\left(\frac{(p-2)\pi}{p}\right) < 2\pi,$$

or, after dividing by $\pi$ and multiplying by $p$,

$$pq - 2q < 2p.$$

Straightforward algebra then gives the following sequence of equivalent inequalities:

$$\begin{aligned} pq - 2p - 2q &< 0 \\ pq - 2p - 2q + 4 &< 4 \\ p(q-2) - 2(q-2) &< 4 \\ (p-2)(q-2) &< 4 \end{aligned} \tag{4}$$

Notice that (4) is symmetric in $p$ and $q$—and thus if you find any values for $p$ and $q$ that satisfy (4) you can find the "complementary" (or dual) solution by exchanging the values for $p$ and $q$. Notice also that since $p \geq 3$ and $q \geq 3$, it follows that $p - 2 \geq 1$ and $q - 2 \geq 1$. Thus if (4) is to be satisfied then the product $(p-2)(q-2)$ must be 1, 2, or 3.

We leave it to the reader to verify that the only possible solutions to (4) are the five shown in the table concerning the Platonic Solids below.[6] These five solids are known as the Platonic Solids. The name of each polyhedron and the corresponding number of vertices ($V$), edges ($E$), and faces ($F$) are also shown in the table. Illustrations of the solids appear in Figure 11.4.

| $p$ | $q$ | Name of solid | $V$ | $E$ | $F$ |
|---|---|---|---|---|---|
| 3 | 3 | Tetrahedron | 4 | 6 | 4 |
| 4 | 3 | Hexahedron (cube) | 8 | 12 | 6 |
| 3 | 4 | Octahedron | 6 | 12 | 8 |
| 5 | 3 | Dodecahedron | 20 | 30 | 12 |
| 3 | 5 | Icosahedron | 12 | 30 | 20 |

## Some Facts about the Platonic Solids

Notice that there is a striking dual relationship between the number of vertices and the number of faces.[7] Important features of this duality are that the $\binom{\text{hexahedron}}{\text{dodecahedron}}$ has the same number of faces as the $\binom{\text{octahedron}}{\text{icosahedron}}$ has vertices and the $\binom{\text{octahedron}}{\text{icosahedron}}$ has the same number of faces as the $\binom{\text{hexahedron}}{\text{dodecahedron}}$ has vertices. So we say that the hexahedron and the octahedron are **duals** of each other; similarly the dodecahedron and icosahedron are duals of each other. The tetrahedron is **self-dual**, since it has the same number of vertices and faces ($V = F = 4$).

These models can be neatly distinguished from each other by what are called **symmetry groups** (see Chapter 8 of [5] for a discussion of symmetry groups), but it will be enough here for the reader simply to observe certain properties about the rotational axes for each of the models. By a **rotational axis** we mean an axis about which the model can be rotated a certain fraction of $2\pi$ and still occupy the same space as in its original position. With the models in hand it is not difficult to see that

[6] The solids are arranged in the table so that the number of faces ($F$) is in ascending order.

[7] We have already observed the dual relationship for the corresponding $p$ and $q$.

(a) The tetrahedron has

  4 axes about which it can be rotated $\pm\frac{2\pi}{3}$
  and 3 axes about which it can be rotated $\frac{2\pi}{2}$.

(b) Both the hexahedron and the octahedron have

  3 axes about which they can be rotated $\pm\frac{2\pi}{4}$
  4 axes about which they can be rotated $\pm\frac{2\pi}{3}$
  and 9 axes (3 joining opposite faces of the cube or vertices of the octahedron; and 6 joining the centers of opposite edges) about which they can be rotated $\frac{2\pi}{2}$.

(c) Both the dodecahedron and the icosahedron have

  6 axes about which they can be rotated $\pm\frac{2\pi}{5}$ or $\pm\frac{4\pi}{5}$
  10 axes about which they can be rotated $\pm\frac{2\pi}{3}$
  and 15 axes about which they can be rotated $\frac{2\pi}{2}$.

We say that the tetrahedron possesses **tetrahedral symmetry**, the hexahedron and the octahedron possess **octahedral symmetry**, and the dodecahedron and icosahedron possess **icosahedral symmetry**. Models belonging to these three symmetry groups are easily distinguished from each other, since octahedral symmetry is the only one allowing rotations of $\frac{2\pi}{4}$, and icosahedral symmetry is the only one allowing rotations of $\frac{2\pi}{5}$.

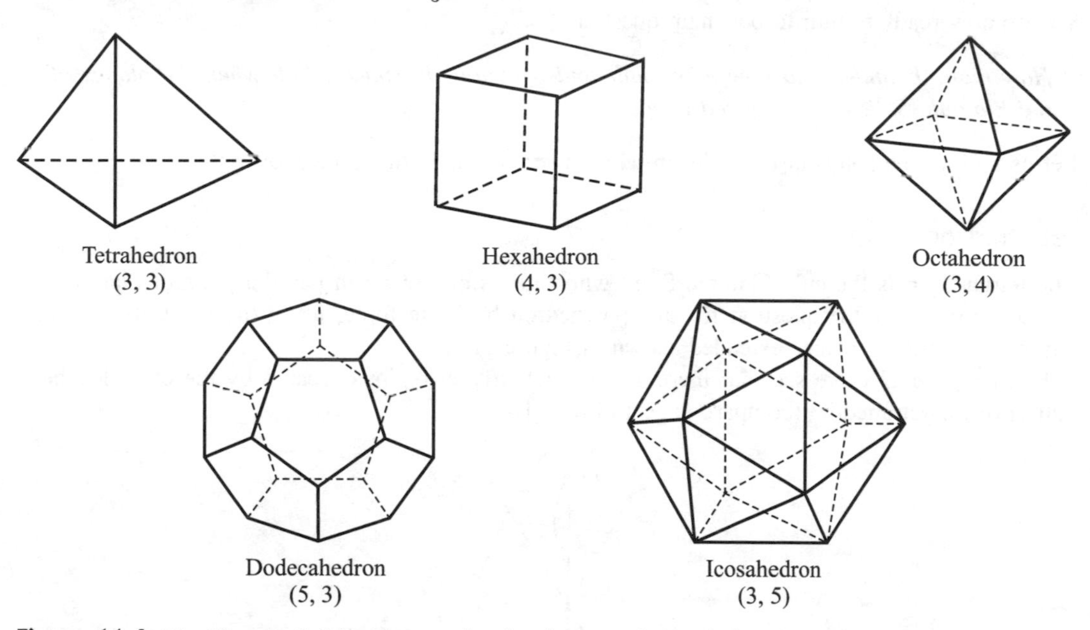


**Figure 11.4.** The Platonic Solids: The notation $(p, q)$ means that each face is a regular $p$-gon and $q$ faces come together at each vertex.

Now, it turns out that there are several classical ways to construct these polyhedra (see [1, 9, 14]). However, one particularly easy way to construct these models, which was discovered by the author, is to braid them from straight strips of paper that have been suitably folded by an iterative folding procedure (see [6] for complete instructions about how to do this). The unexpected payoff is that these braided models can then be used to help answer the main query of this paper. Figure 11.5 shows some of the braided models from [6] that will be playing an important role in the next section.

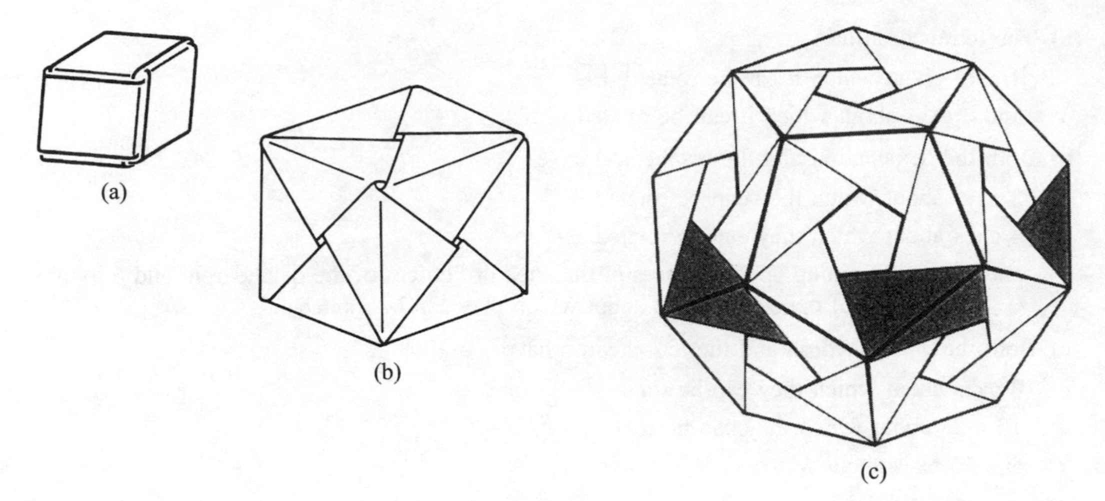


**Figure 11.5.** (a)A cube braided from 3 straight strips (b) A cube (called the diagonal cube) braided from 4 straight strips (c) A dodecahedron (called the golden dodecahedron) braided from 6 straight strips

# 4 Answering the Main Question

We are now ready to turn to our main question:

*How many bounded and how many unbounded regions in space result when the planes of the Platonic Solids are extended in space?*

Let us look at the solids in the order in which they appear in the table above.

### Tetrahedron

The tetrahedron is the only Platonic Solid whose faces do not lie in parallel pairs of planes. We are able to answer the question for this polyhedron by brute force, using the illustration of the tetrahedron with its planes extended shown in Figure 11.6.

From Figure 11.6 we see that there are, in fact, fifteen regions created by the extended face planes of the tetrahedron (compare page 195 of [5]):

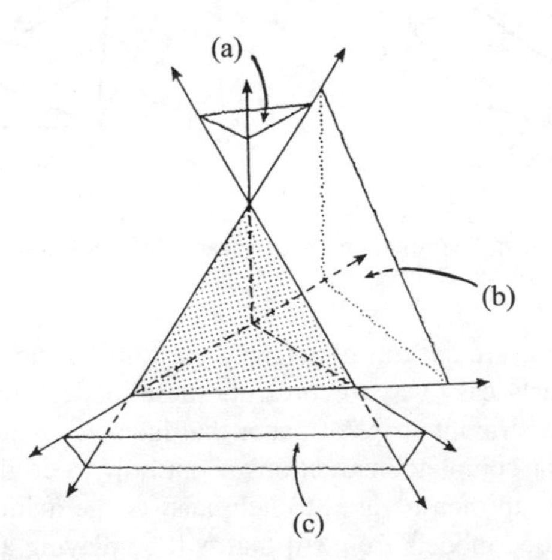


**Figure 11.6.** The extended face planes of a tetrahedron

1 bounded region, the tetrahedron itself with its interior
+ (a) 4 unbounded trihedral regions from its vertices
+ (b) 4 unbounded truncated trihedral regions from its faces
+ (c) 6 unbounded wedges from its edges
= 1 bounded region + 14 unbounded regions.

## Hexahedron (or cube)

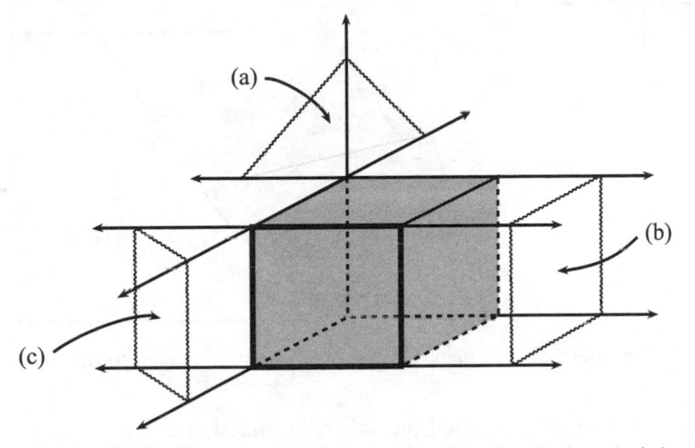


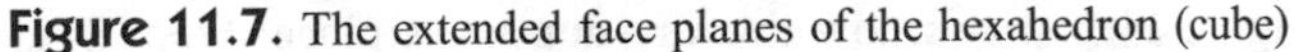
**Figure 11.7.** The extended face planes of the hexahedron (cube)

The extended face planes of the cube partition space into 27 pieces as seen in Figure 11.7:

1 bounded region, the cube itself with its interior
+ (a) 8 unbounded trihedral regions from its vertices
+ (b) 6 unbounded square prisms from its faces
+ (c) 12 unbounded wedges from its edges
= 1 bounded region + 26 unbounded regions.

But now we observe that *because* its faces lie in parallel pairs of planes we are able to make a braided model of it (Figure 11.8). Notice that the edges of the three strips used to create the braided model lie in six planes which intersect each other to form the surface of the cube. The details come later but, meanwhile, whenever braided models appear please compare the various features (holes, 1-thickness regions, and 2-thickness regions) of the braided model with the same features of the associated unbounded regions.

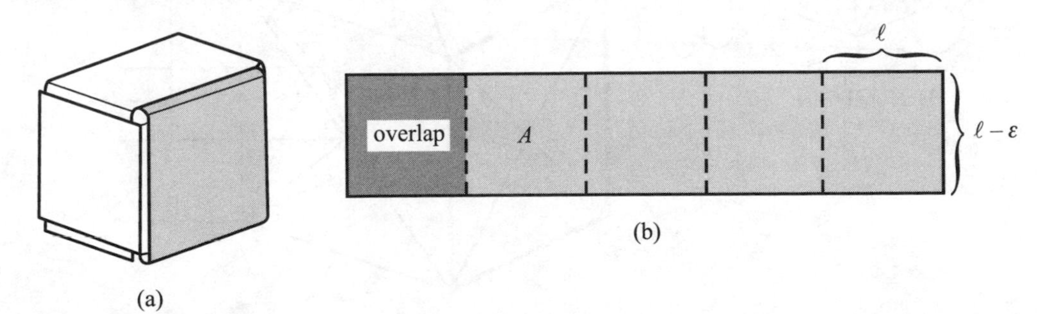


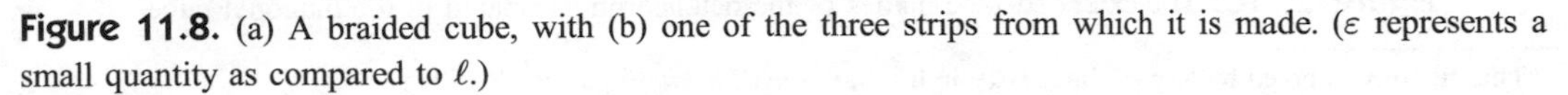
**Figure 11.8.** (a) A braided cube, with (b) one of the three strips from which it is made. ($\varepsilon$ represents a small quantity as compared to $\ell$.)

## Octahedron

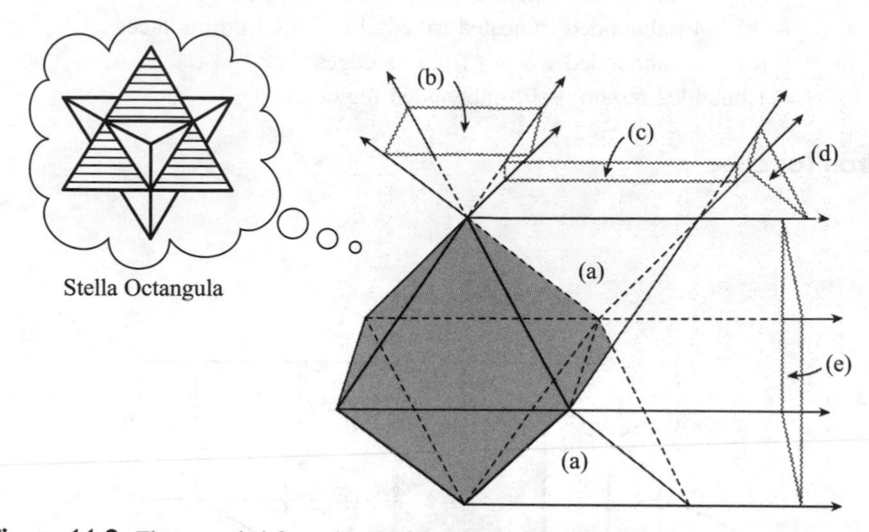


**Figure 11.9.** The extended face planes of the octahedron as related to the Stella Octangula

The face planes of the regular octahedron, when extended, form a slightly more complicated division of space. Figure 11.9 shows how this division takes place. First, 8 tetrahedra, like those marked (a) appear on the octahedron's faces; the visible surfaces of these 8 tetrahedra form the **stella octangula.**[8] Then the unbounded regions are formed. The bubble in Figure 11.9 shows the stella octangula and the main figure shows two of the tetrahedra and one each of the various kinds of unbounded regions. Figure 11.10 shows how the unbounded regions are related to the various regions on the surface of the braided diagonal cube.

A new feature is that not all the unbounded regions grow straight out of a vertex, a face, or an edge of the original octahedron; some grow out of vertices, edges, or faces of one of the tetrahedra. The complete count is:

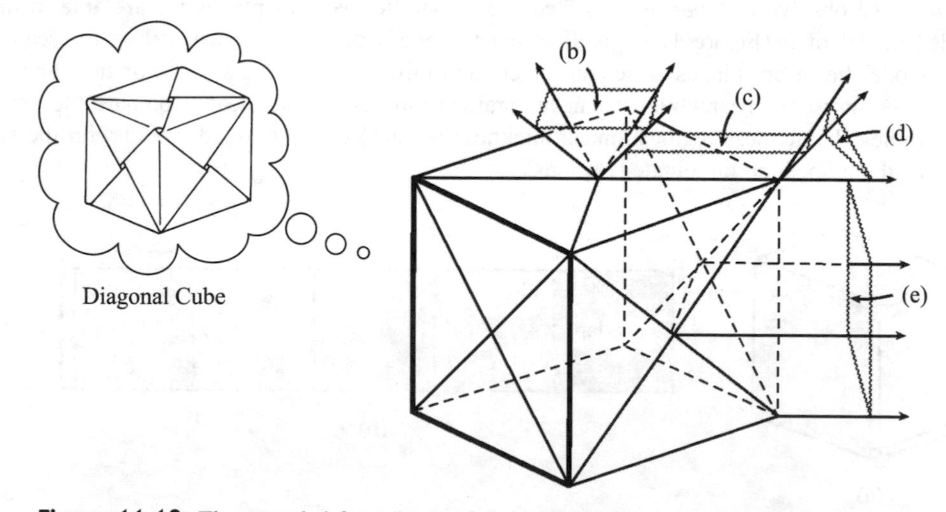


**Figure 11.10.** The extended face planes of the octahedron as related to the diagonal cube

[8]This figure was noted by Kepler about 1619 in his *Harmonia Mundi,* Propositio XXVI.

1 bounded region, the octahedron itself with its interior
+ (a) 8 bounded tetrahedral regions on its faces
+ (b) 6 unbounded tetrahedral regions from its vertices
+ (c) 24 unbounded wedges from the edges of the 8 tetrahedra which do not coincide with edges of the original octahedron
+ (d) 8 unbounded trihedral regions from the outside vertices of the 8 tetrahedra
+ (e) 12 unbounded regions having two finite faces on adjacent tetrahedra and four unbounded faces
= 9 bounded regions +50 unbounded regions.

It may be helpful at this point to construct the diagonal cube shown in Figure 11.11. The procedure is almost self-evident, especially if you remember that every strip must go alternately over and under the other strips on the model. It may help to secure the center square in the original layout with tape that can be removed once the model is finished. All the ends should tuck in neatly on the finished model.

But, in this case, how does the octahedron fit in the diagonal cube? The clue lies in the planes formed by the "open" bases and tops of the four antiprisms made from four strips. These planes define four pairs of mutually intersecting parallel planes positioned symmetrically about a point. Since the braided model has octahedral symmetry, these eight planes form the faces of a regular octahedron. Note that four planes, one from each pair, intersect at the center of each face of the diagonal cube; the six center points are the vertices of the original octahedron, and the eight vertices of the cube are the outside vertices of the associated stella octangula.

Having constructed the diagonal cube you can then read off from its surface the number and nature of the unbounded regions in space defined by the extended face planes of the octahedron. The same is true for the braided cube of Figure 11.8. Notice, for example, that the small square hole shown in the center of each face of the diagonal cube in the bubble of Figure 11.10 represents an

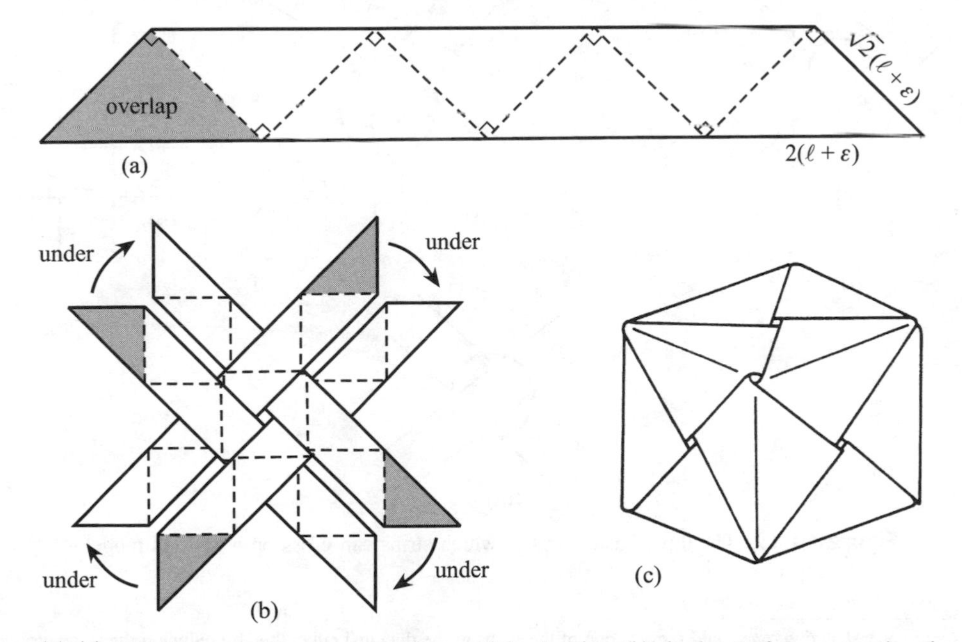


**Figure 11.11.** The diagonal cube showing (a) one of the 4 strips, (b) the layout for constructing the cube, and (c) the completed cube.

unbounded polyhedral region, as does the small triangular hole at the vertex of the same diagonal cube. (The small triangular hole at the vertex of the cube in Figure 11.8(a) also represented an unbounded polyhedral region.) Notice, too, that where there is just one thickness of paper, along the diagonals of the faces of the diagonal cube, each of these regions represents an unbounded wedge. (The same was true for the single thickness that appeared along the edges of the cube in Figure 8(a).) Finally, notice that wherever there is a two-thickness region on the braided model it represents an unbounded region with a polygonal cross section. (A two-thickness region appeared on each face of the cube in Figure 11.8(a).)

Now that we have seen the usefulness of the braided models in the cases of the cube and the octahedron, it should occur to us to take a careful look at the situations that can occur on the surface of any such model. This is, in fact, essential if one wishes to understand similar models for the dodecahedron and the icosahedron. Of course, since in this article we are only concerned with the Platonic Solids, we impose the requirement that the individual strips of our braided models

(a) must have central symmetry;

(b) must cross over every other strip on the model at diametrically opposite places; and

(c) the entire braided configuration must have the symmetry group of the original surface being discussed.[9]

Because of restriction (b) no other lines of intersection between the planes defined by the edges of the strips can occur, and hence there can be no other bounded regions formed by extending the planes *outside* the braided model. Figure 11.12 then shows the three basic situations on the braided model that produce unbounded regions:

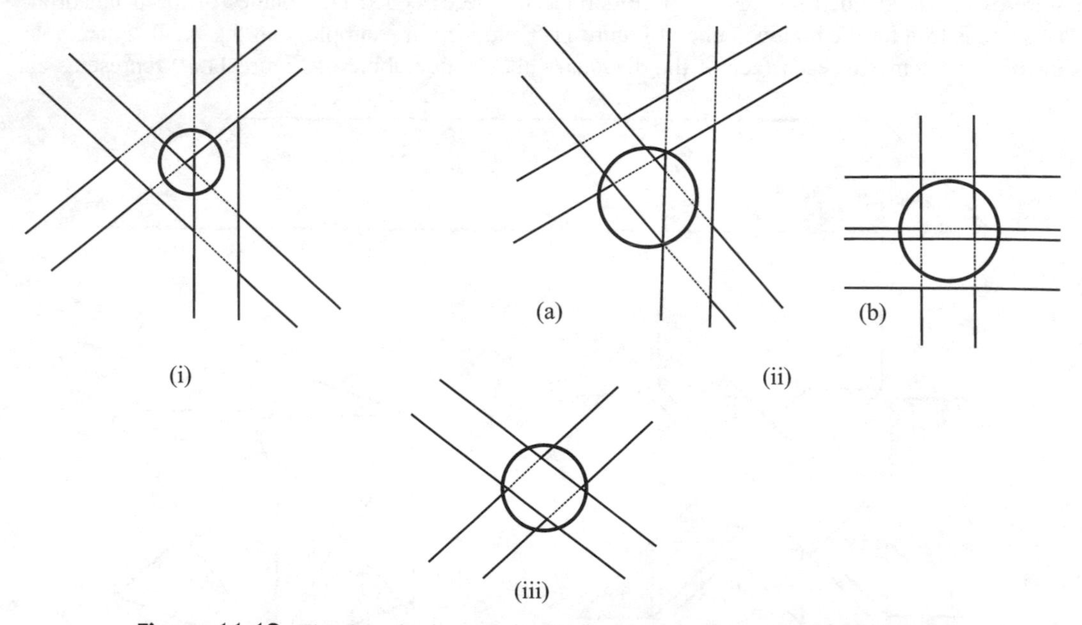


**Figure 11.12.** The three basic ways in which strips can cross on a braided model

[9] In a more general setting, one could remove one of the strips in the diagonal cube, thereby reducing the symmetry group from that of octahedral symmetry to merely dihedral symmetry (see page 248 of [5] for details about this symmetry group) and use the resulting model to discover the unbounded regions of a rhombic hexahedron.

(i) An unbounded polyhedral region occurs whenever there is a hole in the braided model. The number of sides of the polyhedral region is easily determined by looking at the number of sides of the polygon surrounding the hole.

(ii) An unbounded region with 4 unbounded faces and 4 unbounded edges; in the situation shown in (a) the unbounded regions will be "trapezium-like" while in the situation shown in (b) the unbounded regions will be "wedges off edges."

(iii) An unbounded "parallelogram-like" region with 4 unbounded faces and 4 unbounded edges occurs wherever there is a region of 2 thicknesses on the braided model.

## Dodecahedron

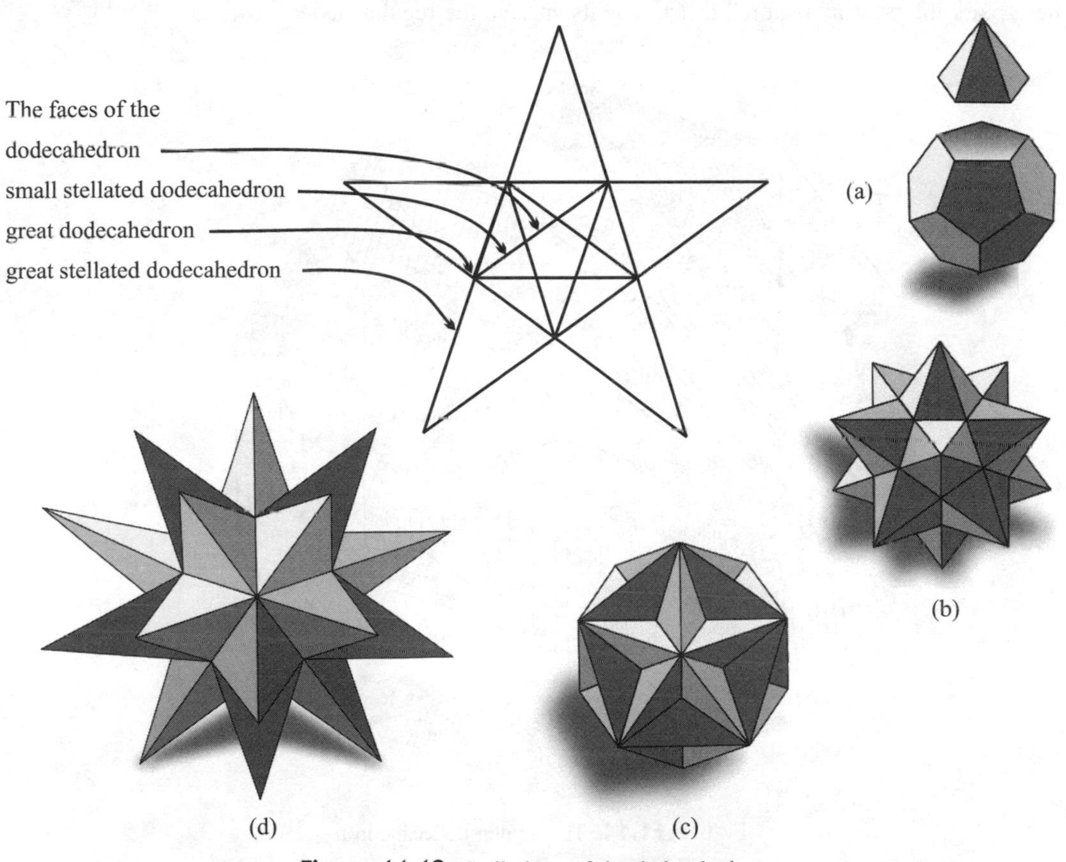


**Figure 11.13.** Stellations of the dodecahedron

Figure 11.13 at the top shows at its center the pentagonal face of the dodecahedron (Figure 11.13(a)). Next, at the top, is the star pentagram, 12 of which interpenetrate each other to form the small stellated dodecahedron of Figure 11.13(b). Then there appears a larger pentagon, 12 of which interpenetrate each other to form the great dodecahedron of Figure 11.13(c). Finally there is the larger pentagram, 12 of which interpenetrate each other to form the great stellated dodecahedron of Figure 11.13(d).[10] These stellations, in fact, produce all of the bounded regions created by extending the face planes of the dodecahedron. We can enumerate them as follows:

[10]In fact, the great stellated dodecahedron fits inside the golden dodecahedron with its outermost vertices touching the vertices of the golden dodecahedron (see [6] for details).

from (a) 1 bounded region, the dodecahedron itself with its interior
from (b) + 12 pentagonal pyramids on the dodecahedron's faces (producing the small stellated dodecahedron)
from (c) + 30 wedge-like tetrahedra on the dodecahedron's edges (producing the great dodecahedron)
from (d) + 20 triangular dipyramids on the dodecahedron's vertices (producing the great stellated dodecahedron)
= 63 bounded regions.

We now use the golden dodecahedron of Figure 11.14 to find the unbounded regions. Notice that this model is formed of 6 strips. There are thus 6 pairs of parallel planes, and the zones between these pairs intersect in a "core" that has as its surface the regular dodecahedron.

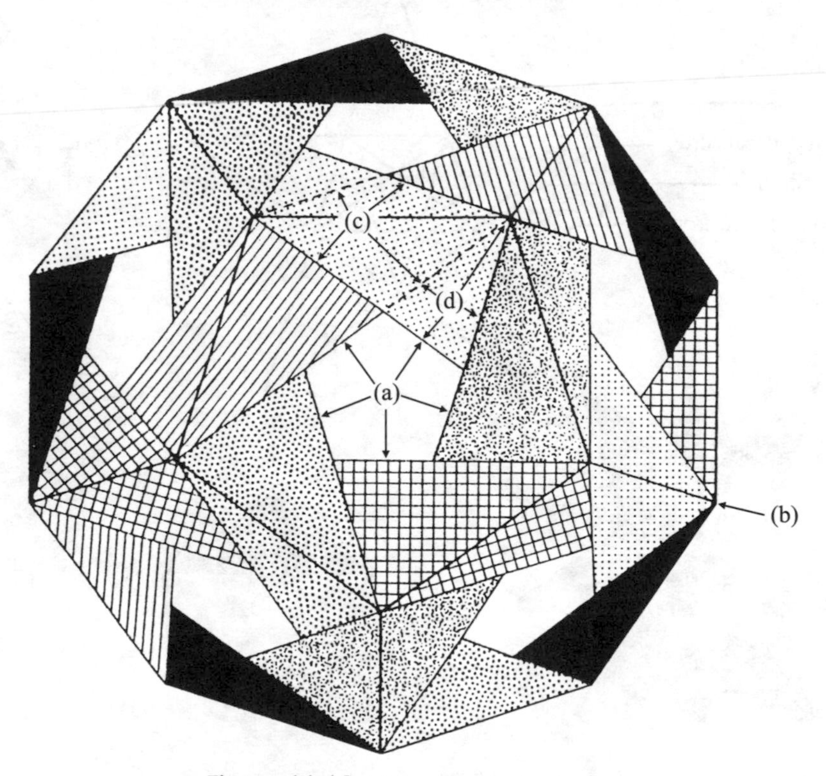


**Figure 11.14.** The golden dodecahedron

Using the surface of the golden dodecahedron, we can count the unbounded regions formed by the face planes of the "core" dodecahedron as follows:

from (a) 12 unbounded pentahedral regions from the pentagonal holes on its faces;
from (b) + 20 unbounded trihedral regions from the triangular holes at its vertices;
from (c) + 30 unbounded regions with "parallelogram-like" cross sections from its edges;
from (d) + 60 unbounded regions with "trapezium-like" cross sections from five regions off the pentagonal holes on each of the twelve faces
= 122 unbounded regions.

## Icosahedron

Not surprisingly, the situation for the icosahedron is the most difficult. The bounded regions are very much more complicated, since there are 59 different kinds of intersections before the 20 face planes of the icosahedron go off into space (see [3]). According to an impressive argument in [7] there are 473 bounded regions for the icosahedron. So let us take this result as proved and try to see how to enumerate the unbounded regions for the icosahedron.

Of course, what is needed is a braided model with 10 strips, having icosahedral symmetry. To see what motivated the discovery of this model the reader is referred to another braided construction of the dodecahedron (see [5] or [6]) in which the regular pentagonal dodecahedron is constructed from straight strips each containing six trapezium sections. What was required, for producing a model to help solve our current problem, was to make each of those six strips narrower so that four more could be introduced into the figure to complete the symmetry group for the dodecahedron.

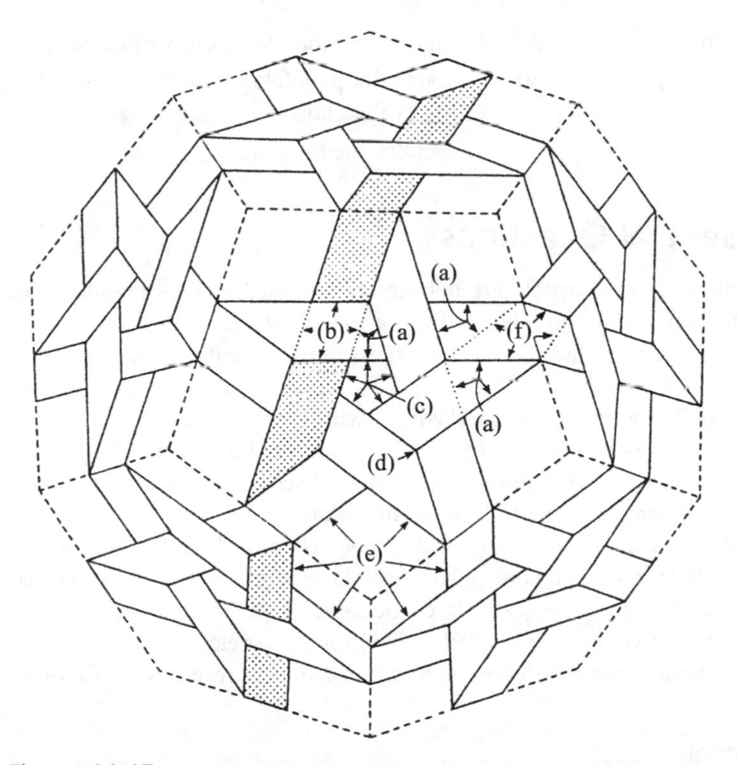


**Figure 11.15.** 10 strips braided together on the ghost of a dodecahedron

The model illustrated in Figure 11.15 can be used, without even constructing it, to obtain the information about the unbounded regions created by extending the face planes of the regular icosahedron. Why is this so? Well, first notice that the model surrounds the "ghost" of a regular dodecahedron. Furthermore, it has 10 strips since each strip crosses 6 of the ghost's faces, and each of these 12 faces must be crossed 5 times:

$$\frac{5 \times 12 \text{ (total number of crossings on the entire ghost)}}{6 \text{ (number of crossings contributed by each strip)}} = 10 \text{ (number of strips on the model).}$$

For clarity, just one of the strips is shown shaded in Figure 11.15; of course, it crosses three other faces of the ghost on the back of the model.

Now notice that the edges of these 10 centrally symmetric strips identify 20 planes which must intersect inside the model to form some kind of centrally symmetric polyhedron. Finally, observe that the braided model has icosahedral symmetry. Thus we are forced to conclude that the polyhedron formed by the intersection of the zones between the pairs of planes determined by the edges of these 10 strips must be the "core" of the icosahedron.

With Figure 11.15 before us we can easily enumerate the unbounded regions made by the extended face planes of the icosahedron.

First, coming from the 12 faces of the ghost we have

| | | |
|---|---|---|
| from (a) | $15 \times 12$ | regions with trapezium-like cross sections; |
| from (b) | $+\ 5 \times 12$ | regions with parallelogram-like cross sections; |
| from (c) | $+\ 1 \times 12$ | pentahedral regions; |
| from (d) | $+\ 5 \times 12$ | trihedral regions; |

then also

| | | |
|---|---|---|
| from (e) | $+\quad 20$ | hexahedral regions from the vertices of the ghost; |
| from (f) | $+\quad 30$ | regions with parallelogram-like sections from the edges of the ghost |
| | $=$ | 362 unbounded regions. |

## 5 More General Questions

The models mentioned in the article are not the only braided models possible. For example, four identical straight strips can be braided to form a model which looks like a truncated octahedron (with holes instead of square faces) and the intersection of the planes determined by the edges of the braided strips is the regular octahedron. Likewise, six identical straight strips can be braided to form what looks like a truncated icosahedron (with holes instead of pentagonal faces) and the intersection of the planes determined by the 12 edges of the anti-prism-like strips is a regular dodecahedron. You may find it challenging to try to discover all the braided models for each of the Platonic Solids having faces that lie in parallel planes.

For a real challenge you might wish to find analogous braided models for the remaining thirteen Archimedean Solids (the semi-regular polyhedra whose edges are of the same length and whose faces are all regular, though not necessarily of the same shape; in addition the faces form the same arrangement about each vertex (see [1], [9] or [14] for more details))—or prove that such a model cannot exist. The author would be very interested in any new results you discover.

### Acknowledgments

The author would like to thank Peter Hilton for his generous and insightful comments, and suggestions, during the preparation of this manuscript. She is also grateful to Sylvie Donmoyer for producing Figures 1, 6, 7, 9, 10, 11, 12 and 13.

The author also wishes to thank John E. Wetzel for bringing this problem to her attention (see [7]) and the authors of [1, 2, 8, 9, 11, 14] for providing the stimulating material in their publications that enabled her to make the connection between her braided models and the main question of this article.

### References

[1] Ball, W. W. Rouse, revised by H. S. M. Coxeter, *Mathematical Recreations and Essays,* (11th edition), Macmillan, 1939.

[2] Coxeter, H. S. M., *Regular Polytopes,* Methuen, 1948.

[3] Coxeter, H. S. M., P. duVal, H. T. Flather and J. F. Petrie, *The Fifty-nine Icosahedra,* Springer-Verlag, 1982.

[4] Hilton, Peter, Derek Holton and Jean Pedersen, *Mathematical Reflections—In a Room with Many Mirrors,* 2nd printing, Springer-Verlag NY, 1998.

[5] ——, *Mathematical Vistas—From a Room with Many Windows,* Springer-Verlag NY, 2002.

[6] Hilton, Peter, and Jean Pedersen, *Build Your Own Polyhedra,* Dale Seymour Publications, 1988.

[7] Kerr, Jeanne W., and John E. Wetzel, Platonic divisions of space, *Mathematics Magazine* 51, No. 4 (1978), 229 – 234.

[8] Cundy, H. Martyn, Antiprism frameworks, *Math. Gaz.* 61, No. 417 (1977), 182–187.

[9] Cundy, H. Martyn, and A. P. Rollet, *Mathematical Models* (2nd edition), Oxford University Press, 1973.

[10] Pedersen, Jean J., Visualizing parallel divisions of space, *Math. Gaz.* 62 (1978), 250–262.

[11] Pólya, George, *Induction and Analogy in Mathematics,* Vol. 1 of *Mathematics and Plausible Reasoning,* Princeton University Press, 1954.

[12] ——, *Mathematical Discovery* (combined edition), John Wiley and Sons, 1981.

[13] Steiner, J., Einige Gesetze über die Theilung der Ebene und des Raumes, *J. Reine Angew. Math.* 1 (1826), 349–364.

[14] Wenninger, Magnus J., *Polyhedron Models,* Cambridge University Press, 1971.

# 12

# Probability by Surprise

Susan Holmes
*Stanford University*

## The Pleasures of Paradoxes and Animations

Probability is involved in many games because they include an element of randomness. It is interesting to experiment with these games as we learn the mathematics behind them. The software tools currently available have made it easy to construct animations which allow us to experiment with games. The original BAMA presentation, which is the basis of this chapter, was about using interactive animations; as such, it is difficult to capture its full nature in writing, but we will attempt that here. Interactive versions of most of the games in this article are available on line at `www-stat.stanford.edu/~susan/surprise/`. [7]

## Aha, Gotcha

The idea to use paradoxes actually preceded the web; it all began when I started teaching statistics and probability to self-declared "math-phobics" in France. This was a course in remedial mathematics for those who had failed to satisfy the high school requirements (a major stigma in France). The students who took this course had decided to go back and get the French equivalent of a GED (General Equivalency Diploma) in order to be able to enter a University to study psychology. I decided to start teaching them by using paradoxes.

There are essentially two ways to teach people who don't "get" math right away. One way is to do things incrementally. I took the opposite approach, which is to say, "Here's a wall, hit the wall and we'll try to find out what the wall is about." This was perfect for psychology students because, for them, what I was saying was: "This is a failure, but this is a psychological failure. You can look at it as a case study of a psychological failure, and so turn yourself into your own first case study." This approach worked very well. Because failing to understand was a success of sorts for the psychological case study, it became something interesting for these students to analyze, and in doing so they also had to become engaged in the mathematics. They could get together and see that their difficulty in understanding the mathematics was, in itself, interesting, and this gave them something psychological to study along with the mathematics. I found this to be very effective.

To teach paradoxes I used a wonderful book by Martin Gardner, called *Aha, Gotcha: Paradoxes to Puzzle and Delight* [5], which is all about mathematical surprises, or "gotchas." Many of these gotchas have to do with probability. There is an example of a three coin puzzle in which one coin has two white sides, another has one white and one black side, while the third has two black sides. Given that you see one side of the coin and it's white, you are asked what are the chances that the second side is also white? What do you guess: one-third? one-half? one-quarter?, or something else?

This is an example of a probabilistic puzzle which many people do not answer correctly. Even some people who are good at mathematics get the wrong answer. It is interesting to find out why some of us are "wired" in such a way that we give the wrong answers to some probability problems.

## How Can We 'Rewire' Ourselves?

Pierre Simon Laplace (1749–1827) wrote, in 1825:

> The Mind, like the sense of sight, has its illusions; and just as touch corrects those of the latter, so thought and calculations correct the former. ... One of the great advantages of probability calculus is that it teaches us to distrust our first impressions.

We will use a number of paradoxes that puzzle a majority of an inexperienced audience (and actually many experienced members as well!). Our intuition is often wrong and we have to go through all the cases in order to eliminate possibilities one by one; in other words, we have to do the math.

## Let's Make a Deal.

The first time I attempted to use a computer in an explanation was with Biology colleagues to whom I was trying to explain the Monty Hall problem (for a description of the problem, see [1]), but to no avail. I looked on the Internet for an example, but they weren't persuaded when they did the simulation that I found, and the probabilities came up one-third or two-thirds. The program just showed the proportion of the simulations that a given strategy would win; the output looked like this:

| For 10,000 simulations | |
|---|---|
| P(win when change door choice) | P(win when don't change) |
| 0.6256 | 0.3132 |

Somehow this was no more convincing than the mathematical argument. Missing from a report of the simulation is the time it takes to follow the probabilities as they converge to their theoretical values, as the number of simulations increases; this is when the law of large numbers starts coming into effect. We need this partial data to update what we think the probabilities are. Simulations that we follow in real time are more persuasive than those where we are presented immediately with a summarizing simulated probability, as in this particular Monty Hall output.

There is, however, a wonderful applet available for playing the Monty-Hall game at: `www.hofstra.edu/~matsrc/MontyHall/MontyHallSim.html`.[1]

So let's try again. This time we will use the simulations to play the game one step at a time, rather than using shortcuts to simply show the final results. Somehow shortcuts are not persuasive. We will spend the time necessary to go through the sequence of actual play many times. This way we can get a genuine feel for what the probability of winning is as we follow the game's evolution, and as the law of large numbers starts to kick in.

Let's look at the animation presented as an interactive game by Constenoble at his website [1]. There are three doors. There are pigs behind two of the doors, and a car behind the third. You choose a door by clicking on it; a green tick mark appears. Then the host opens one of the other doors and a pig (might even be a noisy one!) is shown. Now you have to decide whether you want to switch or not. You can play this game, for real as it were, and experience its wonderful visual (and auditory) effects. You can play it as many times as you wish; you can even change strategies as you play.

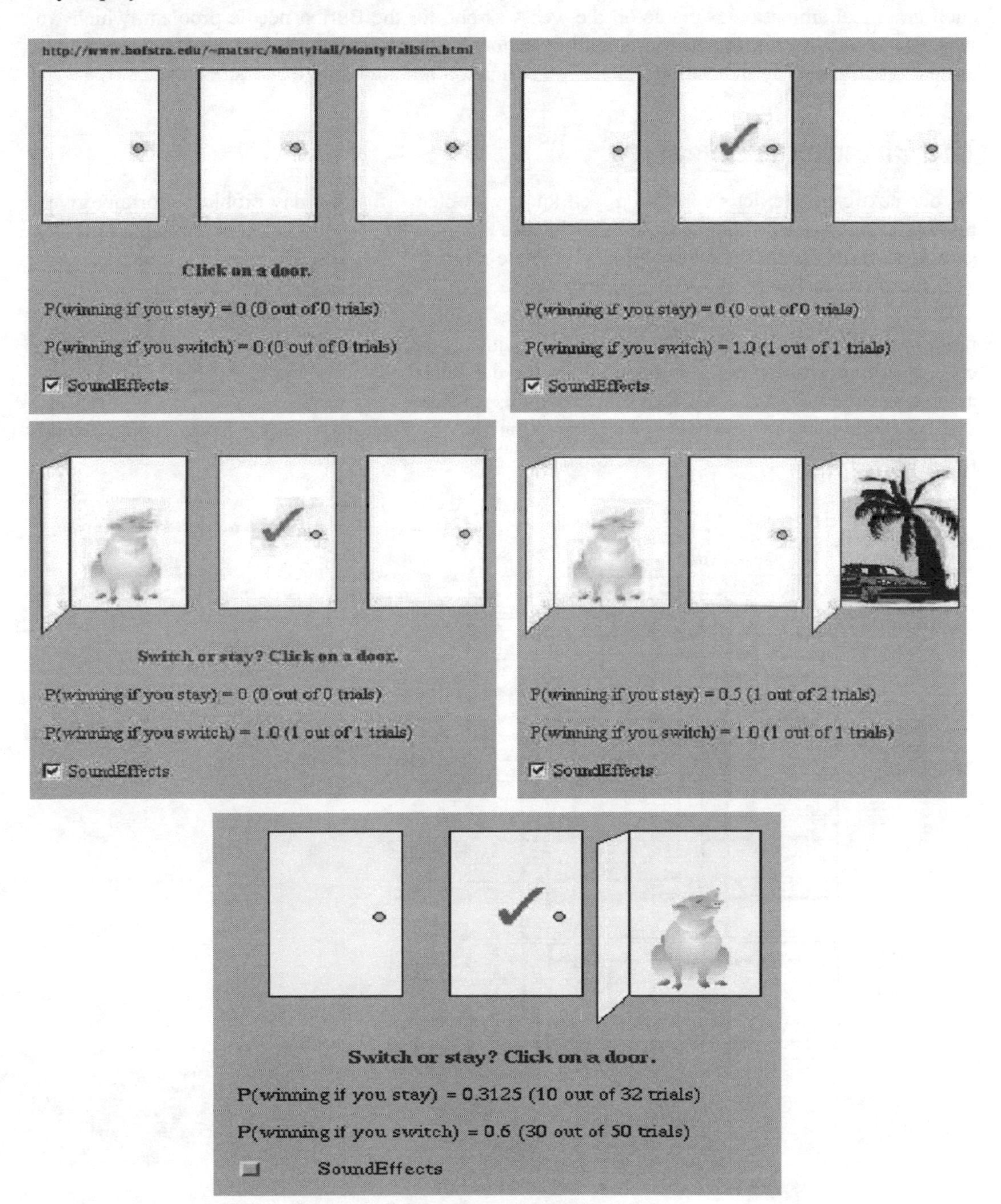

For me this was a perfect animation because you could play it many times and the probability of winning was updated for you. You can see how many times you've won and how many times you've switched. It carefully tracks your winning frequencies for the two strategies and it gives a good feel for how the frequencies converge.

Since this simulation persuaded my colleagues, I was convinced that this was the way to use the Internet. I decided to collect all the graphical programs that worked in this fashion, where the simulation actually runs instances of each "play." Unfortunately, at the time the only other such graphical animation available on the web was one for the Buffon needle problem, which we will talk about in the geometric probability section. As a result, I have had to develop my own animations, funded by the National Science Foundation and aided by many students.

## The Birthday Problem

As our next example, let's consider a well known problem—the birthday problem. For any given number of people in a room, what are the chances that at least two of them have the same birthday (not necessarily in the same year)? Let's simulate this for a group of 50 people. We picture a set of 365 boxes, each box representing a day of the year; then from the rectangle on the right-hand side (see the figure), little balls appear with dates tagging them. The computer chooses a ball at random with probability 1/365, since each ball is just as likely to be chosen as any other one. Each chosen ball goes to the box corresponding to the date attached to that particular ball. When a ball arrives at an already occupied box, that box lights up in red.

50 birthdays is the default for this simulation but the user may choose from among several possible values.

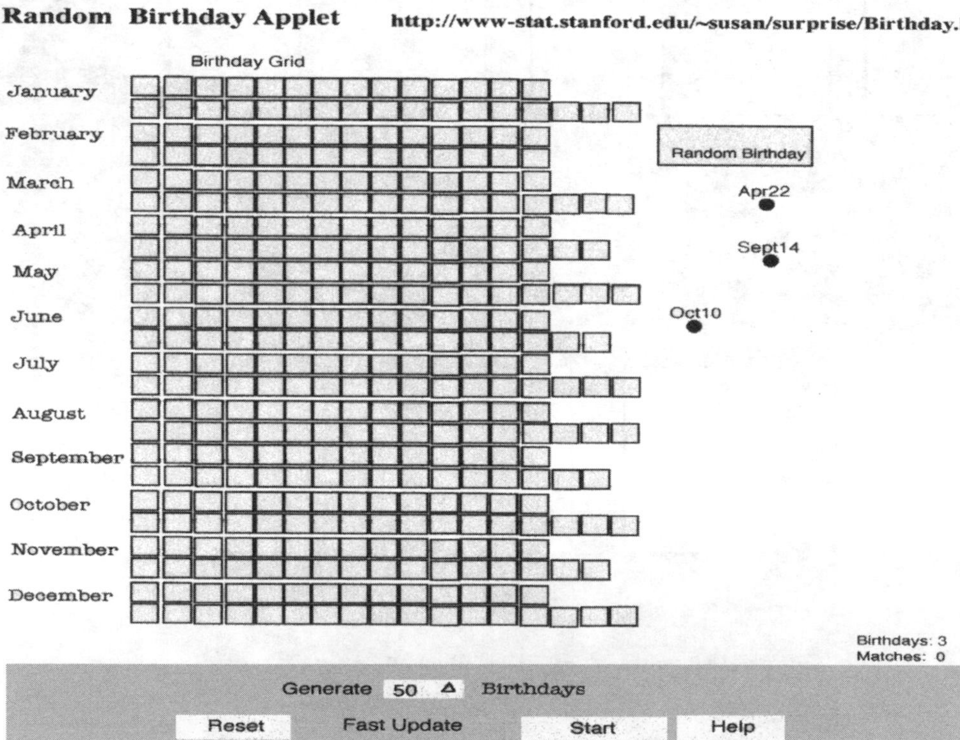

Incidentally, there is a discrepancy between such a simulation and real life. In introductory probability classes of around 20 students, Perci Diaconis and I noticed that we seemed to get a birthday match more than 50% of the time; this is actually more often than we would expect. It can be explained by a number of facts. First, in a class you tend to have more people born in the same year. Second, there is a trend in modern medicine that doctors schedule C-sections or chemically-induced labor on weekdays, thus increasing the probability that babies are born during the week rather than on a weekend. This skews the probability toward more matches than a completely random distribution would predict. One can use a Bayesian probability distribution to account for this by using a prior probability that incorporates this extra information. You'll find that about 16–17 people are enough to get a fifty-fifty chance of a match [2].

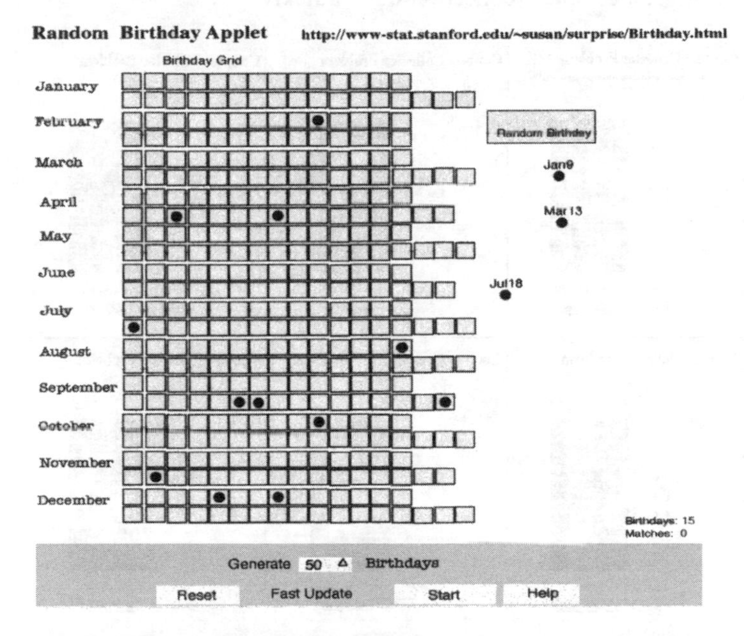

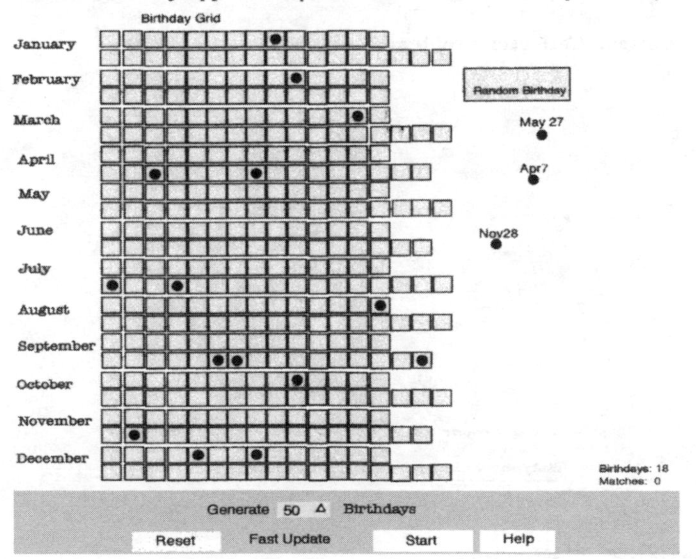

## Coupons, Toys, Baseball Cards. . .

Let us now look at the coupon collecting problem; this is a problem which involves collecting toys, or prizes, from cereal boxes. It is also known as “the fast food toy collection problem”. In the “fast food” version a fast food restaurant includes a toy with each meal and there is a variety of possible toys. If you want to collect all the different toys, how many times do you have to eat at the fast food restaurant?

In the simulation this is illustrated with three different collections: playing cards, Mahjongg tiles, and a die. With 52 playing cards, drawn at random, the simulation takes a long time to get them all. With 42 Mahjongg tiles it’s a bit faster and, of course, collecting all six faces of a die takes much less time. For the die it is possible to get all six faces in as few as six throws; this is a rare event[1]—still you can get a complete collection quite quickly.

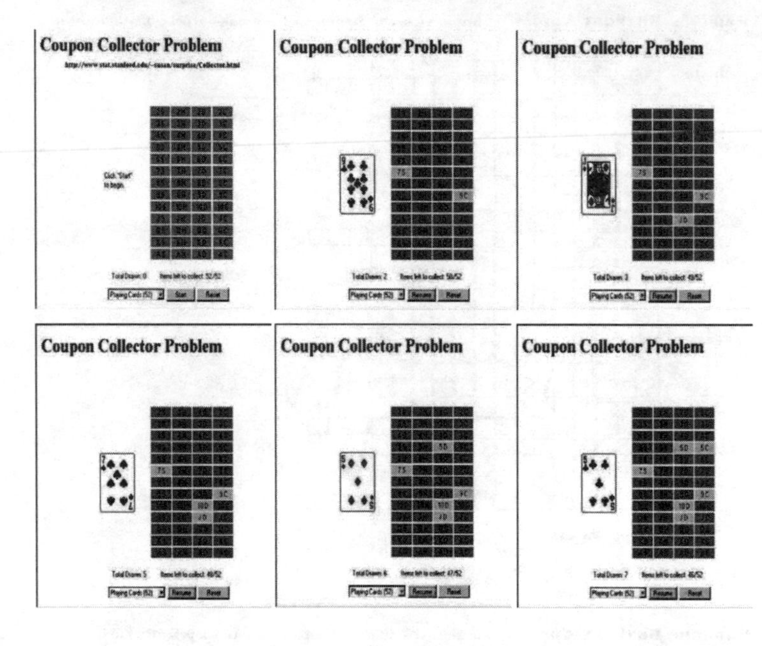


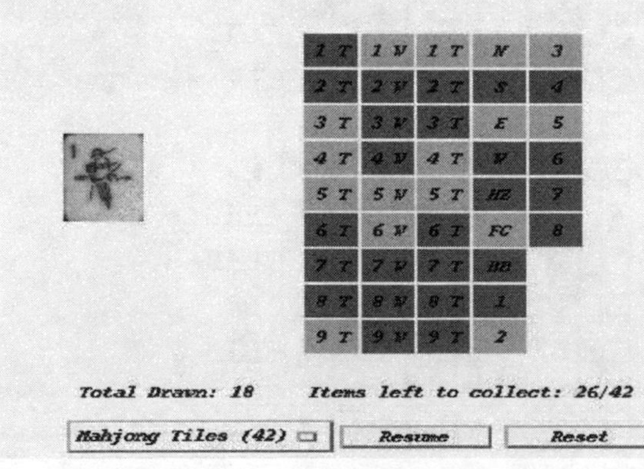


[1]5/324, or about 0.154321; do you see why?

## Matching Problem

A game which was studied in the famous probability classic by W. Feller [4] is a *treize* (French for thirteen), in which there are two stacks of 13 cards each, each stack containing the same set of cards. You and another player turn up one card at a time, and you win if there's a match. We can play different variations of this game, trying to find out what is the probability that this happens at least once with 52 cards in each stack, or 42 Mahjongg tiles each, or 6 faces of a die each. It turns out that regardless of the number of objects you have, whether they're playing cards, Mahjongg tiles, or even dice faces, you'll have about one match on the average if you run through the whole stack or deck; this is especially true as the number of objects grows. This is very surprising! Try this by playing the game 50 times some night.[2]

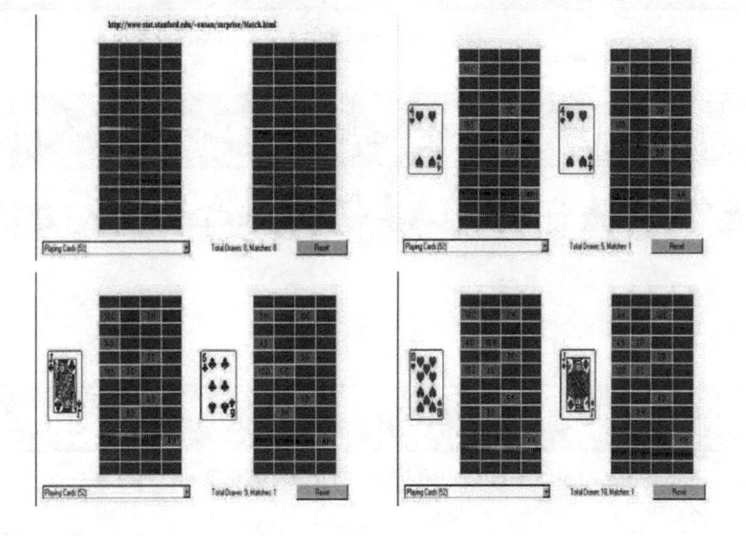


## Pattern Searching

Now there's a completely different feature of paradoxes that actually has a lot to do with psychology. That the human mind searches for patterns is a purely psychological phenomenon. Our intuitive expectations may be different from the results computed by using the laws of probability. This is a direct application of that quote I gave you from Laplace: probabilistic games are interesting because we can try to compute the actual probabilities instead of trusting our intuition. When is our intuition wrong, and in which direction? Our brain is striving to find patterns everywhere it can.

Seeing how the mind interprets things incorrectly can be very interesting. The first example I'd like to show is simply looking at patterns of coin flips. Have you heard of teachers who tell students to go home and flip a coin a hundred times and bring back the results? When the students come back with the results, the teacher can immediately tell who cheated by writing a sequence of heads and tails without actually using a coin: their sequences alternate *too much.* This shows that we have the wrong intuition of what an independent, identically distributed sequence (IID)[3] looks like. It is hard to recognize such a sequence. Let's look at the simplest non-independent case; this

[2] I say night because it is well known that students do their work at night, and actually seem to enjoy spending hours in front of their browser.

[3] A sequence that occurs when every new flip or choice of a color or choice of a shape is independent of the ones that came before it; we also say that it is memoryless.

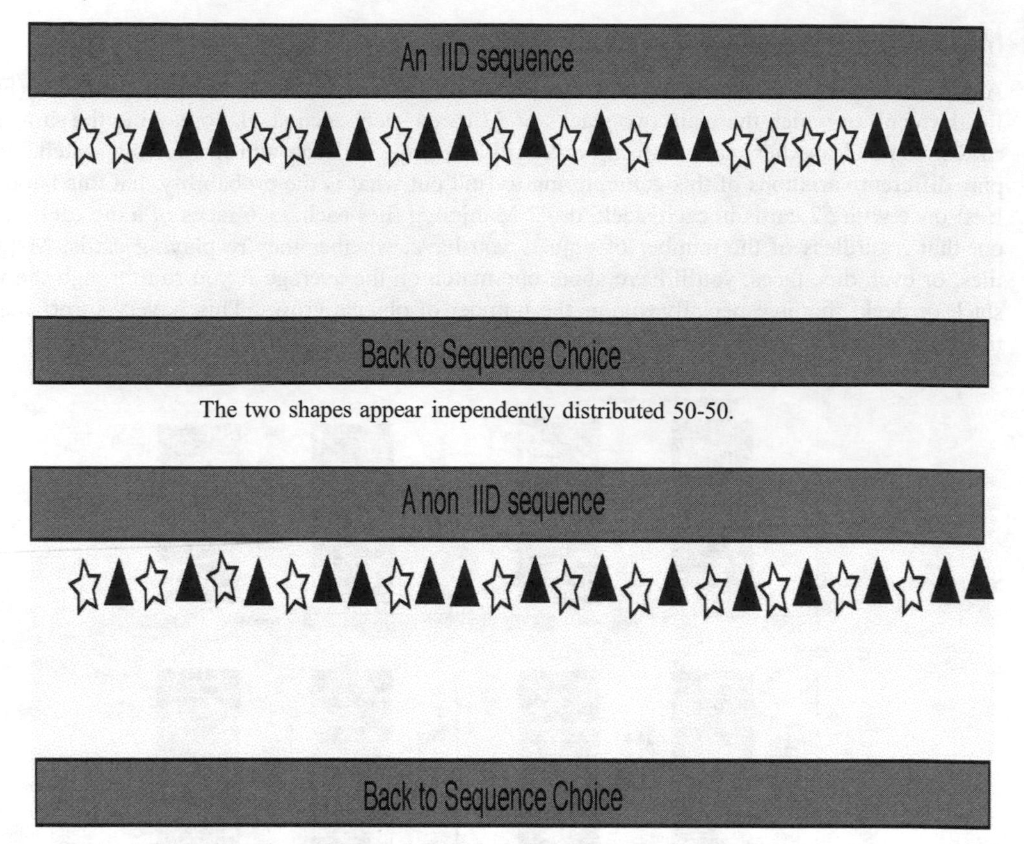


The two shapes appear inependently distributed 50-50.

Each shape has a high probability of changing to a different one.

is a Markov chain in which one shape or color is more often followed by a different one than by the same one. This kind of process tends to produce sequences with alternating shapes. Yet usually we think that this is a better representation of independence than the clumpy patterns. We think wrong!

What I show the students first is the IID case for training and then the non-IID case so they can "see" the difference. Then I show a "mystery sequence", and ask the students, do you believe that you can guess which one this is? IID or not?

By looking at a lot of sequences you can train yourself and get quite good at this game. Playing it will break the intuition that random means alternating a lot. If you have a Markov chain with a very high probability that the color will change—higher than $\frac{1}{2}$, as it would have been in an independent sequence—then you get a nice alternating sequence. It's hard to come to grips with the idea that it's precisely the deviation from IID that makes it alternate.

In the non-IID case, the level of difficulty in recognizing whether it is an IID or not is actually a function of the probabilities involved. There are some Markov chains that are very easy to recognize; one of these is a Markov chain in which the probability of red following blue is 90%. It makes change very probable, and quite easy to recognize once you know that that's the feature. If you make that probability of change smaller, such as 80% or 70%, it's much harder to see. That's the meaning of the level of difficulty. This phenomenon is also closely related to the psychological experiment in which somebody in a booth is shown sequences of red and green lights, and tries to understand how the sequences were built, or guess what the next light will be. People tend to build up stories rather than flip a coin to decide; they want to have a pattern.

## Event Clusters

Our next animation has to do with event clusters. I was quite interested in a series of articles concerning a cluster of autism cases in New Jersey that appeared in the *New York Times* in 1999. It really was about a Poisson process with a focus on the clusters after the fact.

A Poisson process is a sequence of random events that happen independently and homogeneously over an area. If you do not specify ahead of time which area you will study but just allow yourself to pick an area that has the most "hits," then you will be setting yourself up to say, "What are the chances of that many hits? Something terrible must be going on!"

Below is an example of a Poisson process: there are a hundred little bombshells landing on the grid. Some squares have several points and many squares have none. Repeating this many times allows you to see that there usually are areas that are very dense. If you specify ahead of time where you're looking, you won't be surprised. But if you decide to look after the fact at special spots where the bombshells are very dense you may conclude that "something terrible must be going on!"

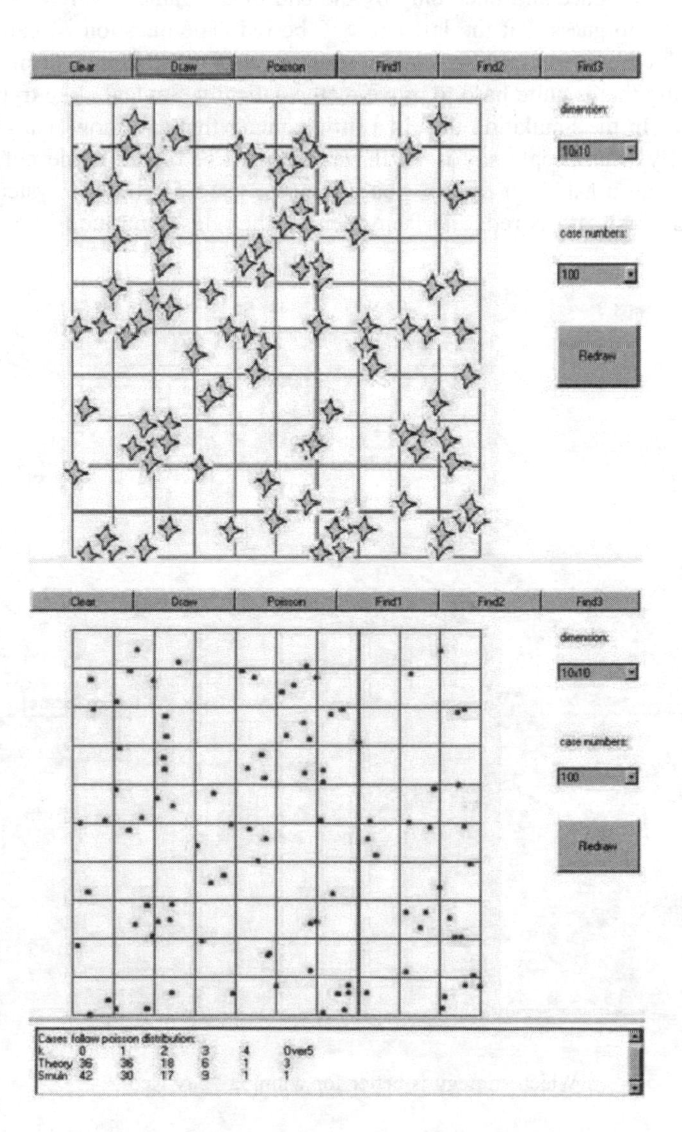

This situation occurred when the British were looking for patterns in the bombing of London, but found that it was completely random (see Feller [4]). Despite all kinds of stories about what the Germans were doing or aiming to do, the bombs weren't really following any particular pattern.

A cluster idea is a very useful one. Instead of "bombshells" these could be cases of illness or something equally undesirable.

A very startling real-life example is given in the article by D. Redelmeier and A. Tversky [10], which provides an analysis of how we explain bouts of arthritis varying with weather changes.

## Say Red

There are lots of games that are interesting to analyze. Here's a game that comes from Bob Connolly, a mathematician at Cornell. It was written up in Martin Gardner's column [6]. With a deck of 52 cards you turn them up one at a time. You say "Red" if you guess that the next card is going to be red, and you say "Red" once and once only by the end of the game. So, if you've seen 51 of the cards, then you have to guess that the last one will be red. The question is, can you improve the odds of 50/50 by having a special strategy? It turns out that no strategy can improve these odds, but this is something that is quite hard to prove mathematically. Students like trying the simulation in the figure below. In the simulation there's a little counter that says how many cards have come out, because usually what people say is: "I'll wait until more blacks come out, then I'll have a better probability than a half." Of course, you can never use this strategy systematically because it could be that the first card is red, and from then on the odds continue to give at most a 50/50

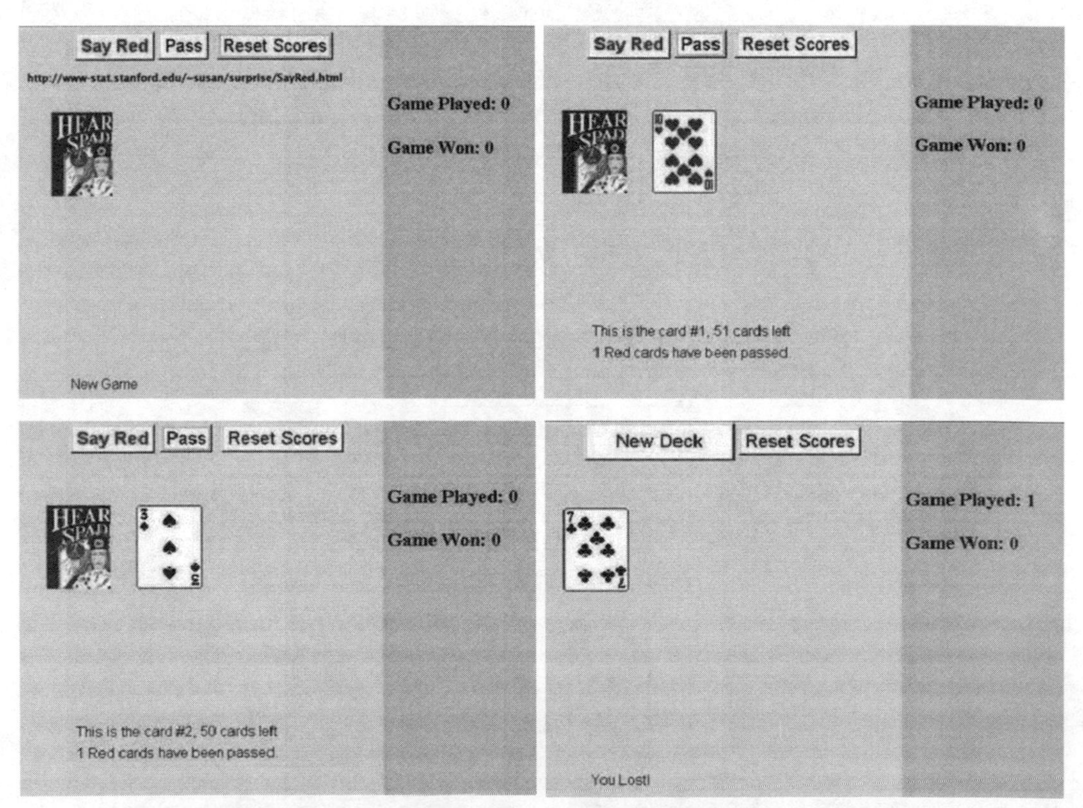

Which strategy is better for winning "Say Red"?

chance of winning. Bob Connolly came up with this game's idea when he was an undergraduate at Harvard, playing a somewhat similar board game. You were dealt a card, face down to begin with so you can't see its color. Then you went around and collected sins and indulgences, but in the end, you turned over the original card that said whether you were saved or condemned. So, in fact, since you had the card at the beginning of the game, the rest didn't matter.

Several strategies are often put forth. Here are some of the suggestions:

1. Say "Red" straight away. (The odds of winning are then seen to be 50/50).

2. Wait for a lead, when more blacks have come up, then say "Red"; but if this never happens, then the last card is black, and you lose. The odds are again 50/50, but this is harder to see.[4]

3. Choose a random time at which you would say "Red"; then the odds of winning are again 50/50.

Probably the best way of seeing this is a thought experiment: The cards are interchangeable, so it doesn't matter whether you guess the first, the last, or the $i$th card; the odds are always 50/50 of getting that card right. Thus, it didn't matter if you had to bet on the top card or the bottom card; and the probability of it being red remains onc half all the way through.

## Everyday Life: Matchboxes

Banach's Matchbox Problem[5] may go away, because there are fewer and fewer smokers, but we could replace it by a healthier version such as the toothpick problem or the chewing gum problem. (This is an example of the kind of silly things mathematicians think about to keep themselves from rcally working).

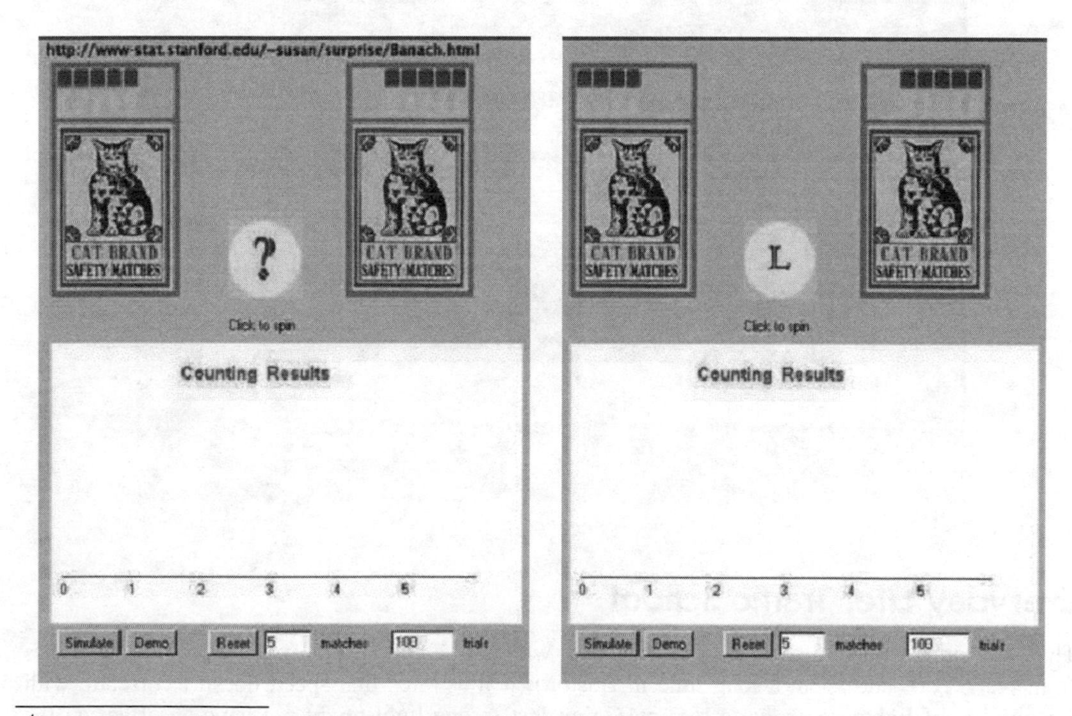


[4]Try to prove that the odds are 50-50 using a deck of 2 red and 2 black cards; a probability tree can help.

[5]Banach invented this problem because he was a pipe smoker.

You have two packets of matches, or gum, or toothpicks; in our simulation it's two boxes of matches, one in each pocket. Each box starts off with the same number of matches. You pick a pocket to get a match from, at random, with a fifty-fifty chance of choosing either pocket; the choice is independent of any previous choice. At some point you are going to draw out an empty box. When you come across your first empty try, you look inside the other box and count the matches.

Banach wondered how many matches would be left in the other box at that time; or rather he wanted to know the distribution of the non-empty box's number of matches. We simulate this problem with a coin flip that is used to choose the left or right matchbox; then you look at the number of matches left in the other box. This kind of distribution is quite a rare one; it is called a negative binomial distribution.

You can play this game interactively, but since it is rather time consuming, an accelerated "simulate" button has been added which takes all the steps at once and shows the results on the histogram. The histogram also shows the expected distribution.

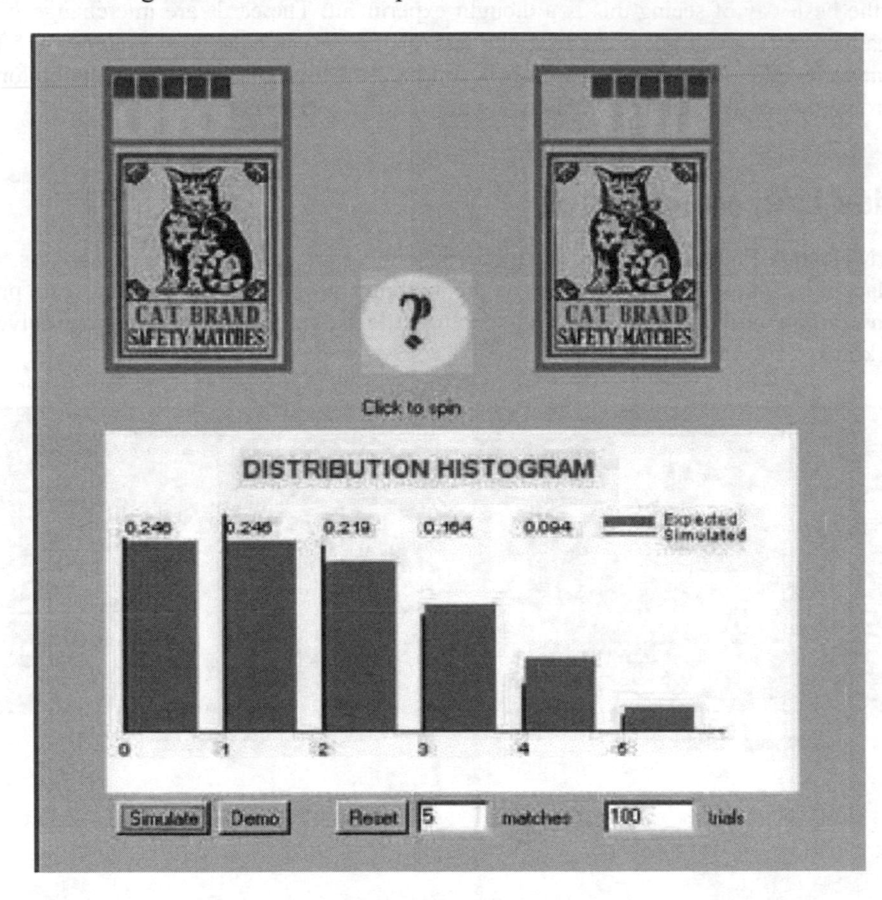


## Everyday Life: Traffic School

This next simulation is dedicated to my husband, who isn't a calm driver (maybe because he comes from New York and spent a long time in Boston); it illustrates that speed doesn't correlate with the number of lights you have to stop at. Going fast to one light might not have an effect on the number of times you have to stop.

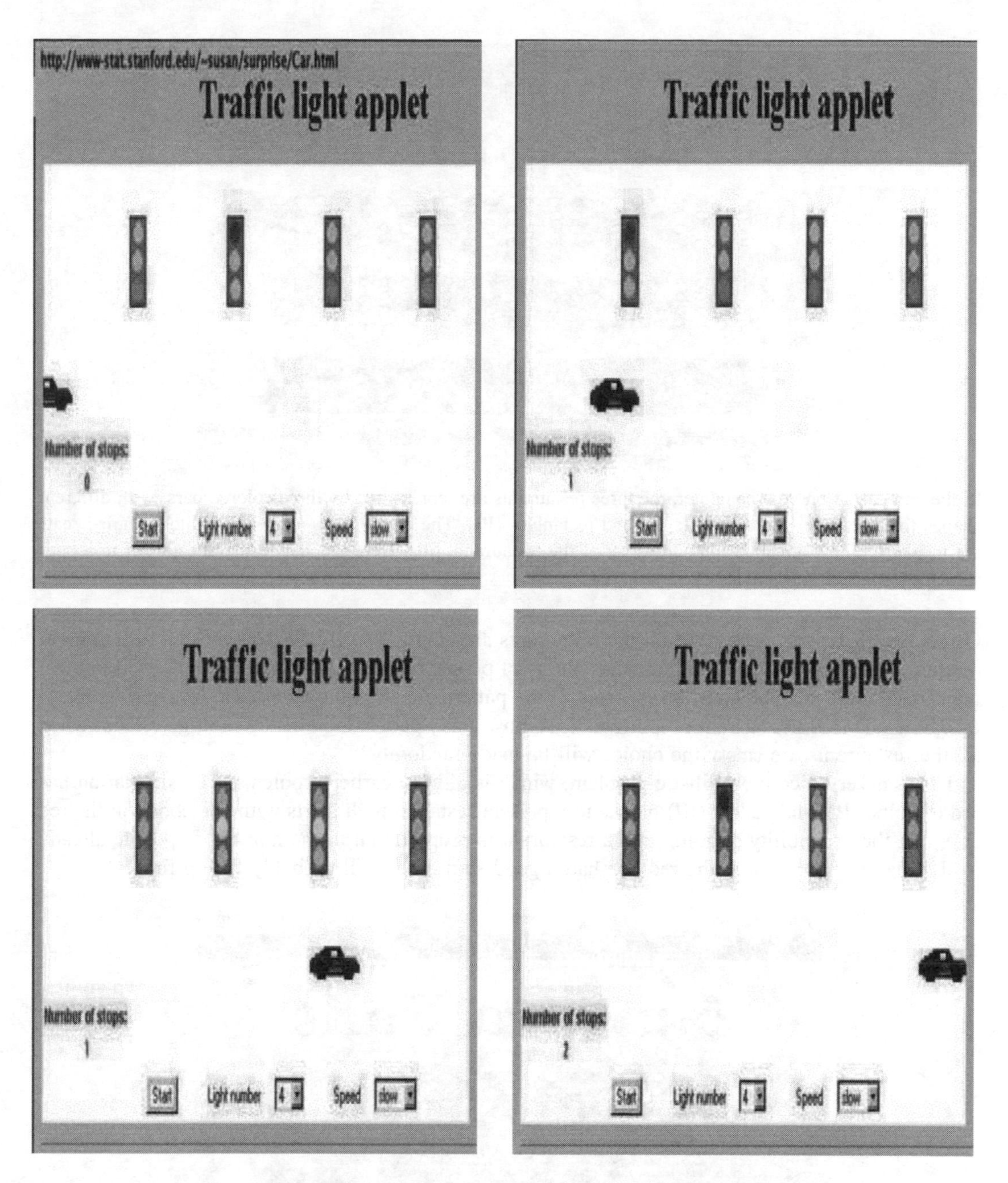


In this simulation the lights change at random, the car's speed is fixed, and you can run the experiment many times to find the distribution of the number of lights that you have to stop at.

## Random Restaurants

Here is another example that has to do with everyday life. You're a tourist somewhere for the first time, for example, at Niagara Falls, and you want to go to a restaurant. You probably won't go to a restaurant that nobody is in. "They must know something to avoid this restaurant," you'll think.

In the simulation on your computer, the three restaurants are represented by three colored bars, with different shapes for the seats: ovals (I), stars (II) and rectangles (III). The empty seats are white while occupied seats are in black. The simulation runs until one of the restaurants fills. Here the star restaurant got off to a poor start.

Of course, if it's very early, and all the restaurants are empty, you'll have to choose a restaurant at random. But if one of the restaurants has plenty of people in it, you'll think that it's a well-known, good restaurant. So, the question is: what is the pattern for the number people in a restaurant, if everyone chooses a restaurant proportionally to how many people are already in it, given that when all the restaurants are empty the choice will be made randomly?

That's a very nice probabilistic problem, which is akin to earlier problems. The simulation has red (I), blue (II), and yellow (III) blocks to represent restaurants. It starts with somebody in the red one, and the probability of going into a restaurant is proportional to the number of people already in it, plus one.[6] Of course, the red one had a good start, so it will probably fill up first.

This time, the star restaurant started off better.

[6]The plus one is to avoid the case where a restaurant is empty and so would never get any customers!

Talking to an econometrician I found out that this is a well-known problem, called the herd phenomenon, although the economists did not know about the underlying probabilistic process. The starting point is random, but once a restaurant has a few customers it has an edge over its competitors, and the other restaurants are not likely to catch up until the game is restarted!

## Geometric Probability

When I started making these applets, I thought that they were a perfect application for geometry. And I indeed found a very nice graphical applet showing the Buffon needle simulation as it actually occurs. This one was written by George Reese[11].

This simulates dropping needles on a hardwood floor with two parallel cracks. The length of the needle is the same as the distance between the cracks. Buffon figured out that the probability of the needle overlapping a crack was $2/\pi$, so he decided to estimate $\pi$ by repeating this experiment hundreds of times.[7] That was the first Monte Carlo simulation ever done.

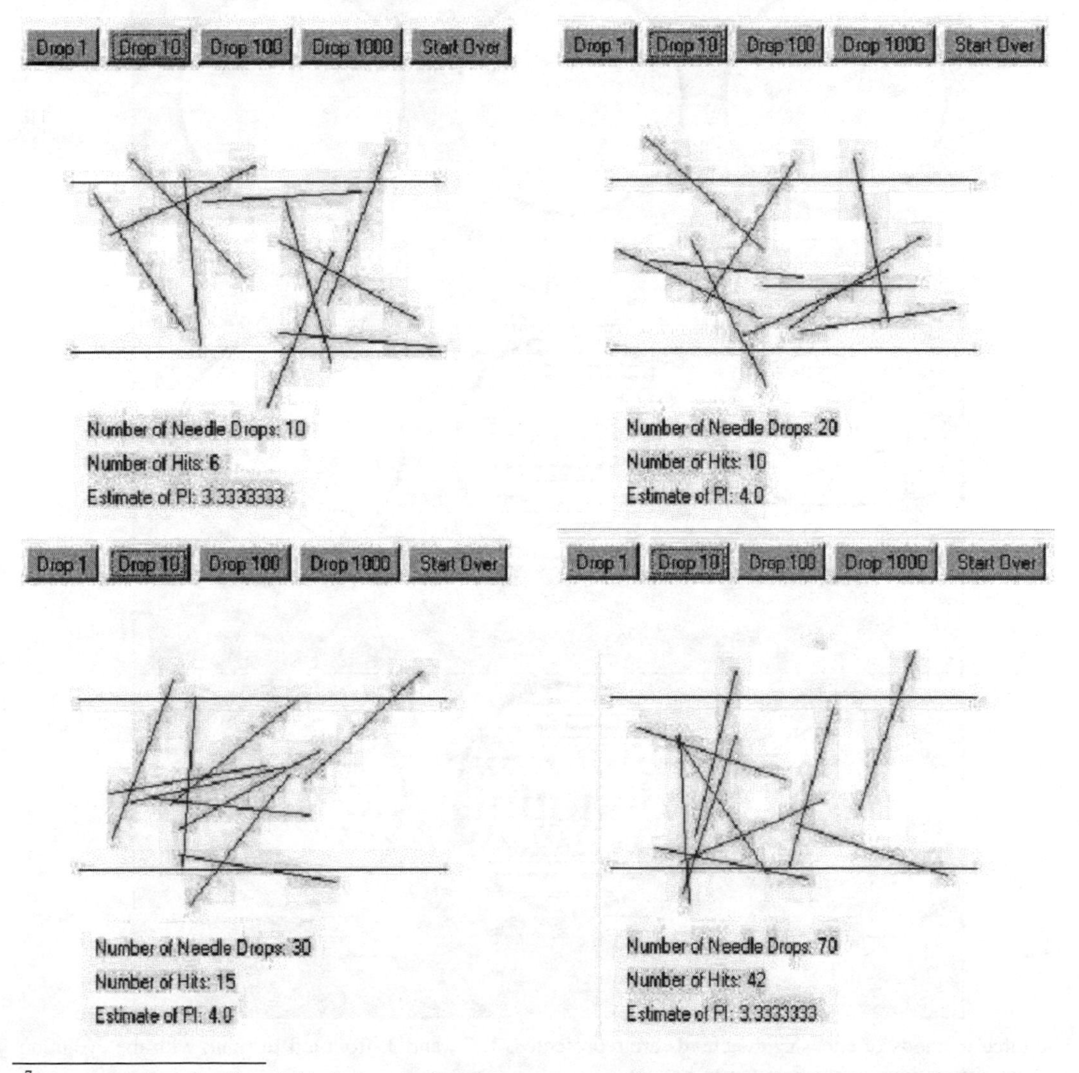


[7]A geometrical explanation of this fact can be found at `www.mste.uiuc.edu/reese/buffon/buffon.html`.

## There's Random and Then There's Random...

Another geometric example is Bertrand's Paradox, where instead of putting down line segments at random, you put down a random chord on a circle; there are various ways to do this. This example illustrates that it is not enough to say "Choose a chord at random": how the chord is chosen must be specified.

We see that a chord can be chosen in three different ways:

1. By choosing the midpoint of the chord as a random point inside the circle.
2. By choosing a coordinate on the $y$-axis, and making the chord parallel to the $x$-axis.
3. By choosing a direction at random from a fixed point.

To illustrate the difference, we could ask what proportion of these random chords will be larger than the edge of an inscribed equilateral triangle. These will be colored in red, the others are blue in the applets; the pictures below show the red lines as dashed and the blue lines as solid.

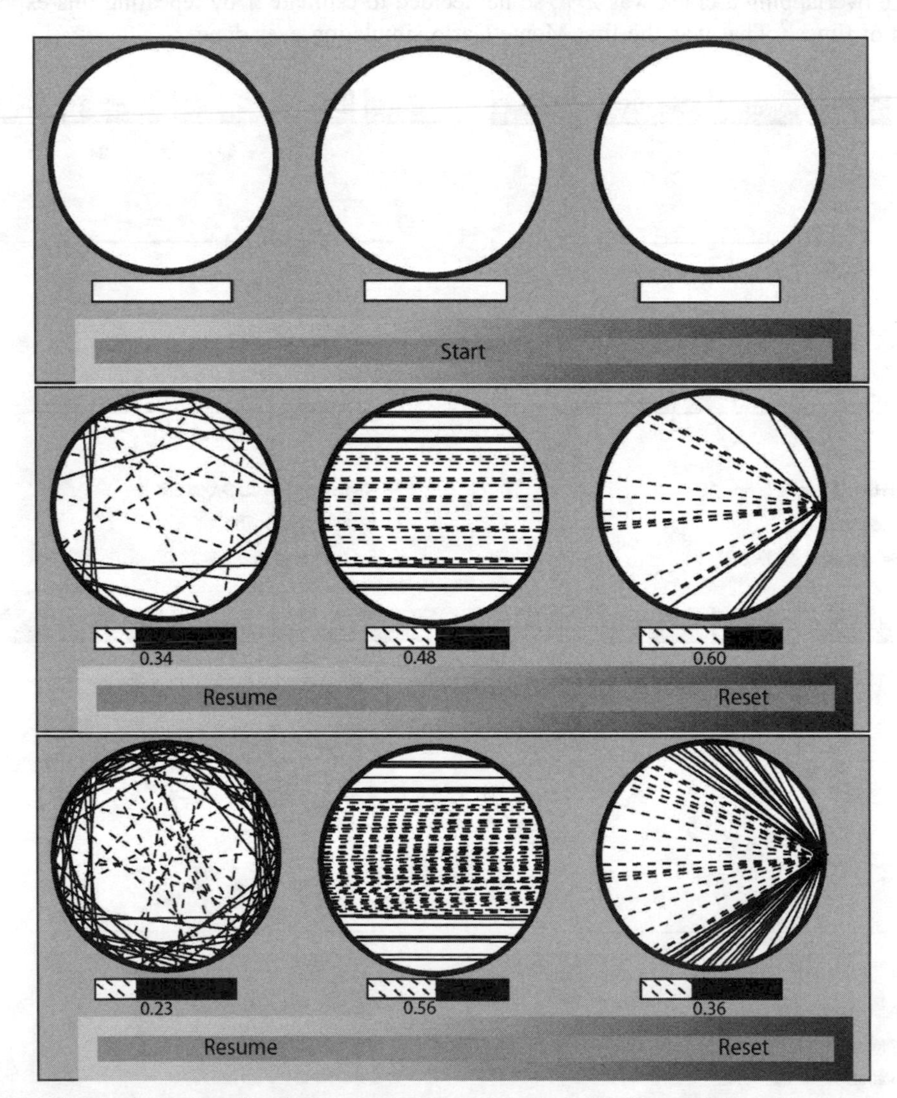


The three methods of choosing the chords are represented, 1, 2 , and 3, from left to right, with the evolution of these simulations going from top to bottom.

It can be shown that the theoretical proportion of dashed lines is a quarter, a half, or a third, depending on the choice of what "random" means to you. If you don't specify what your random variable is, you can come to different conclusions.[8]

## Visualizing Probability

When simple simulation histograms aren't enough to get a feel for a probability, what else can we visualize? Here are two examples of different pictures of probabilities.

### Trees

A common way to explain probabilities of compound events is to break them up and find the probability of a first event, then the probability of a second event given the first, in a hierarchical way. We can build a tree of events and then imagine the simulations as little balls going down the tree. We can even make the size of the tubes the balls go down look proportional to the probability of that branch. Imagine that we have two choices to begin with. The tree has two branches leaving the root, then each branch splits up into as many alternatives as the next experiment offers. Suppose there were two choices again, each one followed by two more possibilities, then there would be eight possible final events.

Here is a picture showing three flips of a biased (80/20) coin:

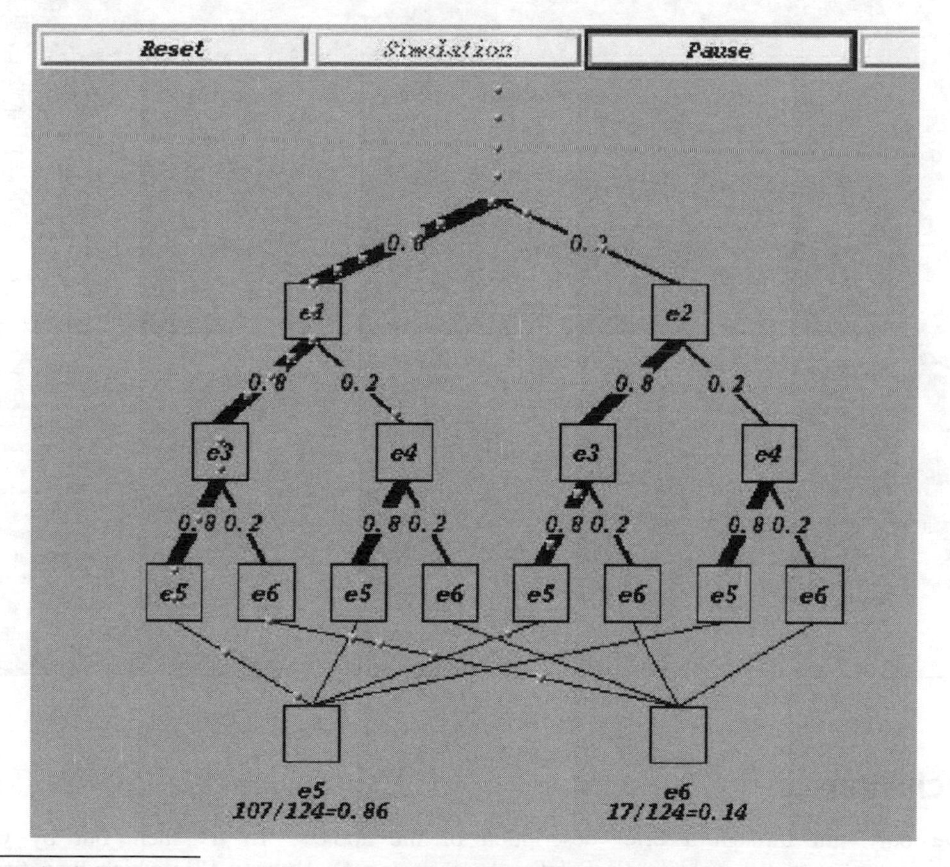


[8]This simulation and geometric explanations for the three different answers may be found at `www.cut-the-knot.com/bertrand.html`.

## Conditional and Marginal Probabilities

Visualizing continuous probabilities is done using densities, but how can we illustrate the difference between the marginal and conditional densities? We need at least a density on a two-dimensional space for this to work. Let's use three-dimensional projections in order to show the difference between conditional and marginal probability, one dimension for the height of densities and two dimensions for the state space. My favorite picture for explaining this came from Jim Pitman's *Introductory Probability* book ([9], page 412).

Three-dimensional applets that are portable have been difficult to create, but I had help from Marc Coram on the applet shown. He found a site in Japan that had a wonderful 3D Java program for showing mesh graphs of 3D pictures. The Surface Plotter was originally written by Yanto Suryono. Marc adapted it to show the difference between the marginal distributions (projected on the sides), and the conditional distribution (the yellow distribution on the side).

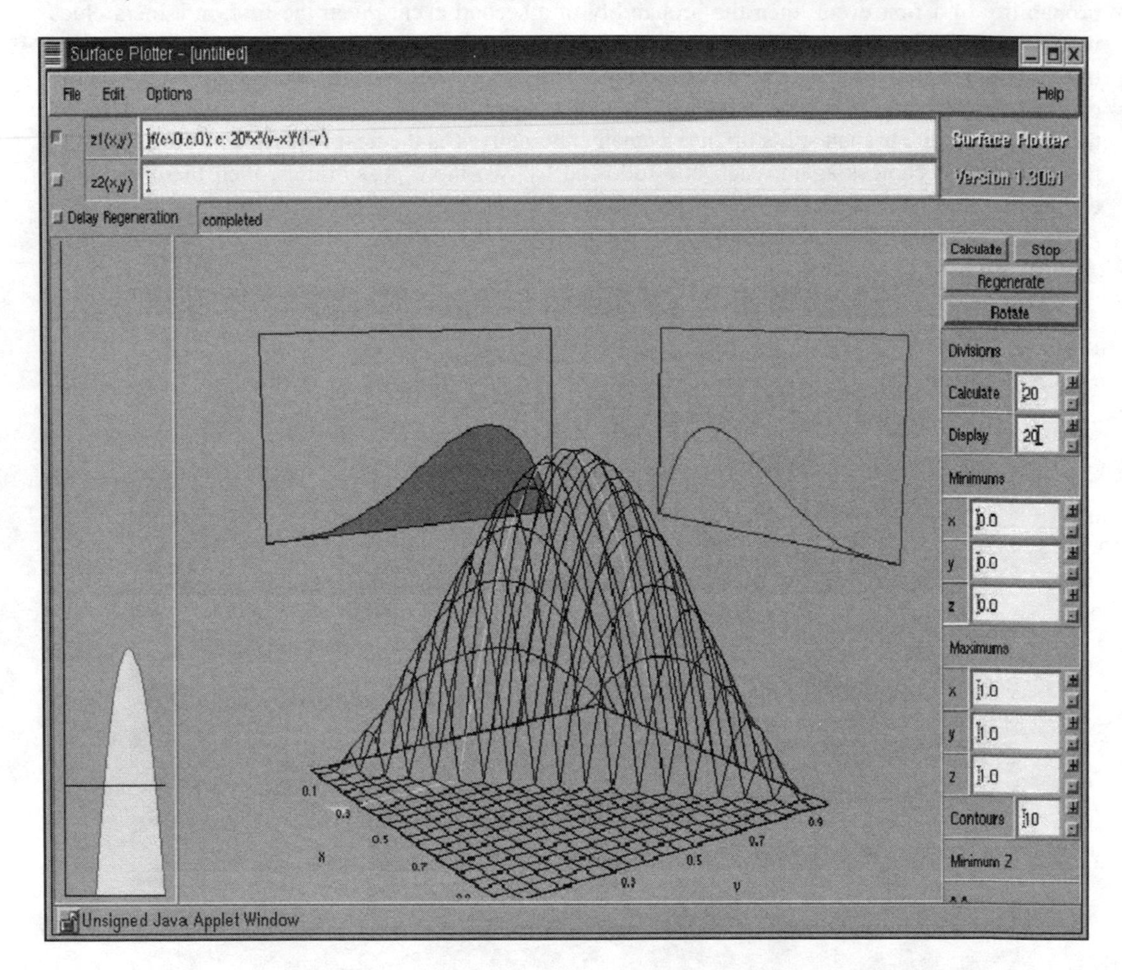


## Conclusion

I have only run through a brief description of the applets. To try them out by yourself you can go to the website *Probability by Surprise,* S. Holmes [7], which was funded by the National Science Foundation (NSF). To understand the underlying probability theory, look at the class notes from the Stanford "Introduction to Probability-Stat116" class notes at `www-stat.stanford.edu/~susan/courses/s116/`.

I have used the same idea of using animations to motivate creative students in a class I taught with Brad Efron, called *Scenarios for Statistics,* B. Efron and S. Holmes [3], where the students wrote their own Macromedia Flash animations. You can look at a presentation of the course on my website [8].

## Acknowledgments

This work was made possible by a grant from NSF-DUE. Collaborators who have written applets include: Brigitte Charnomordic, Marc Coram, Balasubramanuan Narasimhan, Hua Tang, Ying Taur, Jing Zhi, Je Zhu. Consultants who have helped in other ways include Bob Connelly, Persi Diaconis, and Helen Moore. Thanks to Tatiana Shubin, David Hayes, and Peter Ross for setting up the BAMA talk and Laurie Snell and Dan Rockmore for setting up the 2000 Chance lecture where this work was first recorded as an online lecture.

## References

[1] Constenoble, S., Hofstra University, `www.hofstra.edu/~matsrc/MontyHall/MontyHallSim.html`.

[2] Diaconis, P. and S. Holmes, A Bayesian Peek into Feller Volume 1, *Sankhya* (2002).

[3] Efron, B. and S. Holmes, *Scenarios for Statistics,* 2001, `www-stat.stanford.edu/~susan/courses/s30/index.html`.

[4] Feller, W., *An Introduction to Probability,* Volume I, 3rd Edition, Wiley, NY, 1968.

[5] Gardner, M., *Aha, Gotcha! Paradoxes to puzzle and delight,* W. H. Freeman, NY, 1982.

[6] ——, "Psychic Wonders and Probability," chapter 14, 214–27, in *Fractal Music, Hypercards and More...,* W. H. Freeman, NY, 1992.

[7] Holmes, S., *Probability by Surprise,* 1999, `www-stat.stanford.edu/~susan/surprise/`.

[8] ——, *Scenarios for Statistics,* 2001, `www-stat.stanford.edu/~susan/papers/maa.pdf`.

[9] Pitman, J., *Introductory Probability,* Springer Verlag, NY, 1992.

[10] Redelmeier, D., and A. Tversky, "On the belief that arthritis pain is related to the weather," *Proc. Natl. Acad. Sci.* 93(7) (1996), 2895–6.

[11] Reese, G., *Buffon's needle applet,* `www.mste.uiuc.edu/reese/buffon/bufjava.html`.

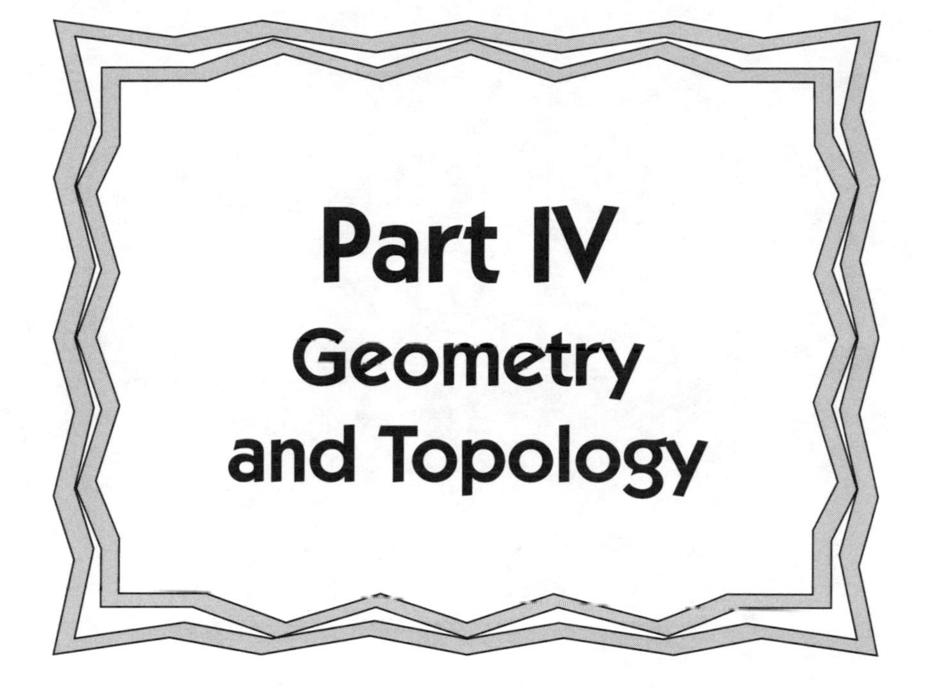

# Part IV
# Geometry and Topology

# 13

# The Rule of False Position

Don Chakerian
*University of California at Davis*

## 1 Introduction

The rule of false position, which is a method for solving linear equations, originated many centuries ago as a sort of "guess and check" technique. Our goal here is to explain how this rule can be viewed as a precursor to what is now called Newton's method for finding solutions of equations, and to see how a geometric interpretation leads to an interesting connection with a famous theorem of Pascal about hexagons inscribed in conic sections.

## 2 Do Two Wrongs Make a Right?

Something called *the rule of single false position* goes back to the ancient Egyptians and amounts to little more than guessing an answer and making an adjustment after substituting the guess. Here we shall restrict ourselves to a method that is sometimes called *the rule of double false position*, involving two guesses. This is *the rule of false position* as presented in many nineteenth century algebra textbooks. A typical example, from *Daboll's Schoolmaster's Assistant* of 1800, is reproduced in the excellent text of Bunt, Jones and Bedient [1, pp.34-35]. The problem, involving those ubiquitous protagonists A, B, and C, is as follows:

> "A, B, and C built a house which cost $500, of which A paid a certain sum, B paid 10 dollars more than A, and C paid as much as A and B both; how much did each man pay?"

The rule of false position asks us to make two guesses for the amount paid by A (note that the cost to A determines everything else) and use the "errors" obtained when checking these guesses to produce the correct answer with formula (M) below. (Thus proving that "two wrongs make a right"?)

Suppose, for instance, we first guess that A pays $80. Then B pays $90 and C pays $170, giving a grand total of $340 paid by all three. This gives an "error" of $500 - 340 = 160$ dollars corresponding to our guess of $80. Next, we might guess that A pays $150. Then B pays $160, C pays $310, giving a total of $620 for all three. This time the "error" is $500 - 620 = -120$ dollars.

The rule of false position consists in calculating

$$\frac{(\text{first guess})(\text{second error}) - (\text{second guess})(\text{first error})}{(\text{second error}) - (\text{first error})}. \tag{M}$$

In our particular case this gives

$$\frac{(80)(-120) - (150)(160)}{-120 - 160} = \frac{-9600 - 24000}{-280} = 120.$$

Now we can check that, indeed, if A pays \$120, then B pays \$130, C pays \$250, and the sum of the three payments is exactly \$500, as we wanted.

You may try other pairs of (distinct) guesses to see that the formula (M) always leads to the correct answer of 120 dollars.

As an aside, note that A gets the best deal here, while C is stuck with the worst part of the bargain. This is consistent with the general plight of poor old C, as detailed by Stephen Leacock in his classic essay [4] dealing with such "word problems."

Let us write (M) in a slightly more formal way. In the sort of problem we just solved, we are given a function $g$ and a number $c$, and we are asked to find a number $t$ such that $g(t) = c$. Our method is to make two "guesses" $x_1$ and $x_2$ and use these to form the "errors" $c - g(x_1)$ and $c - g(x_2)$. Then (M) tells us to take

$$t = \frac{x_1(c - g(x_2)) - x_2(c - g(x_1))}{(c - g(x_2)) - (c - g(x_1))}.$$

There is no loss in generality in defining a function $f$ by $f(x) = c - g(x)$ and trying to find $t$ such that $f(t) = 0$. Then the rule of false position suggests we take

$$t = \frac{x_1 f(x_2) - x_2 f(x_1)}{f(x_2) - f(x_1)}. \tag{1}$$

Will this always produce $t$ such that $f(t) = 0$? If $f$ happens to be a linear function, of the form $f(x) = ax + b$, where $a$ and $b$ are given constants (in which case $g$ is also a linear function), we obtain from (1)

$$t = \frac{x_1(ax_2 + b) - x_2(ax_1 + b)}{(ax_2 + b) - (ax_1 + b)} = \frac{x_1 b - x_2 b}{ax_2 - ax_1} = -\frac{b}{a}.$$

Sure enough, we have

$$f(t) = f(-\frac{b}{a}) = a(-\frac{b}{a}) + b = 0,$$

as desired.

So we find that the rule of false position leads to the correct solution of any linear equation in one variable. This is clearly not the most efficient way to solve a simple linear equation, given the algebraic methods available to us these days! However, you can view this as just a step toward understanding some effective approximation techniques that can be applied in solving much more complicated equations.

To see the geometry that makes the method tick, assuming $f$ is of the form $f(x) = ax + b$, we draw the graph of $y = f(x)$ (a straight line), and we indicate the "guesses" $x_1$ and $x_2$ on the $x$-axis:

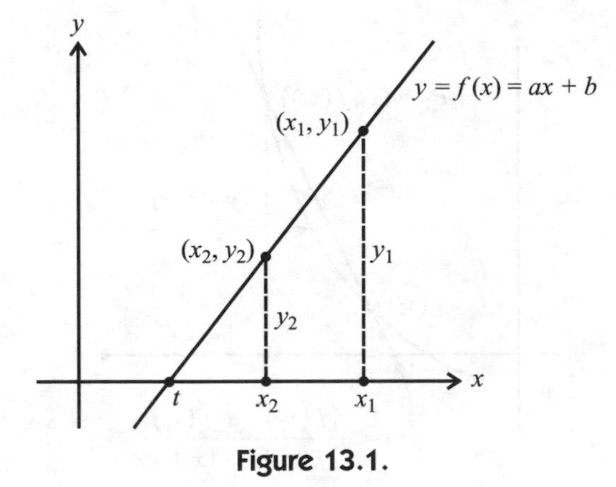


**Figure 13.1.**

On the graph we have the points $(x_1, y_1)$ and $(x_2, y_2)$, with $y_1 = f(x_1)$ and $y_2 = f(x_2)$. We also indicate the abscissa $t$ of the point where the graph intersects the $x$-axis. Similar right triangles in Figure 13.1 tell us that

$$\frac{y_1}{x_1 - t} = \frac{y_2}{x_2 - t}.$$

Solving this equation for $t$ gives

$$t = \frac{x_1 y_2 - x_2 y_1}{y_2 - y_1}, \tag{2}$$

which is exactly the relation (1) prescribed by the rule of false position.

The geometric key to this is that if $P_1 = (x_1, y_1)$ and $P_2 = (x_2, y_2)$ are any points in the $(x, y)$-plane, with $y_1 \neq y_2$, then the expression (2) gives the abscissa $t$ of the point where the line through $P_1$ and $P_2$ intersects the $x$-axis. This observation is crucial for all that follows.

As an exercise, the reader is invited to apply the rule of false position to the problem posed in the following rhyme, which was taken from an old textbook, *The American Tutor's Assistant,* and reproduced in [1, p. 33]:

"When first the marriage knot was ty'd
Between my wife and me,
My age was to that of my bride
As three times three to three.
But now when ten and half ten years
We man and wife have been,
Her age to mine exactly bears
As eight is to sixteen;
Now tell, I pray, from what I've said,
What were our ages when we wed?"

The reader should find that the answer to the question is

"My age when marry'd must have been
Just forty-five; my wife's fifteen."

## 3 Two Wrongs Do Not Make a Right

In Figure 13.2 we have the graph of $y = f(x)$, where $f$ is *not* a linear function, two points $P_1 = (x_1, f(x_1))$ and $P_2 = (x_2, f(x_2))$ on the graph, and the point $(t, 0)$ where the line through $P_1$ and $P_2$ intersects the $x$-axis.

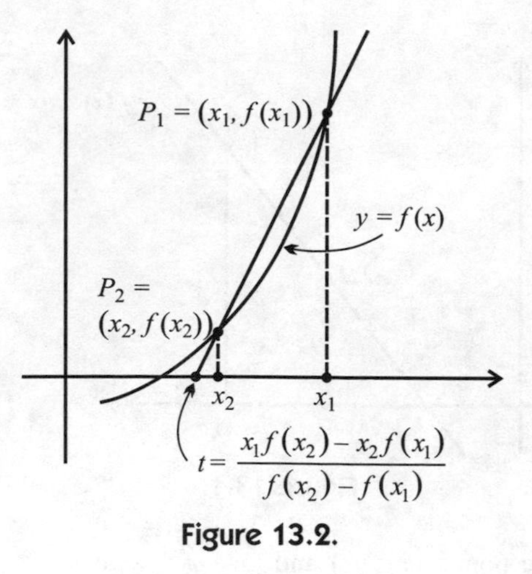


**Figure 13.2.**

By the remark at the end of the previous section, we see that $t$ is given by the expression (1). Since $f$ is not a linear function, the rule of false position does not necessarily give a value of $t$ such that $f(t) = 0$. On the other hand, Figure 13.2 hints that if $x_1$ and $x_2$ are good guesses for the solution of $f(t) = 0$, then the value $t$ given by (1) is liable to be an even better estimate of the solution.

Following up on this hint, we try the example of the quadratic function $f(x) = x^2 - 2$, whose graph is a parabola and whose zeros are $\pm\sqrt{2}$. In this case, for any guesses $x_1 \neq x_2$, the expression (1) gives

$$t = \frac{x_1 f(x_2) - x_2 f(x_1)}{f(x_2) - f(x_1)} = \frac{x_1(x_2^2 - 2) - x_2(x_1^2 - 2)}{(x_2^2 - 2) - (x_1^2 - 2)} = \frac{x_1 x_2^2 - x_2 x_1^2 + 2(x_2 - x_1)}{x_2^2 - x_1^2}$$
$$= \frac{(x_2 - x_1)(x_1 x_2 + 2)}{(x_2 - x_1)(x_2 + x_1)} = \frac{x_1 x_2 + 2}{x_2 + x_1}.$$

Let us denote the expression on the right-hand side of (1) by $F(x_1, x_2)$. This is equal to the abscissa of the point where the $x$-axis is intersected by the line through $(x_1, f(x_1))$ and $(x_2, f(x_2))$. In other words, in Figure 13.2, $F(x_1, x_2) = t$. The above calculation shows that if $f(x) = x^2 - 2$, we have

$$F(x_1, x_2) = \frac{x_1 x_2 + 2}{x_2 + x_1}. \tag{3}$$

If we choose $x_1 = 1$ and $x_2 = 3/2 = 1.5$, we have

$$F(x_1, x_2) = F(1, 3/2) = \frac{(1)(3/2) + 2}{(3/2) + 1} = \frac{7}{5} = 1.4,$$

a better approximation to $\sqrt{2} = 1.41421356237\ldots$ than either 1 or $3/2$.

Thus, starting with the estimates $x_1 = 1$ and $x_2 = 3/2$ for $\sqrt{2}$, we use these to generate the estimate $7/5$, which we shall call $x_3$. Can we iterate the process in some way, obtaining an even better estimate $x_4$, then continuing the process to obtain still better estimates $x_5, x_6, \ldots$ for $\sqrt{2}$? A variety of options are available to us at this stage.

It would be natural to use $x_2$ and $x_3$ to generate $x_4$ in the same manner we obtained $x_3$ from $x_1$ and $x_2$. But we leave this aside for the moment and instead use $x_1$ and $x_3$ to generate $x_4$. Then we use $x_1$ and $x_4$ to obtain $x_5$, and so forth. In other words, we generate a sequence of numbers

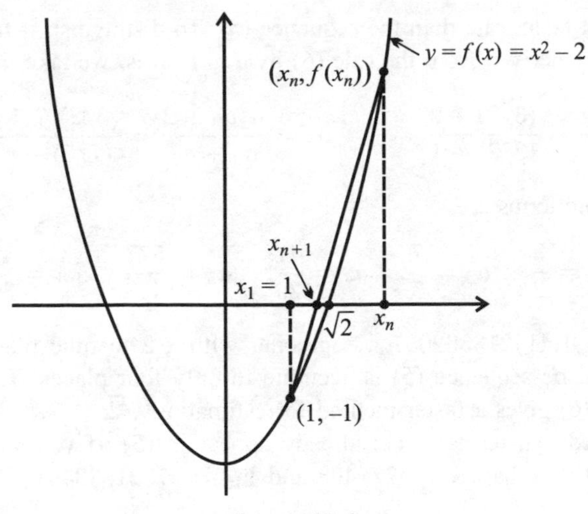


Figure 13.3.

$x_1, x_2, x_3, x_4, \ldots$ according to the rule

$$x_{n+1} = F(x_1, x_n) = \frac{x_1 x_n + 2}{x_n + x_1} = \frac{x_n + 2}{x_n + 1}, \quad n = 2, 3, \ldots, \tag{4}$$

with $x_1 = 1$ and $x_2 = 3/2$. Figure 13.3 exhibits the underlying geometry of this process, showing how $x_{n+1}$ is obtained from $x_1$ and $x_n$ by intersecting the $x$-axis with the line through $(x_1, f(x_1)) = (1, -1)$ and $(x_n, f(x_n))$.

You can check that this procedure produces the sequence

$$x_1 = 1, \quad x_2 = \frac{3}{2}, \quad x_3 = \frac{7}{5}, \quad x_4 = \frac{17}{12}, \quad x_5 = \frac{41}{29}, \quad x_6 = \frac{99}{70}, \ldots, \tag{5}$$

providing successively better approximations of $\sqrt{2}$. The sixth term of the sequence is $99/70 = 1.414285714\ldots$, which agrees with $\sqrt{2}$ to four places after the decimal point. If you are in a mood for calculation, you can check that $x_{11} = 8119/5741 = 1.414213552\ldots$, which is accurate to seven places after the decimal point.

Observe that the terms of the sequence (5) are alternately smaller and larger than $\sqrt{2}$ as they approach $\sqrt{2}$. Examination of Figure 13.3 will help you see geometrically why this occurs.

In the next section we shall see that generating each approximation from the immediately preceding two approximations is considerably more efficient than the process described by (4).

## 4 Better Successive Approximations

We stick with our function $f(x) = x^2 - 2$, but instead of formula (4) to obtain $x_{n+1}$ from $x_n$, we now use

$$x_{n+1} = F(x_{n-1}, x_n) = \frac{x_{n-1} x_n + 2}{x_n + x_{n-1}}, \quad n = 2, 3, \ldots. \tag{6}$$

We shall also choose $x_1 = 1$ and $x_2 = 3/2$. Geometrically, we obtain $x_{n+1}$ by intersecting the $x$-axis with the line through $(x_{n-1}, f(x_{n-1}))$ and $(x_n, f(x_n))$. A moment's thought, with a picture such as Figure 13.3, will convince you that this process should produce a sequence of numbers that approaches $\sqrt{2}$ at a faster rate than the sequence (5). To distinguish it from sequence (5), we shall denote the numbers derived from the rule (6) by $a_n$. That is, we take $a_1 = 1$, $a_2 = 3/2$, and

$$a_3 = \frac{a_1 a_2 + 2}{a_2 + a_1} = \frac{(3/2) + 2}{(3/2) + 1} = \frac{7}{5}, \quad a_4 = \frac{a_2 a_3 + 2}{a_3 + a_2} = \frac{(3/2)(7/5) + 2}{(7/5) + (3/2)} = \frac{41}{29},$$

and so forth. A few more terms are

$$a_1 = 1, \quad a_2 = \frac{3}{2}, \quad a_3 = \frac{7}{5}, \quad a_4 = \frac{41}{29}, \quad a_5 = \frac{577}{408}, \quad a_6 = \frac{47321}{33461}, \ldots. \tag{7}$$

In decimal form, $a_6 = 1.414213562057\ldots$ agreeing with $\sqrt{2}$ to nine places after the decimal point. Recall that $x_6$ in the sequence (5) is accurate to only four places. This is consistent with our suspicion that rule (6) gives a faster method for estimating $\sqrt{2}$.

You may have noticed that terms in (7) already appear in (5). If you were to calculate more terms in (5), you would find that $x_8 = 577/408$ and $x_{13} = 47321/33461$. Consequently we have

$$a_1 = x_1, \quad a_2 = x_2, \quad a_3 = x_3, \quad a_4 = x_5, \quad a_5 = x_8, \quad a_6 = x_{13}, \ldots. \tag{8}$$

In fact, it turns out that all the terms of the sequence $\{a_n\}$ appear somewhere in the sequence $\{x_n\}$. The sequence $\{a_n\}$ turns out to be a *subsequence* of the sequence $\{x_n\}$. Furthermore, notice in (8) that the subscripts of the $x$'s follow the pattern $1, 2, 3, 5, 8, 13, \ldots$, where each integer is the sum of the preceding two integers. This is the Fibonacci sequence. In the next section we will see that this pattern results from a special property (see (11) in Section 5) of the function $F$ given in (3). In Section 7 we will discuss how this property of $F$ is an algebraic form of a famous geometric theorem of Pascal.

## 5 An Important Property of $F$

The function $F$ given in (3) has the form

$$F(x, y) = \frac{xy + 2}{x + y}. \tag{9}$$

It is obvious from (9) that for all $x$, $y$ we have

$$F(y, x) = F(x, y). \tag{10}$$

Another, and more important, property of $F$ we wish to consider in this section is that for all $x$, $y$, and $z$,

$$F(F(x, y), z) = F(x, F(y, z)). \tag{11}$$

One needs to keep in mind that (11) is true only when the expressions involved make sense. Note, for instance, that $F(1, -1)$ is undefined. The proof of (11) is very straightforward. You will find it easy to verify that both sides are equal to

$$\frac{xyz + 2(x + y + z)}{xy + yz + zx + 2}.$$

Now return to the sequence $\{x_n\}$ defined by (4), but allowing $x_1$ to be arbitrary. Thus

$$x_{n+1} = F(x_1, x_n) = \frac{x_1 x_n + 2}{x_n + x_1}, \quad n = 1, 2, \ldots. \tag{12}$$

We will use (11) to prove that

$$x_{m+n} = F(x_m, x_n) = \frac{x_m x_n + 2}{x_m + x_n}, \tag{13}$$

for all integers $m, n \geq 1$. The proof is by induction on $m$. What we want to show is that for each $m \geq 1$ the relation (13) holds for all $n \geq 1$. Now, if we know that for some particular value of $m$ the relation holds for all $n \geq 1$, then certainly

$$x_{m+(n+1)} = F(x_m, x_{n+1})$$

for $n \geq 1$. Therefore

$$\begin{aligned} x_{(m+1)+n} &= x_{m+(n+1)} = F(x_m, x_{n+1}) = F(x_m, F(x_1, x_n)) \\ &= F(F(x_m, x_1), x_n) = F(F(x_1, x_m), x_n) = F(x_{m+1}, x_n), \end{aligned}$$

where we have used (10), (11), and (12). This shows us that (13) holds for all $n \geq 1$ with $m$ replaced by $m+1$. But certainly (13) holds for $n \geq 1$ when $m = 1$, because this is just the way the sequence was defined in (12). By the principle of mathematical induction, we see that (13) holds for all $n \geq 1$ for each $m \geq 1$, as we wanted to prove.

We can use (13) to explain the mysterious appearance of Fibonacci numbers at the end of the preceding section. Let $\{a_n\}$ be the subsequence of those $x_n$ having Fibonacci numbers as subscripts, namely

$$a_1 = x_1, \quad a_2 = x_2, \quad a_3 = x_3, \quad a_4 = x_5, \quad a_5 = x_8, \quad a_6 = x_{13}, \ldots,$$

and so forth. Now observe that (13) gives

$$\begin{aligned} F(a_1, a_2) &= F(x_1, x_2) = x_{1+2} = x_3 = a_3, \\ F(a_2, a_3) &= F(x_2, x_3) = x_{2+3} = x_5 = a_4, \\ F(a_3, a_4) &= F(x_3, x_5) = x_{3+5} = x_8 = a_5, \end{aligned}$$

and so on. In general we have $a_{n+1} = F(a_{n-1}, a_n)$, the rule that gave us the sequence $\{a_n\}$ in section 4.

## 6 Newton's Method Is Even Faster

In Figure 13.2 we saw that for any two numbers $x_1 \neq x_2$, the number given by the rule of false position,

$$t = F(x_1, x_2) = \frac{x_1 f(x_2) - x_2 f(x_1)}{f(x_2) - f(x_1)}, \tag{14}$$

corresponds to the intersection of the $x$-axis and the line through $(x_1, f(x_1))$ and $(x_2, f(x_2))$. In case $f(x) = x^2 - 2$ we found that $F(x_1, x_2)$ is given by (3), an expression that makes sense even when $x_1 = x_2$ although we no longer have two different points on the graph through which to draw a line. What is the geometric interpretation of $F(x, x)$ in this case? Since the line through $(x_1, f(x_1))$ and $(x_2, f(x_2))$ approaches the tangent line at $(x_1, f(x_1))$ as $x_2$ approaches $x_1$, we see that $F(x, x)$ is the abscissa of the point where the tangent line to the graph at the point $(x, f(x))$ intersects the $x$-axis, as indicated in Figure 13.4.

As Figure 13.4 suggests, if $x_n$ is a good approximation of a solution to the equation $f(x) = 0$, then $x_{n+1} = F(x_n, x_n)$ is often a considerably better approximation. This is the intuitive basis for Newton's method of approximation.

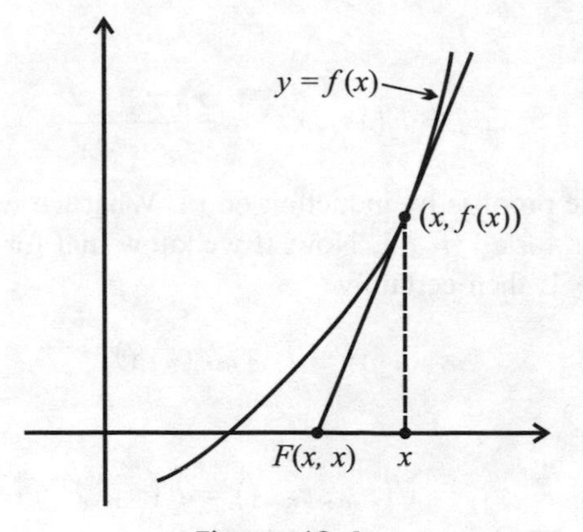


**Figure 13.4.**

As an application of this, we return to our example of $f(x) = x^2 - 2$ and use the rule

$$b_{n+1} = F(b_n, b_n) = \frac{b_n^2 + 2}{2b_n} \tag{15}$$

to generate a sequence $b_1, b_2, b_3, \ldots$, starting with the initial choice $b_1 = 1$. This gives

$$b_1 = 1, \quad b_2 = \frac{3}{2}, \quad b_3 = \frac{17}{12}, \quad b_4 = \frac{577}{408}, \quad b_5 = \frac{665857}{470832}, \ldots. \tag{16}$$

Here $b_5 = 1.4142135623747\ldots$, already agreeing with $\sqrt{2}$ to eleven places after the decimal point. This is better than the sequences we have encountered so far.

You may have noticed that $\{b_n\}$ appears to be a subsequence of the sequence $\{x_n\}$ in (5). In fact, the $b_n$ comprise those $x_n$ whose subscripts are powers of 2. Namely, we have

$$b_1 = x_1, \quad b_2 = x_2, \quad b_3 = x_4, \quad b_4 = x_8, \quad b_5 = x_{16}, \ldots$$

and so on. The validity of this can be shown in a manner similar to our discussion of $\{a_n\}$ in Section 5. We start with $b_1 = 1 = x_1$. Then, using (13), we have $F(x_n, x_n) = x_{n+n} = x_{2n}$ for all $n$. Thus

$$\begin{aligned} b_2 &= F(b_1, b_1) = F(x_1, x_1) = x_2, \\ b_3 &= F(b_2, b_2) = F(x_2, x_2) = x_4, \\ b_4 &= F(b_3, b_3) = F(x_4, x_4) = x_8, \end{aligned}$$

and so forth.

The rule (15) that enables us to approximate $\sqrt{2}$ so efficiently can be rewritten in the form

$$b_{n+1} = \frac{1}{2}(b_n + \frac{2}{b_n}). \tag{17}$$

Thus each term of the sequence is obtained by a "divide and average" technique, where we take the average of the previous term and 2 divided by the previous term. An elaboration of this method, which goes back to the ancient Babylonians, can be found in [3].

We close this section with two exercises for the reader. The first exercise is for those acquainted with the calculus. In the expression (14) for $F(x_1, x_2)$, let $x_1 = x$ and $x_2 = x + h$. Show that this leads to

$$F(x, x+h) = \frac{x(f(x+h) - f(x)) - hf(x)}{f(x+h) - f(x)}.$$

Now, assuming $f$ is a differentiable function, divide numerator and denominator by $h$ and let $h \to 0$ to obtain

$$F(x,x) = x - \frac{f(x)}{f'(x)},$$

where $f'(x)$ is the derivative of $f(x)$. Thus the sequences generated by Newton's method satisfy

$$x_{n+1} = x_n - \frac{f(x_n)}{f'(x_n)}, \qquad n = 1, 2, \ldots.$$

The second exercise extends the results we have obtained for $f(x) = x^2 - 2$ to the more general case $f(x) = ax^2 + bx + c$, where $a$, $b$, and $c$ are constants such that $a \neq 0$. Verify in this case that the rule of false position gives

$$F(x_1, x_2) = \frac{ax_1x_2 - c}{a(x_1 + x_2) + b}. \tag{18}$$

Use this to show that the crucial relation (11) is still valid in this case. It follows that the relationship $x_{m+n} = F(x_m, x_n)$ is also valid for sequences generated by the rule $x_{n+1} = F(x_1, x_n)$.

## 7 Pascal's Theorem Is Behind It All

Figure 13.5 shows us two hexagons $ABCDEG$, one of them "self-intersecting." With the vertices labeled as in the picture, we call $AB$ and $DE$ a pair of "opposite sides." Similarly $BC$ and $EG$ are opposite sides, as are $CD$ and $AG$.

In the following we shall deal with the points of intersection of certain sides of a hexagon; this will always refer to the intersection of the *extended* sides, i.e., the lines containing the segments that are usually called the sides.

Careful drawings of a few examples will show you that the three points obtained by intersecting pairs of opposite (extended) sides of a hexagon do not necessarily lie on a common straight line. However, we have the following **Theorem of Pascal:**

*If a hexagon is inscribed in a conic section, then the intersections of pairs of opposite sides are collinear.*

Figure 13.6 shows two examples of hexagons inscribed in ellipses. $P$, $Q$, and $R$ are the intersections of pairs of opposite sides and are collinear in both examples.

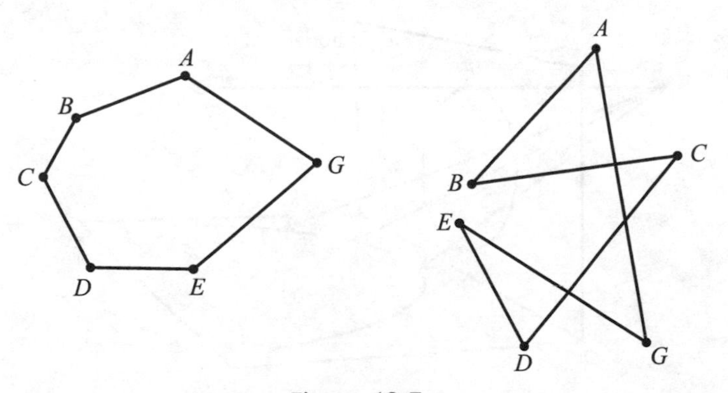


**Figure 13.5.**

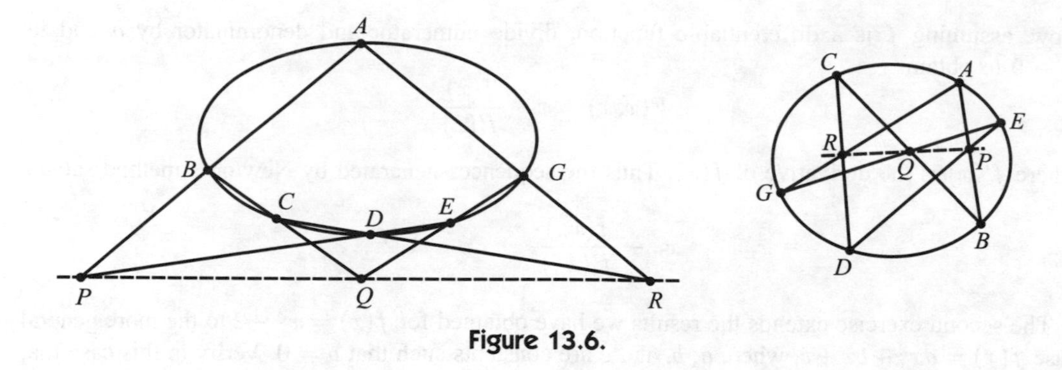


**Figure 13.6.**

Blaise Pascal proved his theorem (presented as "the theorem of the mystic hexagram" in 1640) when he was sixteen years old. He did this for a hexagon inscribed in a circle, but that was enough to establish the general theorem, because any conic section can be transformed to a circle by a projection that maps one plane to another. Such a projection is an example of a *projective transformation,* for which straight lines in one plane correspond to straight lines in the other. That a suitable projection will make a given conic section correspond to a circle boils down to the fact that a conic section is obtained by intersecting a cone whose base is a circle with a suitable plane. For example, in Figure 13.7 the parabola in the horizontal plane corresponds to the circle in the vertical plane under projection from the point $V$.

We have drawn some rays that are generators of a cone with vertex $V$. The vertical plane intersects this cone in a circle. The horizontal plane intersects the cone in a parabola, since the horizontal plane is parallel to a generator of the cone (the ray $e$ in the picture). Had we chosen a higher position for $V$, the circle in the vertical plane would have corresponded to an ellipse in the horizontal plane, while a lower position would have resulted in a branch of a hyperbola.

If we are given a hexagon $ABCDEG$ inscribed in an ellipse as in Figure 13.6, and we want to prove $P$, $Q$, and $R$ are collinear, then we can project the figure in such a way that the ellipse corresponds to a circle and the hexagon corresponds to another hexagon inscribed in that circle. Then the theorem for circles would tell us that the corresponding images of $P$, $Q$, and $R$ are collinear, so the points $P$, $Q$, and $R$ themselves are collinear.

To see how Pascal's theorem is connected to the material in the preceding sections, consider the parabola that is the graph of $y = ax^2 + bx + c$, as in Figure 13.8, where $a$, $b$, $c$ are given constants, with $a > 0$. Let $F$ be the function (18) determined by the rule of false position when applied to this parabola. If $x$ and $y$ correspond to points on the $x$-axis, and $A$ and $B$ are the respective

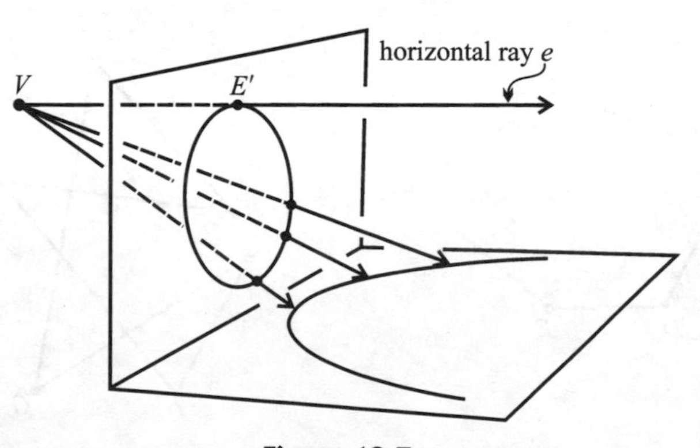


**Figure 13.7.**

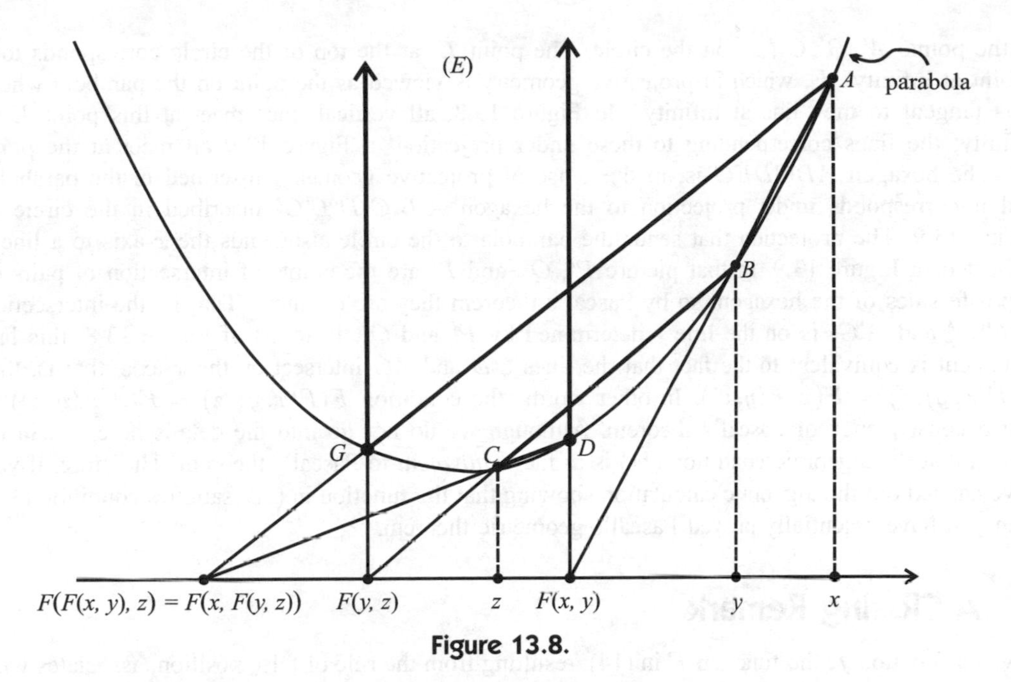


**Figure 13.8.**

points on the parabola above them, as depicted in Figure 13.8, then $F(x, y)$ corresponds to the point where the line $AB$ meets the $x$-axis. In Figure 13.8 we also have $z$ on the $x$-axis and the point $C$ on the parabola above $z$. In this picture $F(y, z)$ corresponds to the point where the line $BC$ intersects the $x$-axis.

Above $F(x, y)$ we have the point $D$ on the parabola and above $F(y, z)$ we have the point $G$. Therefore $F(F(x, y), z)$ corresponds to where the line $CD$ intersects the $x$-axis, while $F(x, F(y, z))$ corresponds to where the line $AG$ intersects the $x$-axis. As shown in the second exercise at the end of the previous section, we have $F(F(x, y), z) = F(x, F(y, z))$. Interpreted geometrically, this says that the intersection of the lines $CD$ and $AG$ is on the $x$-axis, as indicated in Figure 13.8.

Figure 13.9 shows the result of projecting Figure 13.8 (by an appropriate projection as in Figure 13.7) so the parabola corresponds to a circle and the points $A, B, C, \ldots$ on the parabola correspond

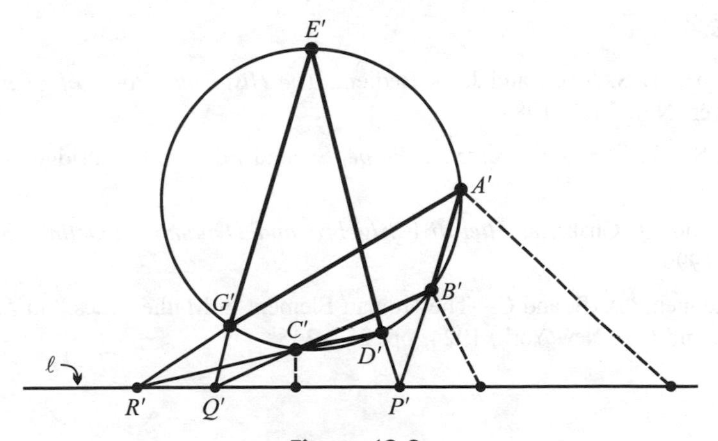


**Figure 13.9.**

to the points $A', B', C', \ldots$ on the circle. The point $E'$ at the top of the circle corresponds to a "point at infinity" $E$, which in projective geometry is viewed as the point on the parabola where it is tangent to the "line at infinity." In Figure 13.8, all vertical lines meet at this point $E$ at infinity; the lines corresponding to these under projection in Figure 13.9 all meet at the point $E'$. The hexagon $ABCDEG$ is, in the sense of projective geometry, inscribed in the parabola, and it corresponds under projection to the hexagon $A'B'C'D'E'G'$ inscribed in the circle in Figure 13.9. The projection that sends the parabola to the circle also sends the $x$-axis to a line $\ell$ indicated in Figure 13.9. In that picture $P'$, $Q'$, and $R'$ are the points of intersection of pairs of opposite sides of the hexagon, so by Pascal's theorem they are collinear. That is, the intersection of $C'D'$ and $A'G'$ is on the line $\ell$ determined by $P'$ and $Q'$. In terms of Figure 13.8, this last statement is equivalent to the fact that the lines $CD$ and $AG$ intersect on the $x$-axis, that is, that $F(F(x,y),z) = F(x,F(y,z))$. In other words, the condition $F(F(x,y),z) = F(x,F(y,z))$ is just a consequence of Pascal's theorem. Although we do not go into the details here, it can be shown that the algebraic condition (11) is in fact *equivalent* to Pascal's theorem. Therefore, if you have carried out the algebraic calculation showing that the function in (18) satisfies condition (11), then you have essentially proved Pascal's geometric theorem.

## 8 A Closing Remark

Given a function $f$, the function $F$ in (14), resulting from the rule of false position, associates with any two numbers $x$ and $y$ another number $F(x,y)$, so it gives us what is called a *binary operation* on the set of real numbers. Let us use the notation $F(x,y) = x \otimes y$ to remind us that this can be viewed as an exotic sort of "multiplication" operation. Then condition (10) says $y \otimes x = x \otimes y$, so this operation is *commutative.* To satisfy (11) means that $(x \otimes y) \otimes z = x \otimes (y \otimes z)$, so the operation is *associative* for the functions we have considered. What we saw in the preceding section is that if we use a parabola and the rule of false position in this way to define a binary operation on the real line, then the fact that this operation satisfies the associative property is equivalent to Pascal's theorem. You can read about the significance of this in a good textbook on projective geometry. See, for example, Coxeter [2], in particular Exercise 3 on page 177 of that book.

As a final exercise, you may want to find the form of the binary operation that results from applying the rule (14) with $f(x) = x/(x-1)$, $x \neq 1$. While $f(x)$ is not defined for $x = 1$, a picture should indicate to you why $F(1,x) = x$ is a natural choice. The graph of $y = f(x)$ is by the way a rectangular hyperbola having the line $x = 1$ as one of its asymptotes.

## References

[1] Bunt, L. N. H., P. S. Jones, and J. D. Bedient, *The Historical Roots of Elementary Mathematics,* Dover, New York, 1988.

[2] Coxeter, H. S. M., *The Real Projective Plane,* Second Edition, Cambridge University Press, Cambridge, 1961.

[3] Kreith, K., and D. Chakerian, *Iterative Algebra and Dynamic Modeling,* Springer-Verlag, New York, 1999.

[4] Leacock, Stephen, "A, B, and C—The Human Element in Mathematics," in *Literary Lapses,* Dodd Mead and Co., New York, 1924, pp. 237–245.

# 14

# Geometric Puzzles and Constructions—Six Classical Geometry Theorems[1]

Zvezdelina Stankova
*Mills College*

## 1 A Warm-up Geometry Problem for Everyone to Play With

We start by presenting a problem for beginners, along with three different solutions. Readers, inexperienced in geometry, are strongly advised to read through the solutions and understand them as much as possible, before moving to Section 3. Advanced readers are encouraged to *solve* this problem without peeking at the solutions below, and then continue with Section 2.

**Problem 1.** *Three congruent squares with bases $AM$, $MH$, and $HB$, are put next to each other to form a rectangle $ABCD$ (see Fig. 14.1). Show that*

$$\angle AMD + \angle AHD + \angle ABD = 90°.$$

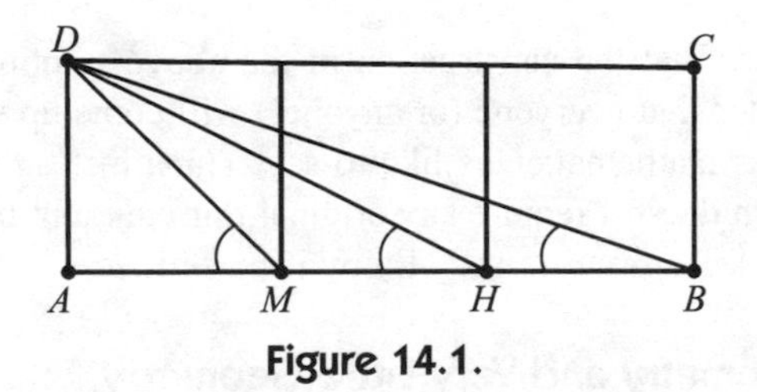


**Figure 14.1.**

This problem was discussed at length at the Math Circle Workshop on June 6th 1998 in the Lawrence Hall of Science, University of California, Berkeley. Let's see if anyone remembers the beautiful solution we saw there! Now, imagine you are in 7–8th grade, and you haven't yet heard of "trigonometry" (oops, that's a hint for the advanced!), and your whole world of geometric tricks

[1]The notes for this paper were assembled from a number of talks on the topic by the author, e.g., Bay Area Math Adventures in September 1999 and the San Jose State University Colloquium in April 2000. For more information and further study, we refer the reader to the bibliography section.

consists of similar and congruent triangles, and, say, you know that the sum of angles in a triangle is 180°. Can you solve the problem with this information? Play with it and see how far you get.

## 1.1 Solution 1: Unexpected Geometric Construction

Reflect the figure across line $DC$, i.e., draw three more squares as shown on Fig. 14.2. Connect the new point $H_1$ with points $D$ and $B$. Note that $\triangle AHD$, $\triangle A_1H_1D$ and $\triangle B_1BH_1$ are congruent to each other. (Why? They are all right triangles with legs of the same length.) This allows us to call three angles on the picture by $\alpha$, and three others by $\beta$. In particular, $\angle AHD = \alpha = \angle B_1BH_1$.

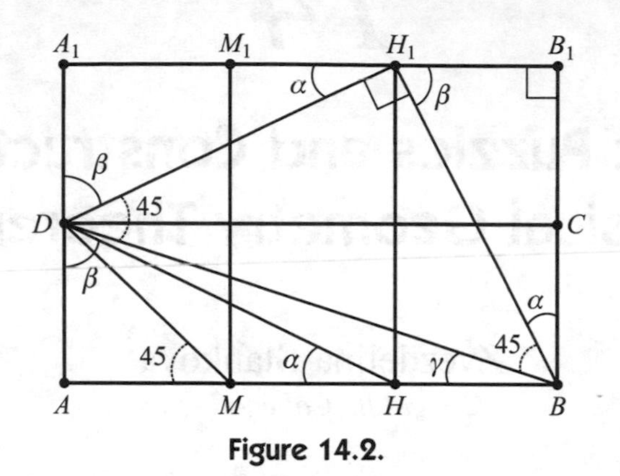


**Figure 14.2.**

Further, note that $\triangle DBH_1$ is a right isosceles triangle. Indeed, $|DH_1| = |BH_1|$ (Why? Because two of the above triangles are congruent. Which triangles do we have in mind?), and

$$\angle DH_1B = 180° - \angle DH_1A_1 - \angle B_1H_1B = 180° - \alpha - \beta = 90°.$$

The last follows from the fact that $\alpha$ and $\beta$ are the two acute angles in right $\triangle AHD$, and hence they sum up to 90°. Thus, in $\triangle DBH_1$: $\angle DBH_1 = 45°$.

To finish the proof, note that the three wanted angles are $\angle AMD = 45°$ (Why? $\triangle AMD$ is also right isosceles) $\angle AHD = \alpha$ and $\angle ABD = \gamma$; and that they appear "miraculously," in the right $\angle ABB_1$:

$$90° = \angle ABB_1 = \gamma + 45° + \alpha. \qquad \square$$

Now, everyone understands that the construction in the above solution is really a very original idea, and there is no guarantee that everyone (or anyone!) will come up with this same idea within a "finite amount of time," as mathematicians like to say. Thus, instead we propose here a simple trigonometric solution, which doesn't require any original thinking, but has the drawback of giving us no idea of *why* these angles sum up to 90°. Enjoy it nevertheless.

## 1.2 Solution 2: Trigonometry and Very Little Geometry

Since $\angle DMA = 45°$ (from right isosceles $\triangle DMA$ as above), it suffices to show that $\angle AHD + \angle ABD = 45°$. Name these two angles by $\alpha$ and $\gamma$ as above. Since they are both acute angles, they cannot sum up to more than 180°; thus, if we show that $\tan(\alpha + \gamma) = 1$, we will be able to conclude that $\alpha + \gamma = 45°$.

The formula for the tangent of a sum comes to the rescue:

$$\tan(\alpha + \gamma) = \frac{\tan\alpha + \tan\gamma}{1 - \tan\alpha \cdot \tan\gamma}.$$

From $\triangle AHD$ and $\triangle ABD$, respectively, we find $\tan\alpha = |AD|/|AH| = 1/2$ and $\tan\gamma = |AD|/|AB| = 1/3$. Substituting $1/2$ and $1/3$ into the above tangent formula yields

$$\tan(\alpha+\gamma) = \frac{1/2 + 1/3}{1 - 1/2 \cdot 1/3},$$

which we leave to the diligent reader to check that it equals 1. □

**Question:** Why did we use tangents? Would it be easier to use sines or cosines, or some other trigonometric function of the angles?

### 1.3 Solution 3 via Geometry without Extra Constructions

This solution was suggested by Professor Peter Ross (Santa Clara University) after the BAMA'99 lecture.

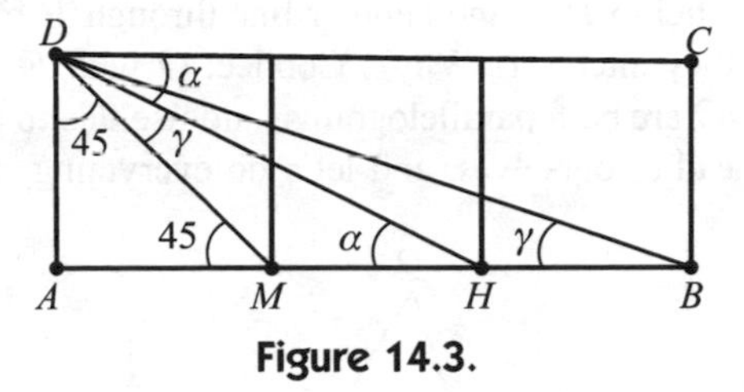


**Figure 14.3.**

With the notation in Fig. 14.3, we must show that $\angle AMD + \angle AHD + \angle ABD = 90°$. But $90° = \angle ADM + \angle MDH + \angle HDC$, where $\angle ADM = 45° = \angle AMD$, and $\angle HDC = \angle AHD = \alpha$ as alternate interior angles for the parallel lines $AB$ and $DC$. Thus, it remains to show that $\angle MDH = \angle ABD = \gamma$. If this were true, then $\triangle DMH$ and $\triangle BMD$ would be similar by (AA) because they also share the common $\angle DMH$. Now, to prove this similarity, we note this common angle, and the ratio of adjacent sides $MD/MH = \sqrt{2}/1 = 2/\sqrt{2} = BM/MD$. Thus, $\triangle DMH \sim \triangle BMD$ and hence $\angle MDH = \gamma$, which completes the proof. □

## 2 A Warm-Up Problem for the Die-Hards

Don't try this problem at home unless you really know what you are doing, for it is really hard. When you see the solution you will be surprised that it doesn't require any advanced mathematical tools; **but** how one can come up with such a solution—that's where the mystery is! So, good luck.

**Problem 2.** *Let $ABCDEF$ be a convex hexagon (see Fig. 14.4). Let $P$, $Q$, and $R$ be the intersections of the lines $AB$ and $EF$, $EF$ and $CD$, $CD$ and $AB$, respectively. Let $S$, $T$, $U$ be the intersections of the lines $BC$ and $DE$, $DE$ and $FA$, $FA$ and $BC$, respectively. Show that*

$$\frac{AB}{PR} = \frac{CD}{RQ} = \frac{EF}{QP} \Rightarrow \frac{BC}{US} = \frac{DE}{ST} = \frac{FA}{TU}.$$

(Math Olympiad Summer Program '98, Homework Assignment.)

**Solution to Problem 2.** The given triple ratios remind us suspiciously of a criterion for similar triangles (SSS): as if someone wants to tell us that $\triangle PRQ$ is similar to another triangle with sides $AB$, $CD$ and $EF$, but no such similar triangle can be found on the given picture. Let's construct it!

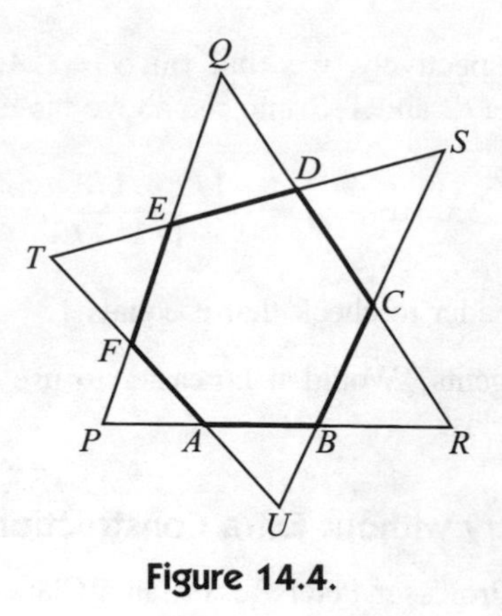


**Figure 14.4.**

Draw a line through $A$ parallel to $PQ$, and another line through $B$ parallel to $RQ$, and let them intersect in point $O$ (Would they intersect? Why?) Connect $O$ with $E$ and with $D$. Our goal is to prove that $AOEF$ and $BCDO$ are both parallelograms, and use this to prove what is wanted in the problem, but let's not get ahead of ourselves, and let's do everything step by step (see Fig. 14.5).

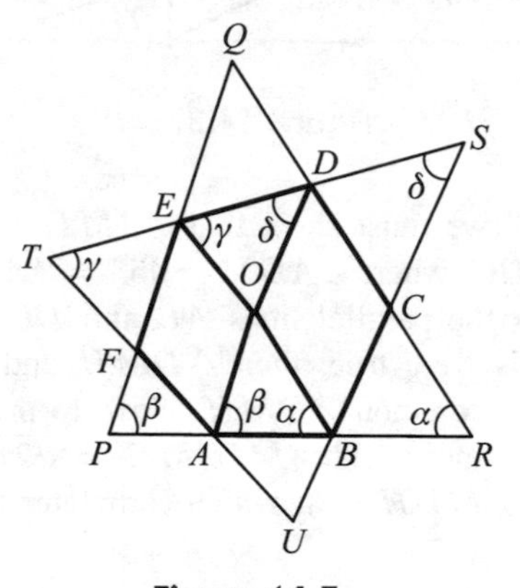


**Figure 14.5.**

For starters, do you see any similar triangles? By construction, $\triangle ABO$ and $\triangle PQR$ are similar: check out their equal angles $\alpha$'s and $\beta$'s from the parallel lines in our construction. Therefore, the sides of these two triangles are proportionate, i.e.,

$$\frac{AB}{PR} = \frac{BO}{RQ} = \frac{OA}{QP}.$$

But we have by hypothesis that

$$\frac{AB}{PR} = \frac{CD}{RQ} = \frac{EF}{QP}.$$

Since the first ratio is the same in both equations, all those five ratios are equal, in particular, $\frac{BO}{RQ} = \frac{CD}{RQ}$ and $\frac{OA}{QP} = \frac{EF}{QP}$. We conclude that $BO = CD$ and $OA = EF$.

Recall now that by construction $BO$ is parallel to $CD$, and $OA$ is parallel to $EF$. Therefore, indeed we do have parallelograms $AOEF$ and $BCDO$.

Now, we can play the same game for $\triangle TSU$ and $\triangle EOD$, by reversing the above argument. Are they similar? Since $EO$ and $TU$ are parallel, and $DO$ and $SU$ are parallel (from the parallelograms above) we conclude that the two triangles have equal angles $\gamma$'s and $\delta$'s, and therefore they are similar.

Thus, the sides of $\triangle TSU$ and $\triangle EOD$ are proportionate:

$$\frac{OD}{US} = \frac{DE}{ST} = \frac{OE}{TU}.$$

But $OD = BC$ and $OE = FA$ (again from the parallelograms), thus

$$\frac{BC}{US} = \frac{DE}{ST} = \frac{FA}{TU}. \qquad \square$$

## 3 Six Related Classical Theorems in Geometry

In this section we state six famous classical geometry theorems: Desargues', Menelaus', Pascal's, Brianchon's, the Poncelet–Brianchon, and Pappus'. Readers are encouraged to familiarize themselves with these statements and to attempt to prove them from scratch. Do not be disappointed if your attempts fail—these theorems are hard, and the purpose of this article is to present you with different ways of thinking and proving them, along with the necessary theory and problem-solving techniques. Another task that we shall be interested in is the connections between the six theorems: does one theorem imply another? What mathematical areas are evoked when proving each theorem?

In what follows, we shall say that several points are *collinear* if they lie on a line. Similarly, several points are *concyclic* if they lie on a circle; an *inscribed* (cyclic) polygon has its vertices lying on a circle. If three distinct points $A$, $B$, and $C$ are collinear, then the *directed ratio* $\overline{AB}/\overline{CB}$ is the ratio of the lengths of segments $AB$ and $CB$, taken with a sign "+" if the segments have the same direction (i.e., $B$ is *not* between $A$ and $C$), and with a sign "−" if the segments have opposite directions (i.e., $B$ is between $A$ and $C$). Several objects (lines, circles, etc.) are *concurrent* if they all intersect in some point.

**Problem 3 (Desargues' Theorem)** *$\triangle ABC$ and $\triangle A_1B_1C_1$ are positioned in such a way that lines $AA_1$, $BB_1$, and $CC_1$ intersect in a point $O$. If lines $AB$ and $A_1B_1$, $AC$ and $A_1C_1$, $BC$ and $B_1C_1$ are pairwise not parallel, prove that their points of intersection, $L$, $M$ and $N$, are collinear (see Fig. 14.6).*

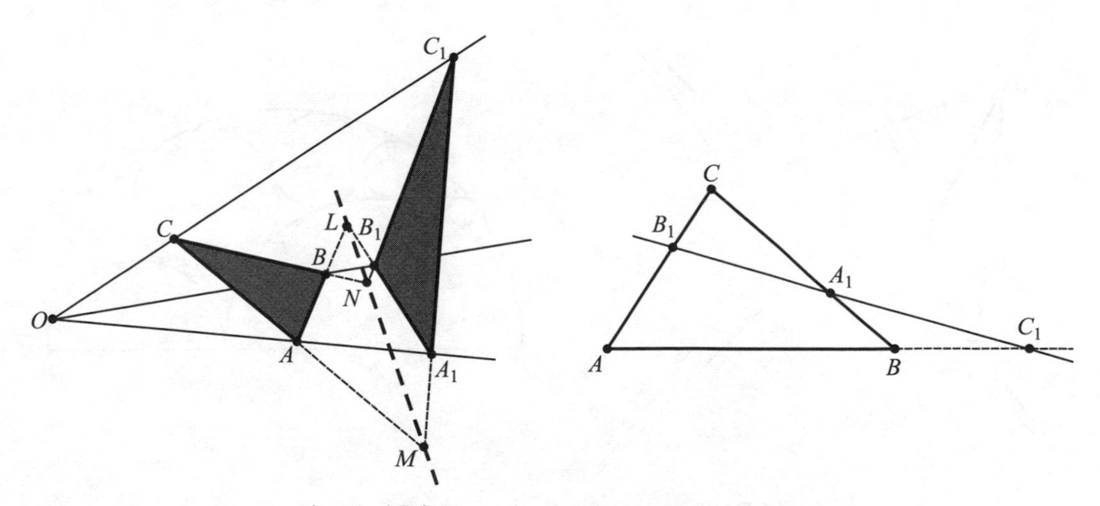


**Figure 14.6.** Desargues' and Menelaus' Theorems

**Problem 4 (Menelaus' Theorem)** *Let $A_1$, $B_1$, and $C_1$ be three points on the sides $BC$, $CA$, and $AB$ of $\triangle ABC$. Then they are collinear if and only if*

$$\frac{\overline{AB_1}}{\overline{CB_1}} \cdot \frac{\overline{CA_1}}{\overline{BA_1}} \cdot \frac{\overline{BC_1}}{\overline{AC_1}} = 1.$$

**Problem 5 (Pascal's Theorem)** *If the hexagon $ABCDEF$ is cyclic and its opposite sides, $AB$ and $DE$, $BC$ and $EF$, $CD$ and $FA$, are pairwise not parallel, prove that their three points of intersection, $X$, $Y$, and $Z$, are collinear (see Fig. 14.7.)*

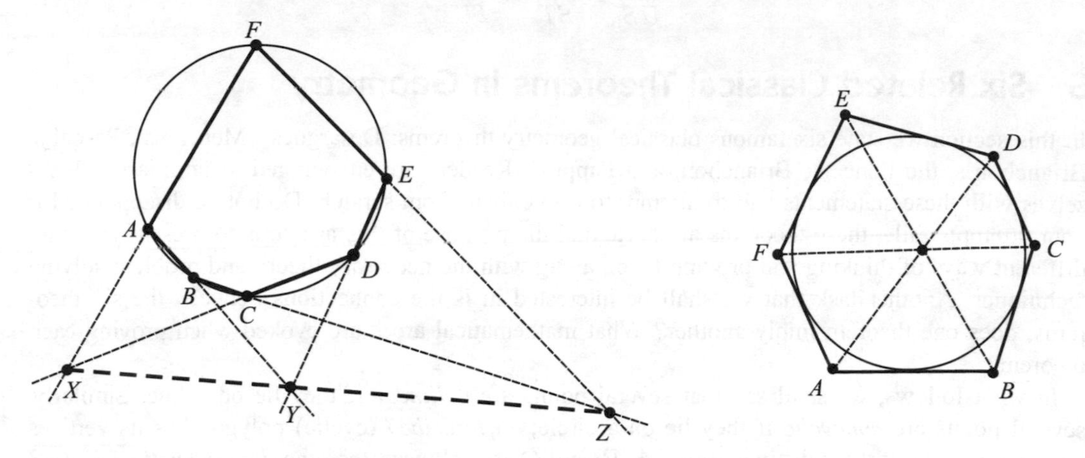


**Figure 14.7.** Pascal's and Brianchon's Theorems

**Problem 6 (Brianchon's Theorem)** *If the hexagon $ABCDEF$ is circumscribed around a circle, prove that its three diagonals $AD$, $BE$, and $CF$ are concurrent (see Fig. 14.7.)*

**Problem 7 (The Poncelet–Brianchon Theorem)** *Let $A$, $B$, and $C$ be three points on a rectangular hyperbola (i.e., a hyperbola with perpendicular asymptotes.) Prove that the orthocenter of $\triangle ABC$ also lies on the hyperbola (see Fig. 14.8.)*

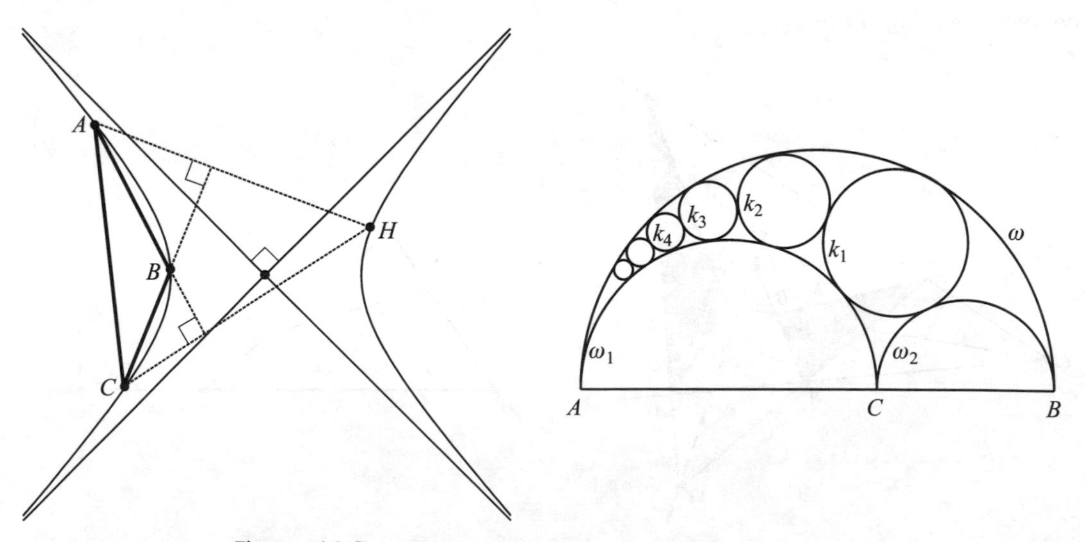


**Figure 14.8.** The Poncelet–Brianchon and Pappus' Theorems

**Problem 8 (Pappus' Theorem)** *Let $\omega$ be a semicircle with diameter $AB$. Let $\omega_1$ and $\omega_2$ be semicircles externally tangent to each other at $C$, and internally tangent to $\omega$ at $A$ and $B$, respectively. Let $k_1$, $k_2$,... be a sequence of circles, each tangent to $\omega$ and $\omega_1$, such that $k_1$ is tangent to $\omega_2$, and $k_{n+1}$ is tangent to $k_n$ for all $n \geq 1$. Let $r_n$ be the radius of $k_n$, and $d_n$ the distance from the center of $k_n$ to $AB$. Prove that $d_n = 2nr_n$ for all $n$ (see Fig. 14.8.)*

# 4 First Proof of Desargues' Theorem via Menelaus'

In this section we see how assuming Menelaus' Theorem implies Desargues' Theorem. The solution is simple since it doesn't require too much thinking, just concentrating on the technical details and trying not to make a calculation error. A serious drawback of this solution is that it doesn't give us a clue *really why* Desargues' Theorem works. On the other hand, it shows that Menelaus' is a stronger theorem than Desargues'—a relation frequently desired in mathematics. But first, let's briefly discuss the idea for the proof of Menelaus' Theorem.

## 4.1 Sketch of Proof of Menelaus' Theorem

In one direction, suppose that the 3 points are collinear. Drop the perpendiculars from the vertices $A$, $B$, and $C$ to line $A_1B_1C_1$, and consider three pairs of similar triangles (which ones?). For the other direction, suppose that the given equality is satisfied, and let line $A_1B_1$ intersect $AB$ in point $C'$. Using the proven direction above for the points $A_1$, $B_1$, and $C'$, conclude that $C' \equiv C_1$. □

## 4.2 Proof of Desargues' Theorem via Menelaus'

Apply Menelaus' Theorem three times to, respectively, $\triangle OBC$ and line $NB_1C_1$, $\triangle OAB$ and line $LB_1A_1$, and $\triangle OAC$ and line $MA_1C_1$:

$$\frac{\overline{CN}}{\overline{BN}} \cdot \frac{\overline{BB_1}}{\overline{OB_1}} \cdot \frac{\overline{OC_1}}{\overline{CC_1}} = 1$$

$$\frac{\overline{BL}}{\overline{AL}} \cdot \frac{\overline{AA_1}}{\overline{OA_1}} \cdot \frac{\overline{OB_1}}{\overline{BB_1}} = 1$$

$$\frac{\overline{AM}}{\overline{CM}} \cdot \frac{\overline{CC_1}}{\overline{OC_1}} \cdot \frac{\overline{OA_1}}{\overline{AA_1}} = 1$$

Now we multiply the three equalities and cancel out everything we can. We are left with

$$\frac{\overline{AM}}{\overline{CM}} \cdot \frac{\overline{CN}}{\overline{BN}} \cdot \frac{\overline{BL}}{\overline{AL}} = 1$$

which again by Menelaus (the reverse direction of the theorem) implies that points $M$, $N$, and $L$ are collinear. □

**Question:** What happens if some of the pairs of lines in the problem (or in the solution) do not intersect, i.e., they are parallel? Can you still solve the problem using a modification of the above method? The advanced reader should investigate these cases and come up with some resolution.

# 5 Second Proof of Desargues' Theorem via Projective Geometry

It turns out that it is not so bad to let some of the pairs of lines in the setting of Desargues' Theorem be parallel. In fact, making *all* such lines parallel pairwise is the basis for the following *Projective Geometry* proof.

## 5.1 A Short Discussion of Projective Geometry

There are certain nice transformations in the plane, called *projective,* which send lines to lines—nothing really surprising here: say, reflections across a point or across a line, rotations and parallel translations are examples of such transformations. However, the "magic" of projective transformations works when we are able to "separate" intersecting lines, i.e., to make them parallel without changing too much the structure of the original picture. This is possible because we add one extra "line" to the usual plane, called the *line at infinity*. For every family of parallel lines in the usual plane there is a (different) point on the line at infinity $l$. Conversely, any point on $l$ is "born" by a (unique) family of parallel lines.

Note that it is very hard to imagine exactly the picture of this augmented plane, called the *projective plane*. This is because we are used to thinking in 3 dimensions, and the projective plane is simply too complex to be "fitted" in 3D space. Instead of trying to imagine it, think of the projective plane as an *abstract construction* with some useful applications. When you take an introductory course in algebraic geometry, you will see various descriptions of the projective plane. These will, we hope, help you construct a satisfactory mental image of the projective plane.

## 5.2 Projective Proof of Desargues' Theorem

For now, let's glimpse the magic performed by a well-chosen projective transformation. In the setting of Desargues' theorem, consider points $L$ and $N$. If they exist, it means that the pairs of corresponding lines intersect, i.e., $AB \cap A_1B_1 = L$ and $BC \cap B_1C_1 = N$. The idea is to apply a projective transformation to the plane, sending points $L$ and $N$ to the line $l$ at infinity, and thus, in effect making line $AB$ parallel to $A_1B_1$ (they will intersect at a point "at infinity"), and similarly $BC$ parallel to $B_1C_1$.

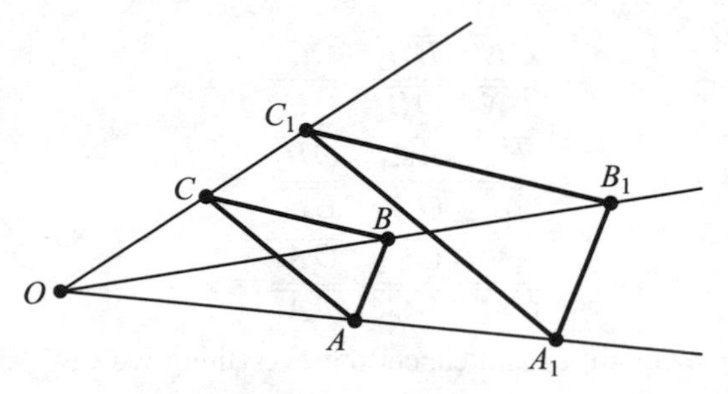


**Figure 14.9.** Projective Proof of Desargues' Theorem

It is now not hard to prove that $AC$ and $A_1C_1$ are also parallel: use similar triangles $\triangle OAB \sim \triangle OA_1B_1$ (why?), and $\triangle OBC \sim \triangle OB_1C_1$ (why?), to conclude that $OA/OA_1 = OB/OB_1 = OC/OC_1$. This in its turn implies that $\triangle OCA \sim \triangle OC_1A_1$ (why?), and therefore $AC$ is parallel to $A_1C_1$ (why?). We leave justification of the "whys" to the dedicated reader.

So what? The fact that $AC$ and $A_1C_1$ are parallel means that they intersect at a point at infinity, namely, $M$. The nice thing about the projective plane is that no matter what point you choose from which to view it, the picture you see will be essentially the same—you will see the usual (called "finite") plane, and whichever line you won't see, you can think of it as the "line at infinity." In particular, all lines are "created" equal, regardless of whether they are usual lines or the "line at infinity." In other words, the fact that all three points $L$, $M$, and $N$ happen to lie on the "line at infinity" makes them *collinear*.

To finish the proof, one has to apply the inverse of whatever projective transformation was applied in the beginning in order to obtain the original picture of Desargues' setting. In the process, the "line at infinity" $LMN$ will be sent to some line in the usual plane, on which our original points $L$, $M$, and $N$ must have been lying. □

## 6 Third Proof of Desargues' Theorem via Exit into 3D Space

It is very counterintuitive to attempt to solve a (plane) 2D problem by a 3D solution. That is, to cook up an argument in 3D space which somehow implies our 2D version. This type of reasoning is called an *exit into 3D*.

In the setting of Desargues' Theorem, imagine that everything originally lies in some plane $\gamma$, but we "lift" the ray $OCC_1$ vertically from the plane in 3D space, keeping all lines, triangles and intersection points the same as before. The goal is then to show that the "new" points $L$, $M$, and $N$ lie on a line in 3D space; we then project our new 3D picture back to the original 2D picture in the plane $\gamma$, and necessarily the "space" line $l = LMN$ will project onto another line $l_1$ in $\gamma$. This line $\gamma$, we conclude, must have contained our original points $L$, $M$, and $N$, so we will be done.

So, what are we waiting for? The 3D picture appears as shown in Fig. 14.10. Note that we have created the three planes $\gamma = (OAB)$, $\alpha = (OBC)$, $\beta = (OCA)$, which can be thought of as forming part of the pyramid $OABC$, and the two planes formed by the two new triangles: plane $P = (ABC)$ and plane $P_1 = (A_1B_1C_1)$.

Then point $L$ is the intersection of lines $AB$ and $A_1B_1$; but line $AB$ is the intersection of planes $P$ and $\gamma$, while line $A_1B_1$ is the intersection of planes $P_1$ and $\gamma$. In short:

$$L = AB \cap A_1B_1 = (P \cap \gamma) \cap (P_1 \cap \gamma) = P \cap P_1 \cap \gamma.$$

The serious reader will also verify similarly that

$$M = P \cap P_1 \cap \beta, \quad \text{and} \quad N = P \cap P_1 \cap \alpha.$$

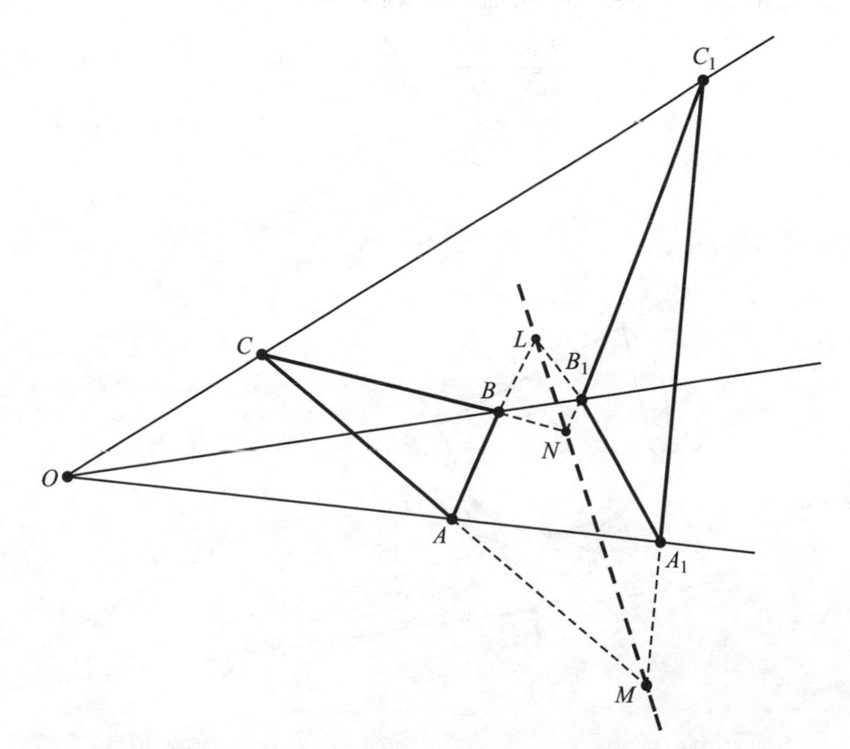


**Figure 14.10.** Proof of Desargues' Theorem via Exit into 3D Space

But planes $P$ and $P_1$ intersect in some line (why?), which we call on purpose $l$. Thus, we have seen above that all three points $L$, $M$, and $N$, lie on the line $l = P \cap P_1$, i.e., they are collinear. Projecting line $l$ onto the original plane $\gamma$ yields the wanted line. □

## 7 Proof of Pascal's Theorem via Menelaus' Theorem

In order to use Menelaus' Theorem in the setting of Pascal's Theorem, we need to recreate Menelaus' setting, i.e., we need to draw some extra line intersections. At first it is not clear which lines to intersect, but follow closely the solution below and you will see the reasoning behind our choice of extra constructions.

Create $\triangle PQR$ by intersecting the following lines: $AB \cap CD = \{R\}$, $CD \cap EF = \{P\}$, and $EF \cap AB = \{Q\}$. Then apply Menelaus' Theorem three times to $\triangle PQR$ and lines $XAF$, $CBZ$, and $DYE$, respectively:

$$\frac{\overline{PX}}{\overline{RX}} \cdot \frac{\overline{RA}}{\overline{QA}} \cdot \frac{\overline{QF}}{\overline{PF}} = 1$$

$$\frac{\overline{PC}}{\overline{RC}} \cdot \frac{\overline{RB}}{\overline{QB}} \cdot \frac{\overline{QZ}}{\overline{PZ}} = 1$$

$$\frac{\overline{PD}}{\overline{RD}} \cdot \frac{\overline{RY}}{\overline{QY}} \cdot \frac{\overline{QE}}{\overline{PE}} = 1$$

Now we multiply the three equalities and cancel out everything we can. In particular, note that $RA \cdot RB = RC \cdot RD$, $QF \cdot QE = QA \cdot QB$, and $PC \cdot PD = PF \cdot PE$, by the *Power of Point Theorem* [8, p. 246] applied consecutively to points $R$, $Q$, and $P$, and circle $k$ circumscribed around the hexagon $ABCDEF$. Thus, we are left with

$$\frac{\overline{PX}}{\overline{RX}} \cdot \frac{\overline{RY}}{\overline{QY}} \cdot \frac{\overline{QZ}}{\overline{PZ}} = 1,$$

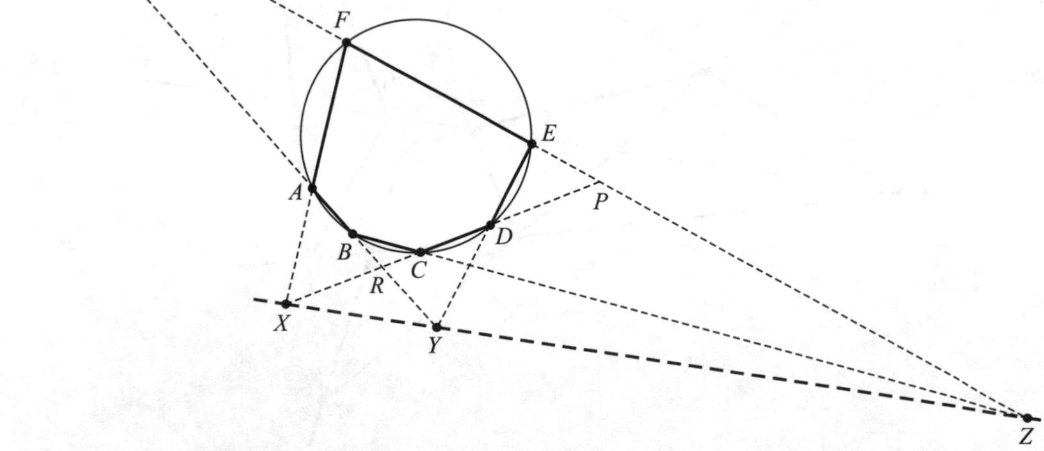

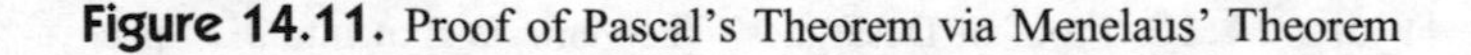


**Figure 14.11.** Proof of Pascal's Theorem via Menelaus' Theorem

which again by Menelaus (the reverse direction of the theorem) implies that points $X, Y$, and $Z$ are collinear. □

## 8 Proof of Brianchon's Theorem via Exit into 3D

The proof requires several steps and some spacial (3D) reasoning. We leave the details to the curious reader to complete.

*Step 1.* Create a spacial hexagon $A_1B_1C_1D_1E_1F_1$ which projects onto the given planar hexagon, as shown in Fig. 14.12. (Why does such a hexagon exist?) Start with point $A_1$ in space, projecting onto $A$, and then construct the remaining 5 points one by one; use six pairs of similar triangles to prove that you will eventually come back to $A_1$ in your construction.

*Step 2.* Note that to prove that diagonals $AD$, $BE$ and $CF$ meet in a point, it will suffice to show that $A_1D_1$, $B_1E_1$ and $C_1F_1$ meet in a point $X_1$ (in space)—projecting $X_1$ onto the plane will yield the required intersection point of the original diagonals.

*Step 3.* To show that $A_1D_1$, $B_1E_1$, and $C_1F_1$ intersect in space, it suffices to show that every two of them intersect in space. Indeed, if $X, Y$, and $Z$ are the pairwise intersections of the three segments, **and** we suppose by contradiction that $X, Y$, and $Z$ are distinct, this implies that $A_1D_1$, $B_1E_1$, and $C_1F_1$ all lie in a plane (together with $X, Y$, and $Z$). Now that's a contradiction since $A_1B_1C_1D_1E_1F_1$ is not planar, but spacial by construction.

*Step 4.* To show that, say, $A_1D_1$ and $B_1E_1$ intersect in space, it is sufficient to show that lines $A_1B_1$ and $D_1E_1$ lie in a plane (why?), or equivalently, to show that $A_1B_1$ and $D_1E_1$ intersect.

*Step 5.* Show that all of the 12 marked angles are equal. (Use again the 12 triangles as above, and "equal tangents" from a point to a circle.) This means that all six lines formed by the sides of the spacial hexagon $A_1B_1C_1D_1E_1F_1$ form the same angle with the original plane.

*Step 6.* Show that, say, lines $A_1B_1$ and $D_1E_1$ intersect by using two facts: they form the same angle with the original plane, and equal tangents are obtained after extending $DE$ and $AB$ until they meet. ($A_1B_1$ and $D_1E_1$ will be parallel if $DE$ and $AB$ are parallel.)

*Step 7.* Put together all pieces above to conclude that the diagonals of the spacial hexagon are concurrent, and hence the diagonals of the original planar hexagon are also concurrent. □

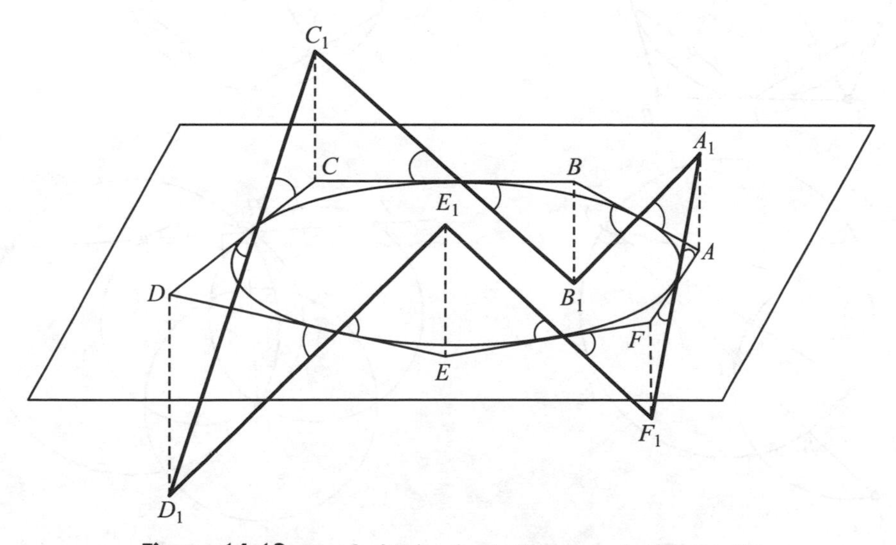


**Figure 14.12.** Proof of Brianchon's Theorem via Exit into 3D

# 9 Pascal's and Brianchon's Theorems are Dual Statements

We can find even further connections between our six classical theorems, by considering the relation between Pascal's and Brianchon's Theorem. We have already presented proofs of them, but now we show that these two theorems directly imply each other via inversion in the plane [8, p. 243], i.e., they are *dual statements*.

## 9.1 Inversion of Brianchon's Theorem yields Pascal's Theorem

We present here a sketch of the argument. The reader is encouraged as usual to fill in the details.

*Step 1.* Apply inversion with respect to the circle $k$ inscribed in the hexagon $ABCDEF$. Then the vertices of the circumscribed hexagon $ABCDEF$ map to the midpoints $A_1, \ldots, F_1$ of the inscribed hexagon $IJKLMN$.

*Step 2.* The diagonals $AD$, $BE$ and $CF$ intersect (Brianchon) if and only if their images intersect in a point other than $O$. The images are three circles $k_1$, $k_2$ and $k_3$ passing through $O$ and respectively through $A_1$ and $D_1$, through $B_1$ and $E_1$, and through $C_1$ and $F_1$.

*Step 3.* The circle $k_2$ through $O$, $B_1$ and $E_1$ has as diameter $OY$, where $Y$ is the intersection point of sides $LM$ and $IJ$ of the inscribed hexagon. Similarly for the other two circles $k_1$ and $k_3$ and the intersection points $X$ and $Z$ of the other pairs of opposite sides of $IJKLMN$.

*Step 4.* The circles $k_1$, $k_2$ and $k_3$ intersect in another point $H_1$, besides $O$, if and only if there is a point $H_1$ such that $\angle OH_1X = \angle OH_1Y = \angle OH_1Z = 90°$, i.e., $X$, $Y$ and $Z$ are collinear.

We conclude that via inversion fixing the original circle, Brianchon's theorem for the circumscribed hexagon $ABCDEF$ is equivalent to Pascal's theorem for the inscribed hexagon $IJKLMN$.

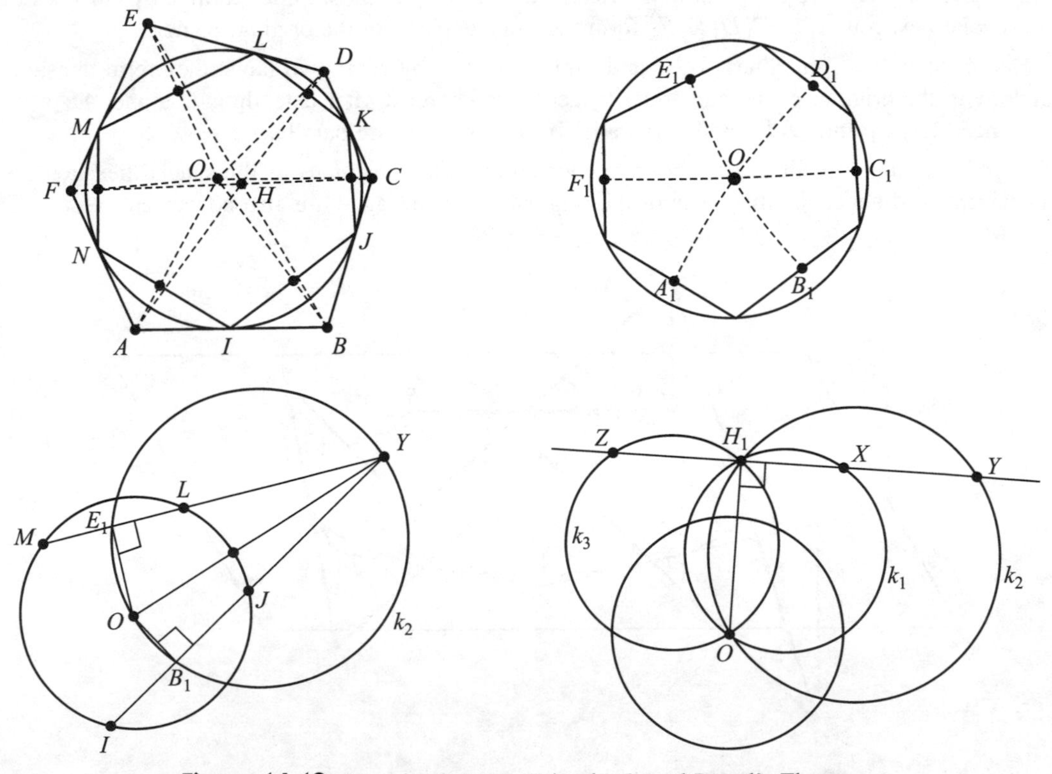


**Figure 14.13.** Inversion between Brianchon's and Pascal's Theorems

Under this correspondence, the intersection point $H$ of the diagonals of $ABCDEF$ is mapped to the orthogonal projection $H_1$ of $O$ onto line $XYZ$. □

### 9.2 Application of Brianchon's Theorem

The reader is challenged to find a direct application of Brianchon's Theorem to the following problem from the St. Petersburg Mathematical Olympiad.

**Problem 9.** *Point $I$ is the incenter of $\triangle ABC$. Some circle with center $I$ intersects side $BC$ in $A_1$ and $A_2$, side $CA$ in $B_1$ and $B_2$, and side $AB$ in $C_1$ and $C_2$. The six points obtained in this way lie on the circle in the following order: $A_1$, $A_2$, $B_1$, $B_2$, $C_1$, $C_2$. Points $A_3$, $B_3$, and $C_3$ are the midpoints of the arcs $A_1A_2$, $B_1B_2$, and $C_1C_2$ respectively. Lines $A_2A_3$ and $B_1B_3$ intersect in $C_4$, lines $B_2B_3$ and $C_1C_3$ in $A_4$, and lines $C_2C_3$ and $A_1A_3$ in $B_4$. Prove that the segments $A_3A_4$, $B_3B_4$, and $C_3C_4$ intersect in one point.*

## 10 Proof of the Poncelet-Brianchon Theorem via Pascal's Theorem

The idea of the proof is to apply the converse of Pascal's Theorem, and thus to reduce the problem to a simple geometry problem (Bay Area Mathematical Olympiad (BAMO) 2001).

One of the many startling properties of projective geometry is the fact that in the projective plane all the conic sections—circles, ellipses, parabolas, hyperbolas—are essentially the same object. Thus the Poncelet–Brianchon theorem, as well as Pascal's theorem and its converse, apply to all conics. This also allows us to use the terms ellipses, hyperbolas, and conics interchangeably.

*Proof of the Poncelet–Brianchon Theorem.* In the projective plane, let $D$ and $E$ be the two points of intersection of the line at infinity $l_\infty$ with the two asymptotes of the hyperbola $\Lambda$. Note that since $\Lambda$ is a conic, $D$ and $E$ are also the points of intersection of $\Lambda$ with $l_\infty$. We apply:

**Converse of Pascal's theorem.** *If the three pairs of opposite sides in a hexagon intersect in collinear points, then the hexagon in inscribed in a conic (see Fig. 14.14.)*

Note that through any 5 points in general position (i.e., no 3 of them are collinear), there passes a unique conic. In our situation, the points $A$, $B$, $C$, $D$, and $E$ lie on a hyperbola, and hence they are in general position.

To show that the orthocenter $H$ of $\triangle ABC$ also lies on this hyperbola, it suffices to verify that the points $X, Y$, and $Z$ are collinear, where $X = AB \cap HD$, $Y = BC \cap HE$, and $Z = AE \cap CD$. Thus, the problem is equivalent to the following

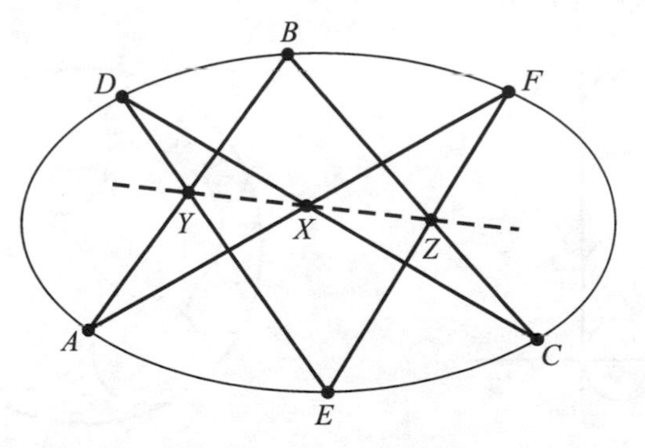


**Figure 14.14.** Converse of Pascal's Theorem

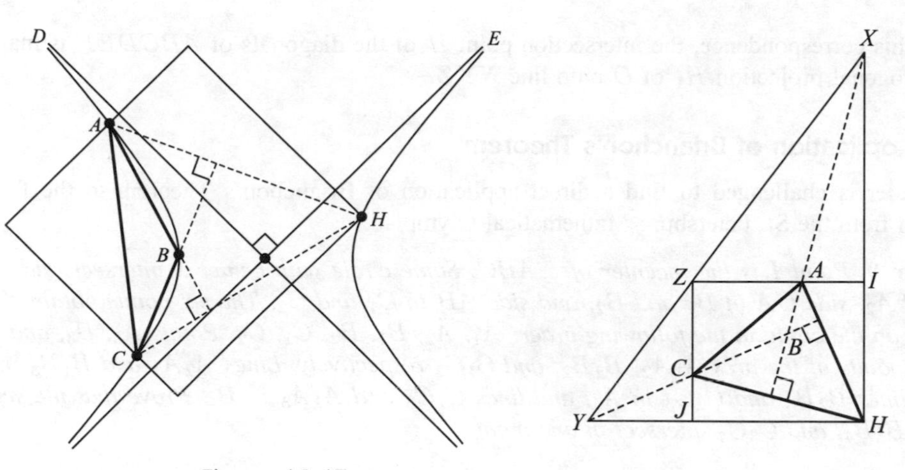


**Figure 14.15.** The Poncelet–Brianchon Theorem

**Problem 10 (BAMO 2001)** *Let $JHIZ$ be a rectangle, and let $A$ and $C$ be points on sides $ZI$ and $ZJ$, respectively. The perpendicular from $A$ to $CH$ intersects line $HI$ in $X$, and the perpendicular from $C$ to $AH$ intersects line $HJ$ in $Y$. Then $X$, $Y$, and $Z$ are collinear (see Fig. 14.15).*

*Proof.* $\angle XAI = \angle XHC = \angle HCJ$, hence $\triangle XAI \sim \triangle HCJ$, and thus $\frac{XI}{HJ} = \frac{AI}{CJ}$. Similarly, $\frac{YJ}{HI} = \frac{CJ}{AI}$. Putting these together yields $\frac{XI}{HJ} = \frac{HI}{YJ}$.

$$\Rightarrow \frac{XI}{ZI} = \frac{ZJ}{YJ} \Rightarrow \triangle XZI \sim \triangle ZYJ.$$

Since $\angle JZI = 90°$, this immediately implies $\angle YZX = 180°$, and $X$, $Y$, $Z$ are collinear. □

## 11 Inversion on Pappus' Theorem

Inversion is a very versatile tool, applicable to a wide range of problems. We already saw earlier how it makes Brianchon's and Pascal's Theorems equivalent. Here we apply it to prove Pappus' Theorem easily.

*Proof of Pappus' Theorem.* Apply inversion with respect to a circle with center $A$ perpendicular to $k_n$. This fixes the circle $k_n$ and the line $ABC$ as sets, and sends $\omega$ and $\omega_1$ to two parallel lines $l$ and $l_1$, both perpendicular to $ABC$. Now $k_n$ and all other circles are each tangent to $l$ and $l_1$, forming a chain of tangent circles that ends with $\omega_2$ centered on line $ABC$. From here the desired identity is obvious (see Fig. 14.16.) □

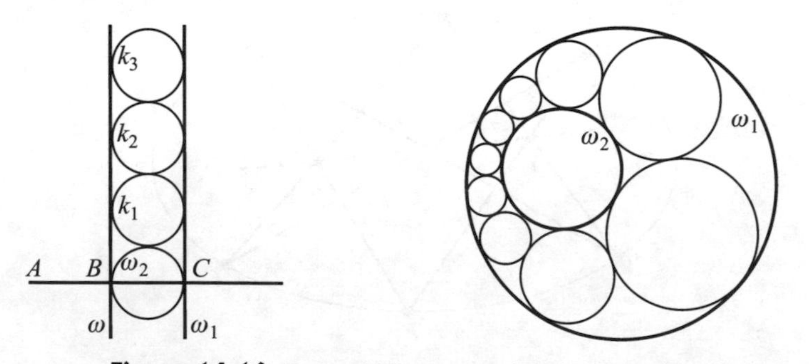


**Figure 14.16.** Pappus' Theorem and Steiner's Porism

A further application of inversion is needed in the following

**Problem 11 (Steiner's Porism)** *Suppose that two nonintersecting circles have the property that one can fit a "ring" of circles between them, each tangent to the next. Then one can do this (fit such a ring of circles) starting with any circle tangent to both given circles (see Fig. 14.16.)*

*Proof of Steiner's Porism.* There is an inversion which sends the two circles into concentric ones. The problem is now formulated exactly as before, except that $\omega_1$ and $\omega_2$ are concentric, and the conclusion follows trivially. □

## 12 Summary of Relations Between Our Classical Theorems

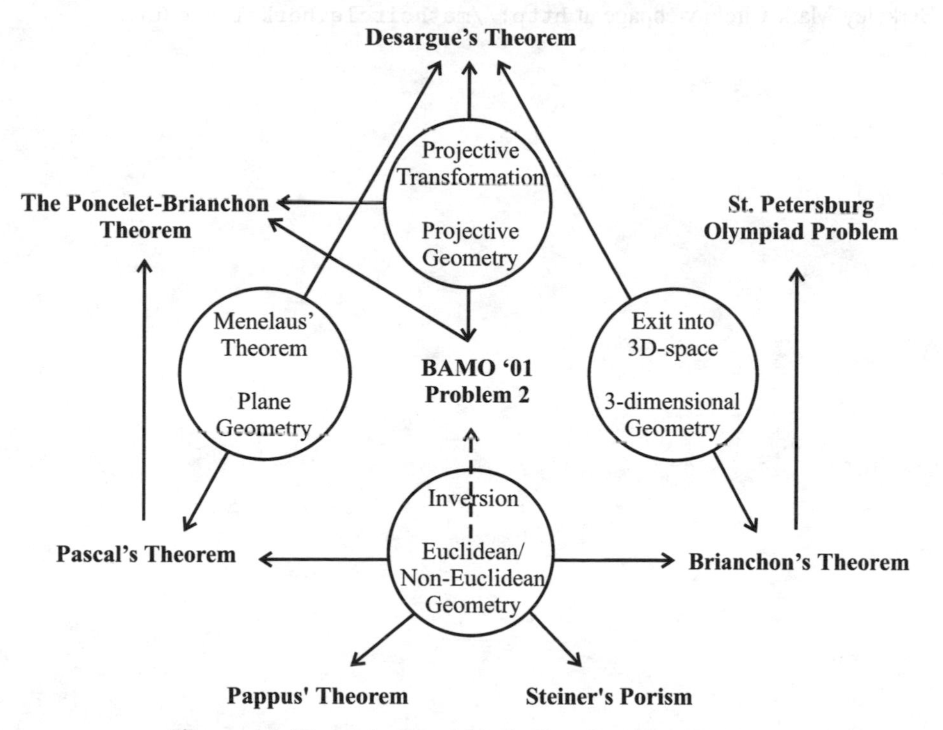


**Figure 14.17.** Methods and implications between theorems

## References

[1] Zvezdelina Stankova, "Inversion in the Plane. Part I," Berkeley Math Circle, (http://mathcircle.berkeley.edu/).

[2] ——, "Classical Theorems in Plane Geometry", Berkeley Math Circle, (http://mathcircle.berkeley.edu/).

[3] ——, "Introduction to Algebraic Geometry. Projective Geometry," lecture notes, Mathematical Olympiad Summer Program, Lincoln, Nebraska, 1999 (http://mathcircle.berkeley.edu/).

[4] Joe Harris, *Algebraic Geometry*, Springer-Verlag, 1992.

[5] Dmitri Fomin, Sergey Genkin, and Ilia Itenberg, *Mathematical Circles (Russian Experience)*, American Mathematical Society 1996.

[6] *Kvant Selecta: Algebra and Analysis, I and II*, American Mathematical Society 1999.

[7] George Lenchner, *Math Olympiad Contest Problems for Elementary and Middle Schools*, Glenwood Publications, 1997.

[8] Marvin Greenberg, *Euclidean and Non-Euclidean Geometries*, W.H. Freeman & Company, 1997.

[9] Robin Hartshorne, *Companion to Euclid*, American Mathematical Society, 1997.

[10] Berkeley Math Circle webpage at `http://mathcircle.berkeley.edu/`.

# 15

# Cusps

Dmitry Fuchs
*University of California at Davis*

Among the graphs that calculus teachers love to assign their students, there are curves containing sharp turns, which mathematicians call *cusps.* A characteristic example is given in Figure 15.1 below. This is a *semicubic parabola*, a curve given by the equation $y^2 = x^3$.

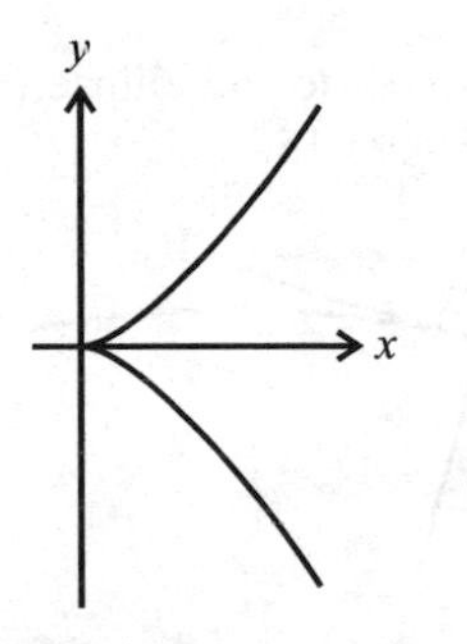


**Figure 15.1.** Semicubic parabola

The next example is the famous *cycloid* (Figure 15.2). You will observe it if you make a colored spot on the tire of your bike and then ask your friend to ride the bike; the spot will trace out a cycloid.

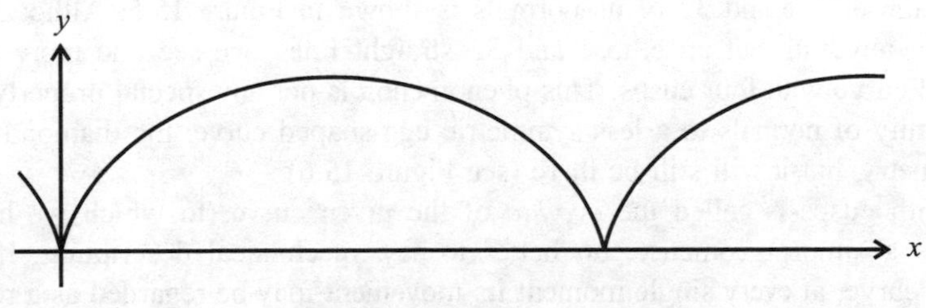


**Figure 15.2.** Cycloid

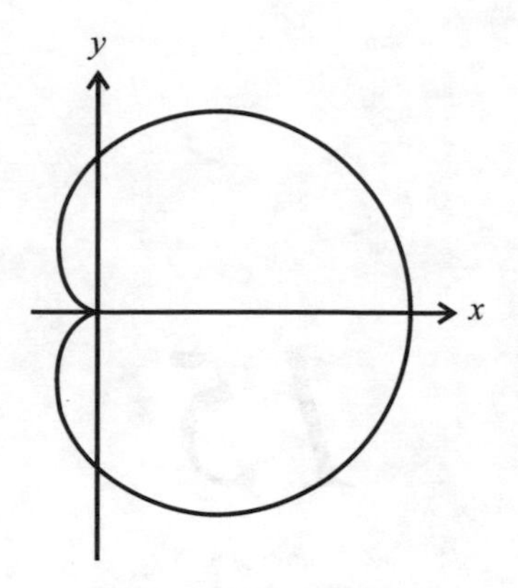


**Figure 15.3.** Cardioid

Our last example is the so-called *cardioid* (Figure 15.3), a curve whose name reflects its resemblance to a drawing of a human heart. Mathematicians usually represent this curve by the polar equation $\rho = 1 + \cos\theta$.

Certainly, the cusps on these graphs may seem to be something occasional, accidental: there are so many curves without cusps. But do not make premature conclusions. My goal is to convince you that cusps appear naturally in so many geometric or analytic contexts that we can justly say: *cusps are everywhere!*

Let us draw an ellipse, the one given by the equation

$$\frac{x^2}{4} + y^2 = 1$$

and a sufficiently dense family of *normals* to the ellipse (a normal is a line perpendicular to the tangent at the point of tangency, see Figure 15.4).

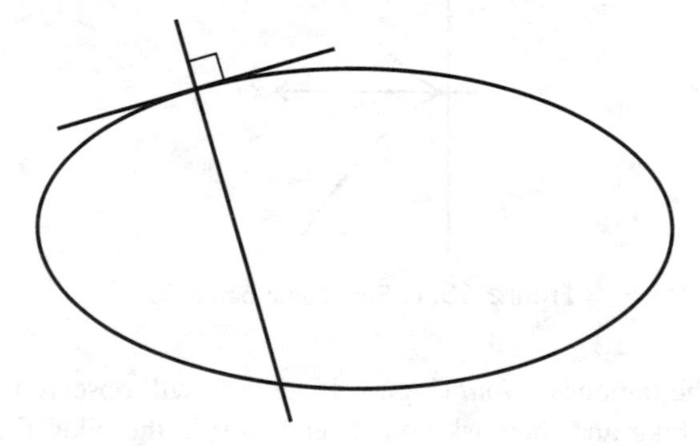

**Figure 15.4.** A tangent and a normal to an ellipse

A picture of the ellipse and 32 of its normals is shown in Figure 15.5. Although Figure 15.5 does not contain anything but an ellipse and 32 straight lines, we see one more curve on it: a diamond-shaped curve with four cusps. This phenomenon is not any special property of an ellipse. If we take a family of normals to a less symmetric egg-shaped curve, the diamond will also lose its perfect symmetry, but it will still be there (see Figure 15.6).

The curve with cusps is called the *evolute* of the given curve (to which we have taken the normals). It has a simple geometric, or, better to say, mechanical description. If a particle is moving along a curve, at every single moment its movement may be regarded as a rotation around a certain center. This center changes its position at every moment, thus it also traces a curve. It is

**Figure 15.5.** An ellipse with thirty-two normals

**Figure 15.6.** An egg-shaped curve with normals

this curve that we see in Figures 15.5 and 15.6. Evolutes always have cusps. Moreover, the celebrated *Four Vertex Theorem* (proved more than 150 years ago, but still appearing mysterious) states that if the given curve is non-self-intersecting (like an ellipse or the egg-shaped curve of Figure 15.6), then the number of cusps on the evolute is at least four.

For self-intersecting curves this is no longer true; Figure 15.7 shows a family of normals to a self-intersecting curve; the evolute is clearly visible in this picture, and it has only two cusps.

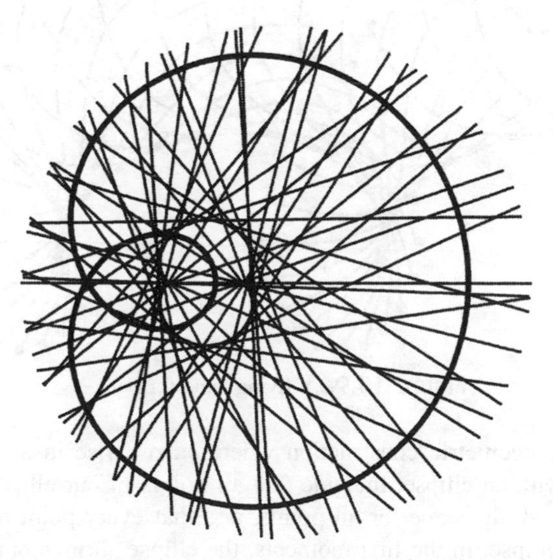

**Figure 15.7.** A self-intersecting curve with normals

To be honest, this seemingly spontaneous appearance of a curve with cusps in a picture of a family of normals is not directly related to normals. You will see something very similar if you take a "sufficiently arbitrary," or "sufficiently random," family of lines. Imagine an angry professor who throws his cane at his students. The cane flies and rotates in its flight. If you draw a family of subsequent positions of the cane in the air, you will see something like Figure 15.8.

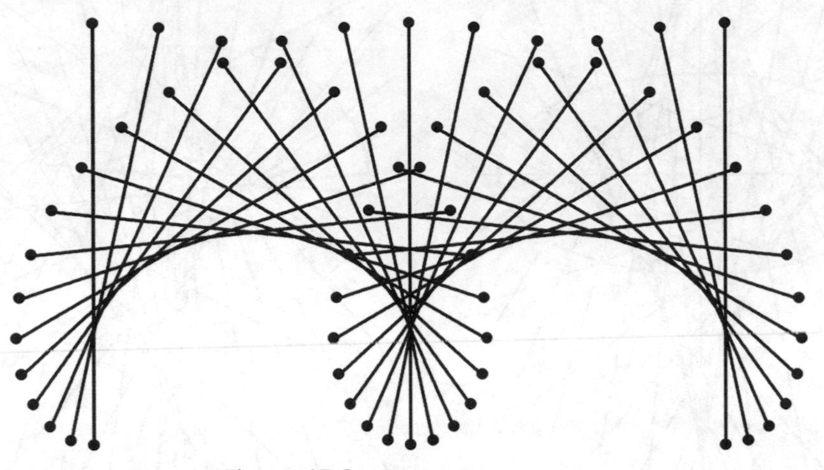

**Figure 15.8.** A flying cane (straight)

You see here 32 subsequent positions of the cane, but also a curve looking a bit like the cycloid (Figure 15.2), with cusps (one of them is clearly seen in the middle of the drawing). And straight lines do not play any special role, it is simply more convenient to draw them. If the professor is old and heavy, and his cane has long lost its linearity—say, it is shaped like a sinusoid—then the picture of Figure 15.8 will look different, but the cusps will remain (see Figure 15.9).

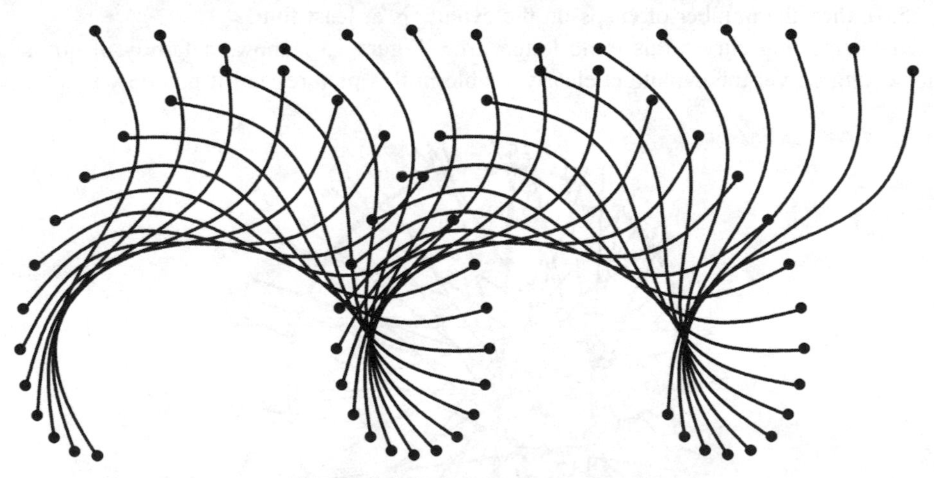

**Figure 15.9.** A flying cane (curvy)

Let us turn to another geometric construction where cusps arise in an even more unexpected way. Let us again start with an ellipse. Imagine that all points of our ellipse simultaneously begin moving at a constant speed, the same for all points, and that every point moves inside the ellipse along the normal to the ellipse. In the first moments, the ellipse shrinks but still retains its smooth oval shape (Figure 15.10).

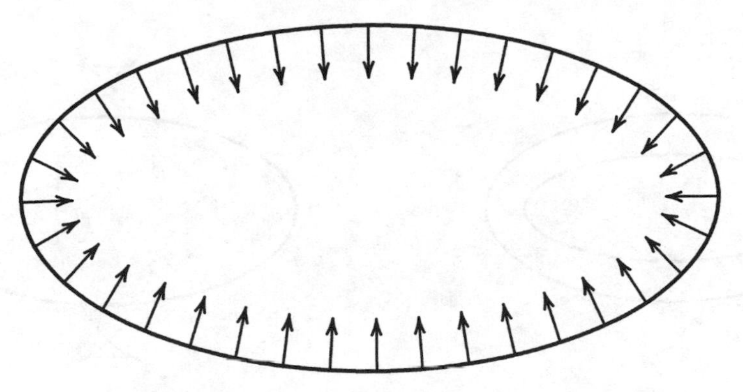

**Figure 15.10.** First, the ellipse retains its oval shape

Then the points begin forming sorts of crowds at the left-hand and right-hand extremities of the curve (Figure 15.11). Next the trajectories of the points cross each other (no collisions, they pass through each other), and, believe it or not, the curve acquires four cusps (Figure 15.12).

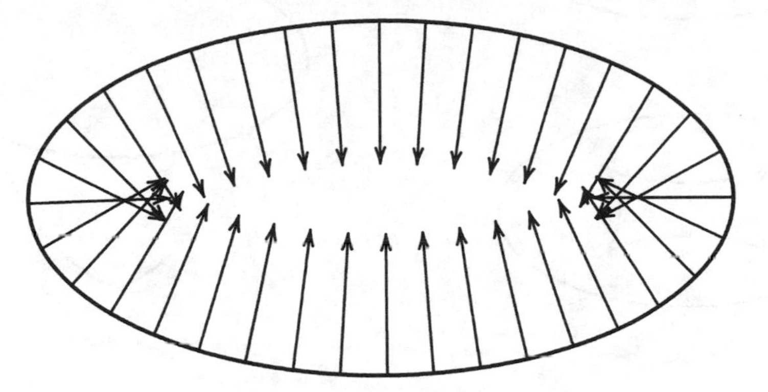

**Figure 15.11.** Then the points begin forming crowds

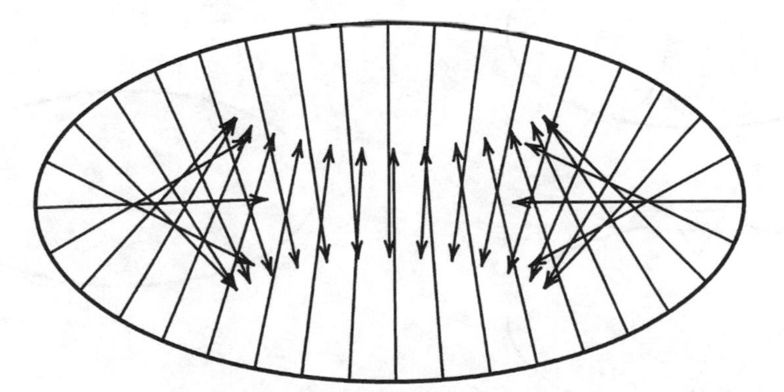

**Figure 15.12.** Eventually, the curve acquires cusps

The evolution of the moving curve, which is conveniently called a front, can be seen in Figure 15.13. We see that after the appearance of the four cusps, the curve consists of four sections between the cusps, two short and two long, and the long sections cross each other twice. Then the short sections become longer, and the long sections become shorter. At some moment, the "long"

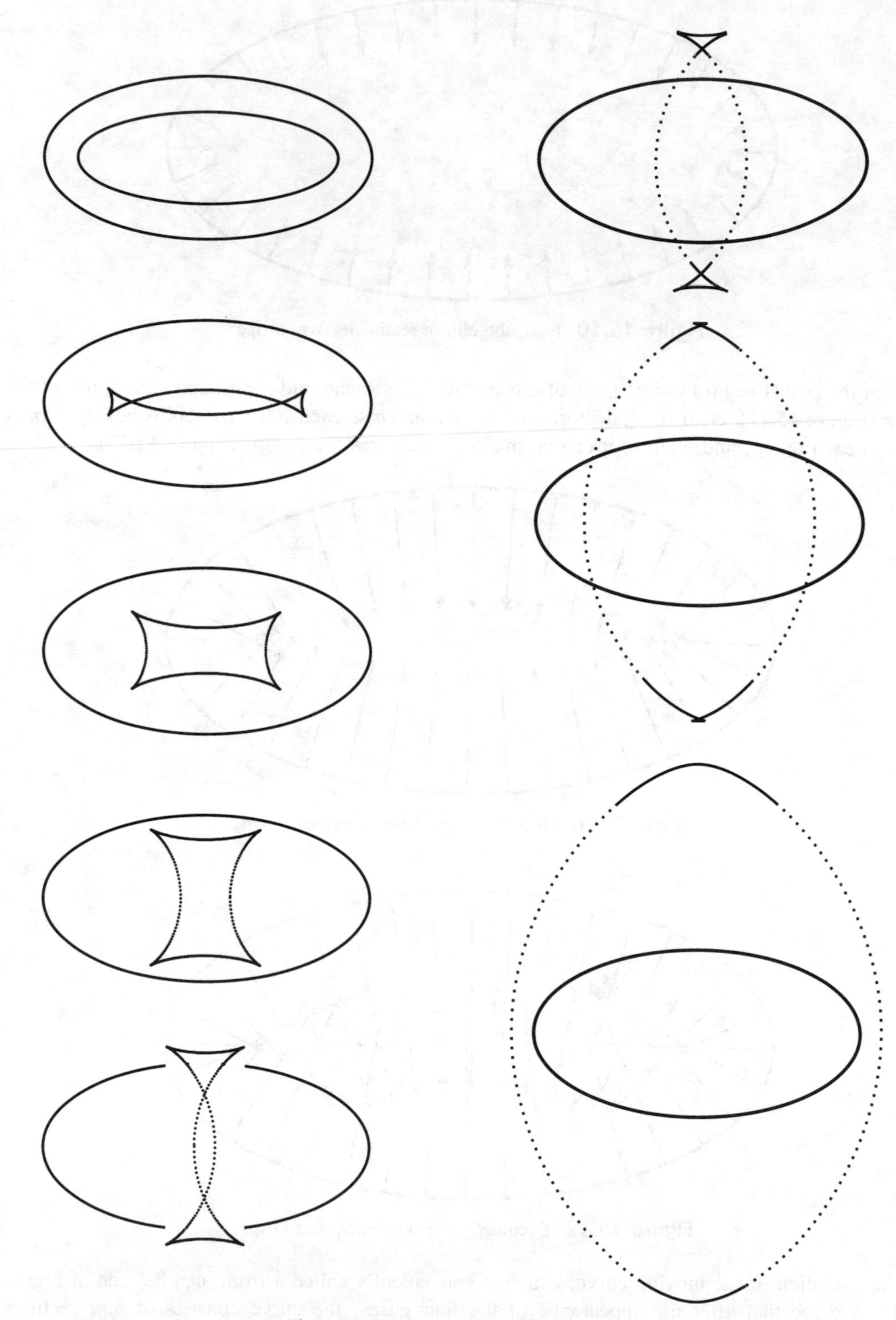

**Figure 15.13.** The evolution of a front

sections (which are not so long at this moment) go apart; then the "short" sections (which are quite long at this moment) meet and form two crossings. Then the cusps bump into each other and disappear, and the curve becomes again more or less elliptic.

It is interesting to draw all the fronts of Figure 15.13 in one picture. The cusps of the fronts form a curve themselves (Figure 15.14), and if you compare Figure 15.14 with Figure 15.5, you will see that our curve is nothing but the evolute of the ellipse.

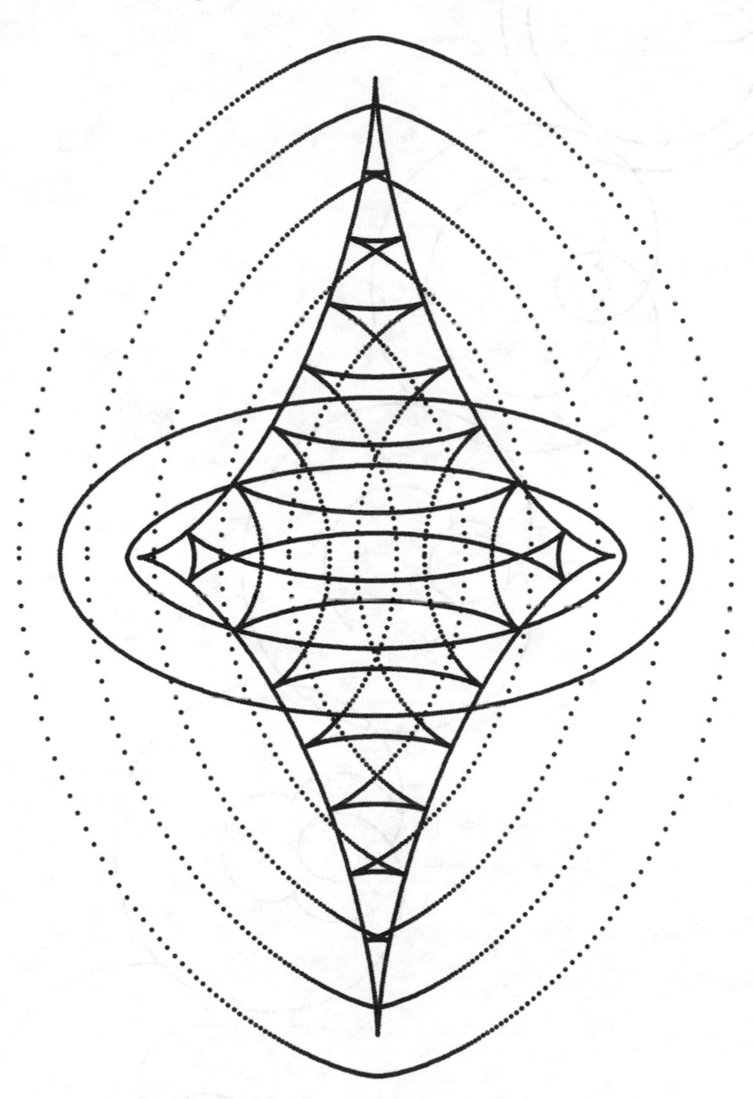

**Figure 15.14.** The fronts and the evolute

Similarly, the movement of the fronts of the self-intersecting curve of Figure 15.7 is shown in Figures 15.15–15.16. The drawing in Figure 15.17 presents the whole family in one picture; if you trace, mentally, the curve of cusps, you will get the two-cusp evolute visible in Figure 15.8. If you want to have more examples, observe the family of fronts of a sinusoid (Figure 15.18). You can guess how the evolute of a sinusoid looks. (The evolute of a curve with *inflection points* always has asymptotes, which are the normals to the curve at the inflection points. If the words "inflection points" and "asymptotes" do not say much to you, forget about them.)

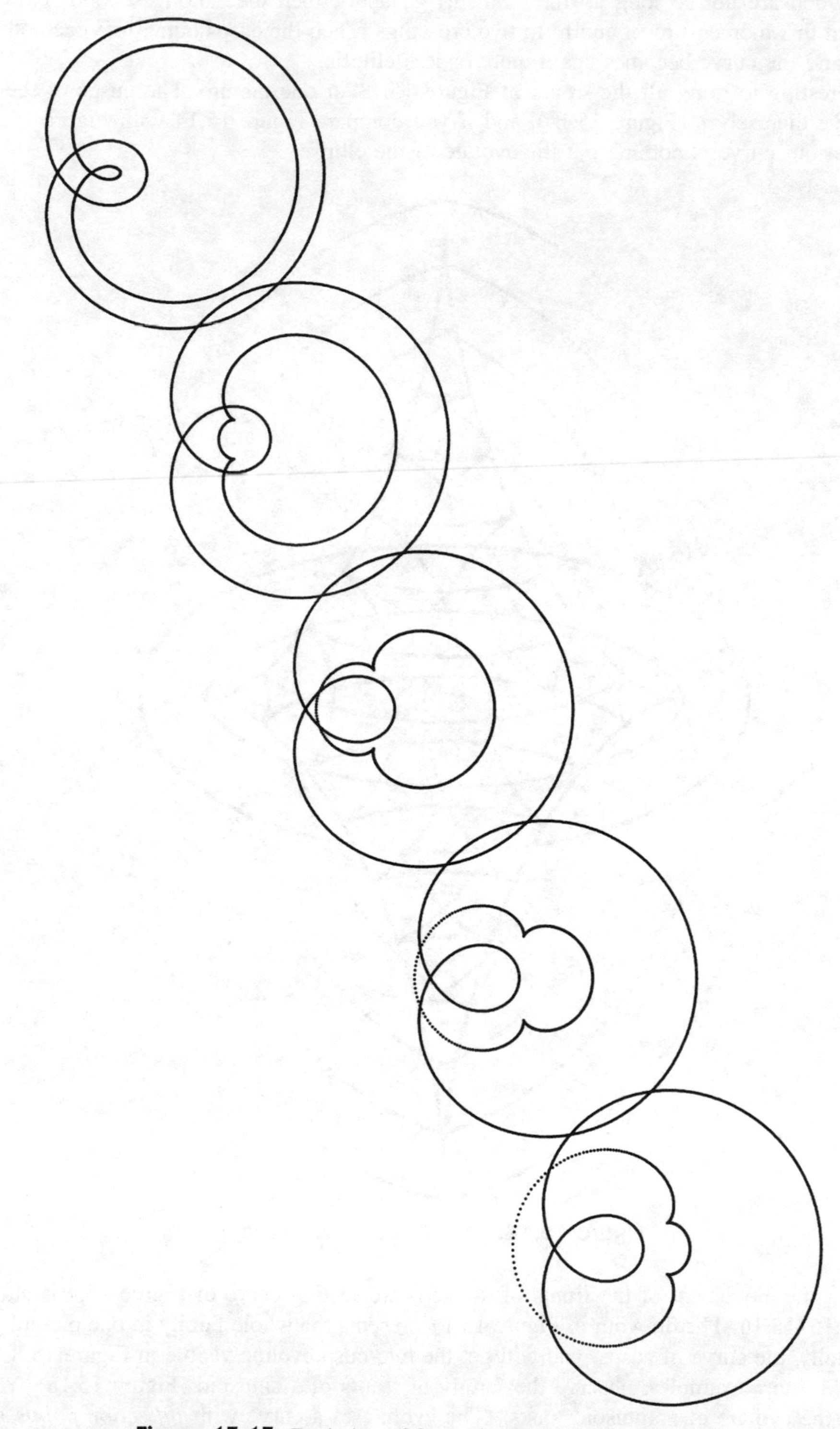

**Figure 15.15.** Evolution of fronts for a self-intersecting curve

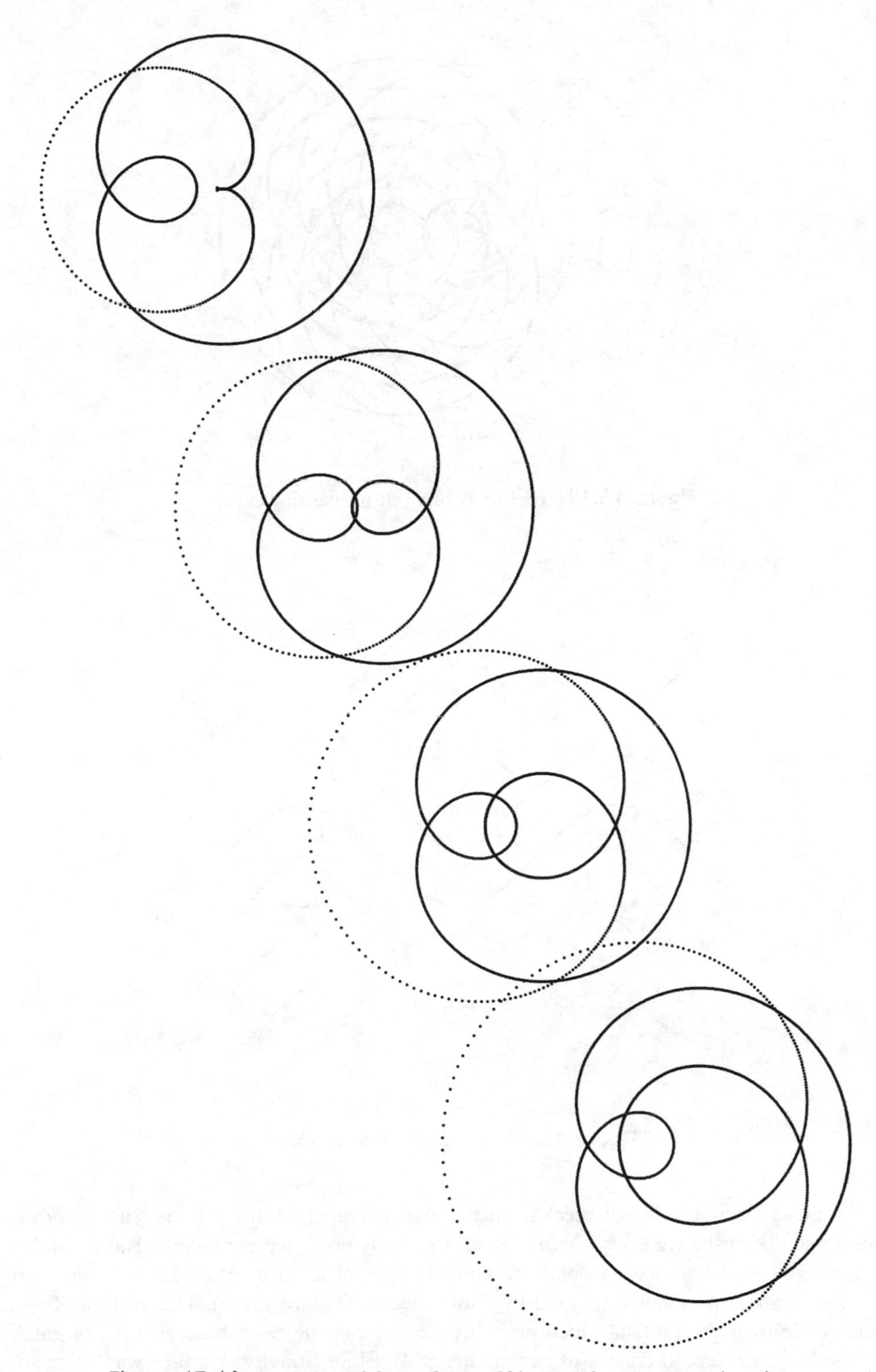

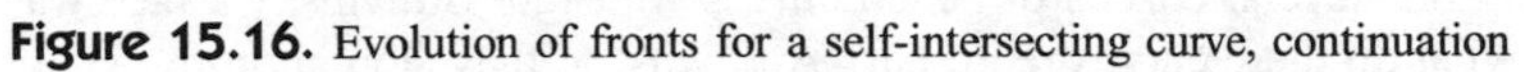

**Figure 15.16.** Evolution of fronts for a self-intersecting curve, continuation

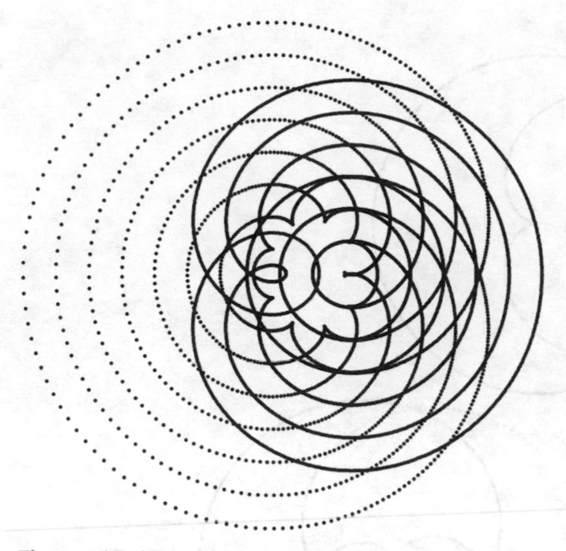

**Figure 15.17.** The fronts of a self-intersecting curve

**Figure 15.18.** The fronts of a sinusoid

Still, all these examples do not seem to justify the statement "*cusps are everywhere around.*" One can argue, "if cusps are everywhere around, then why don't we see them?" But we do!

To be convinced, let us look around through the eyes of a great artist. Look at the famous portrait of Igor Stravinsky drawn in 1932 by Pablo Picasso (Figure 15.19). This picture of a great composer made by a great artist, which probably bears a notable resemblance to the original, is, actually, nothing but several dozens of pencil curves. But Igor Stravinsky's face was not made of curves! Then, what do these curves represent? And why do they stop abruptly without any visible reason?

**Figure 15.19.** Pablo Picasso. *Portrait of Igor Stravinsky* (1932)


Certainly, this drawing is too complicated to begin thinking of such things. Let us consider a simpler drawing. Imagine young Pablo Picasso first entering an art school in his native Malaga, or, maybe, later, in Barcelona. It is very probable that his teacher offered him a jug to draw (art students often begin their studies with jugs).

It is very unlikely that Pablo's picture of a jug, even if it ever existed, can be still found anywhere. But maybe it looked like one of the drawings below.

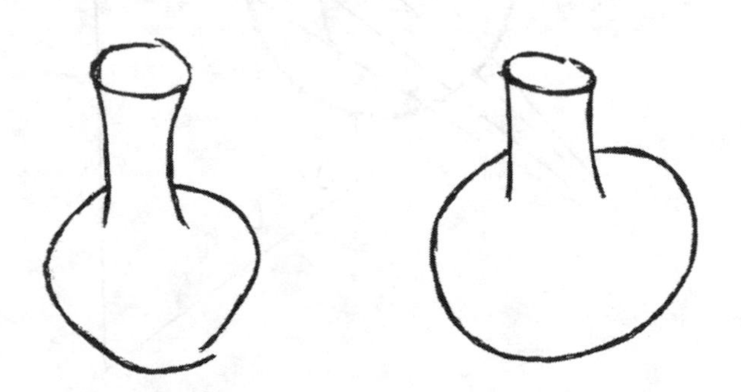

**Figure 15.20.** Two jugs as young Pablo might have drawn them.

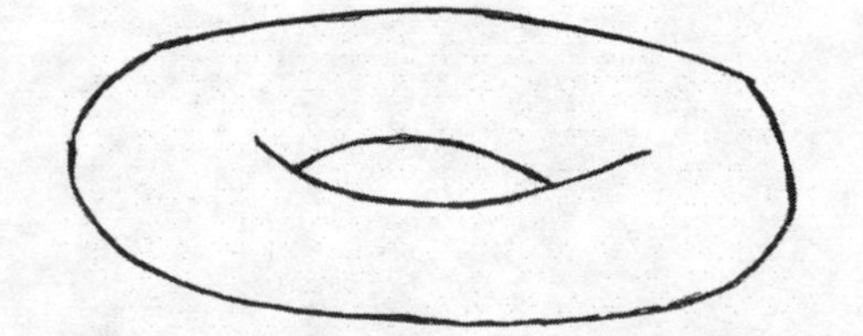

**Figure 15.21.** Young Pablo's possible rendition of a torus

Or, maybe, the art teacher was a geometry lover, and Pablo's first assignment was a torus (the surface of a bagel, if you do not know what a torus is). Then Pablo's first drawing could look like Figure 15.21.

In these simple drawings, we see the same things as in the masterpiece: there are curves—some of them end abruptly, either when they meet other curves, or without any visible reasons.

Let us think about the reasons. The curves we see (and draw) are boundaries of visible shapes, or, in other words, they are made of points of tangency of the rays starting from our eyes with the surface we look at. Let us denote this surface by $S$ and the curve made of tangency points by $C$ (see Figure 15.22). If we place, mentally, a screen behind the surface, then our rays will trace a curve on the screen, and this curve $C'$ looks precisely like the contour of the surface that we see. If the shape of the surface $S$ is more complicated, then some parts of the curve $C$ may be hidden from our eye by the surface (geometrically this means that the ray crosses the surface, maybe more than one time, before the tangency). This is what happens when a curve stops upon meeting another curve: if the things we are drawing were transparent, the curve would not have stopped, it would have gone further as a smooth curve.

The second case, when the curve stops without meeting another curve is more interesting. As I have said before, the tangency points of the rays of our vision form a curve $C$ on our surface $S$. A

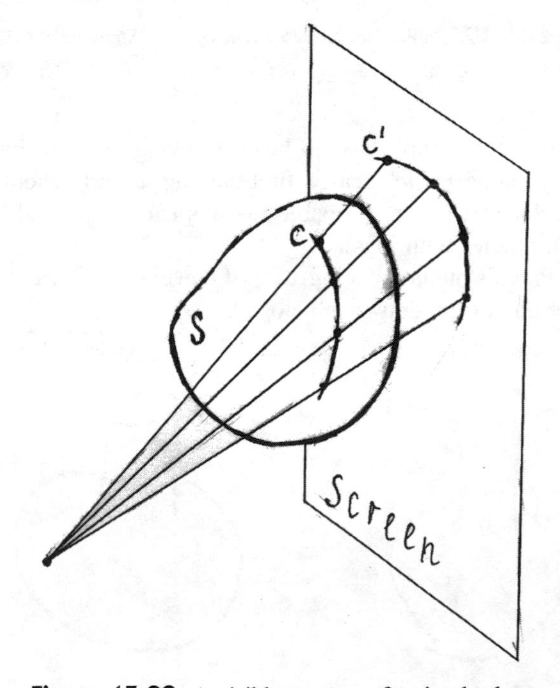


**Figure 15.22.** A visible contour of a simple shape

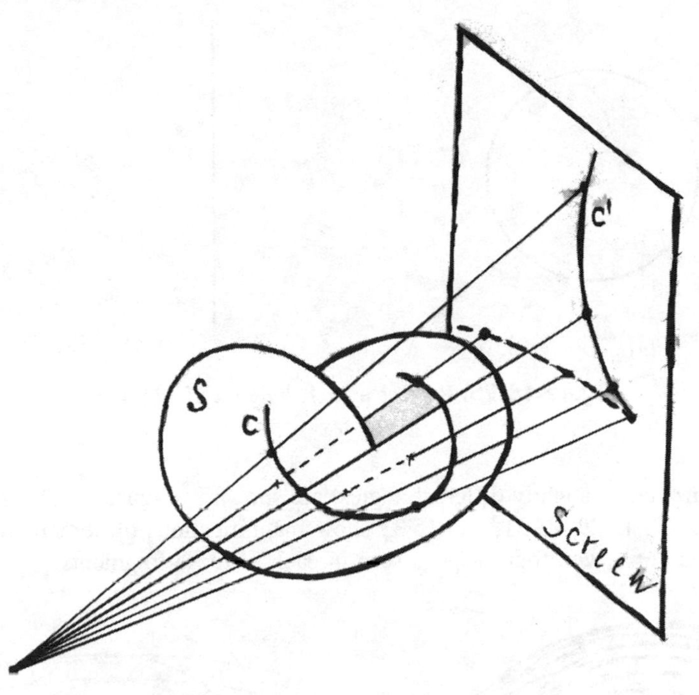


**Figure 15.23.** A visible contour of a complicated shape

simple analytic argument, which I skip here, shows that this curve $C$ is always smooth. However, the ray may be tangent not only to the surface $S$, but also to the curve $C$. In this case, the image of this curve on the screen, and, hence, inside our eye, or in our drawing, forms a cusp (see Figure 15.23). But, if the things we are looking at or drawing are not transparent, we see only one half of this cusp, while the second half is hidden behind the shape. Thus, if the things around us were transparent (sounds Nabokovian!), we would never have seen stopping curves; but we would have seen a lot of cusps, which in real life are visible only by half.

For example, if Pablo's jugs and torus were transparent, he would have supplemented his drawing by the curves shown (dotted) in Figure 15.24.

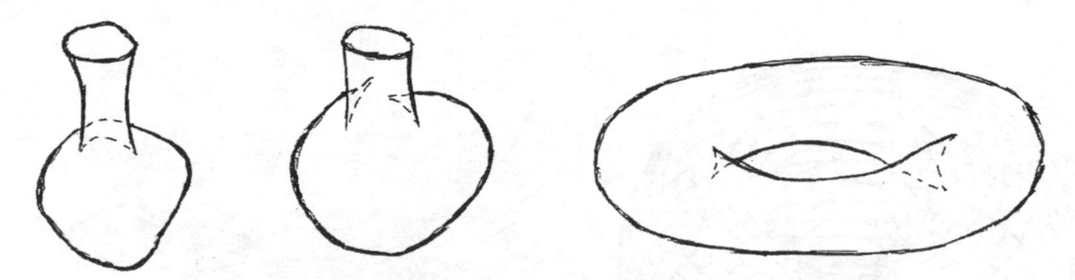

**Figure 15.24.** Transparent things

In conclusion, let us look at the projections of a transparent torus. To make it transparent, we replace it by a dense family of circles in parallel planes. More precisely, a torus is a surface of revolution of a circle around an axis not crossing the circle (see Figure 15.25a). We replace the circle by a dense set of points, in our example by the set of vertices of an inscribed regular 32-gon (Figure 15.25b).

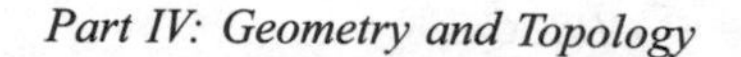

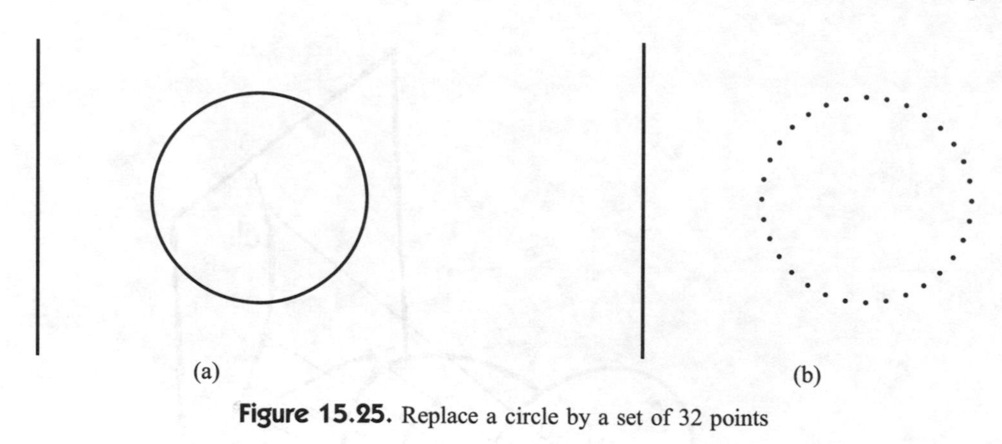


**Figure 15.25.** Replace a circle by a set of 32 points

Four projections (under slightly different angles) are shown in Figures 15.26, 15.28, 15.30, and 15.32. Figures 15.27, 15.29, 15.31, and 15.33 show magnifications of central fragments of these four drawings; the curve with four cusps is seen in each of these fragments.

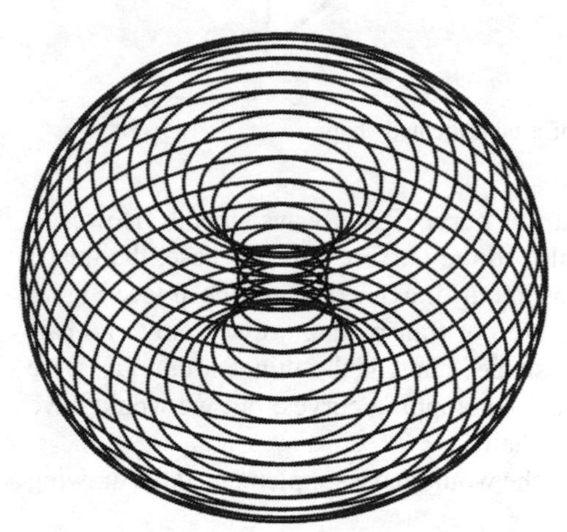

**Figure 15.26.** A projection of a torus

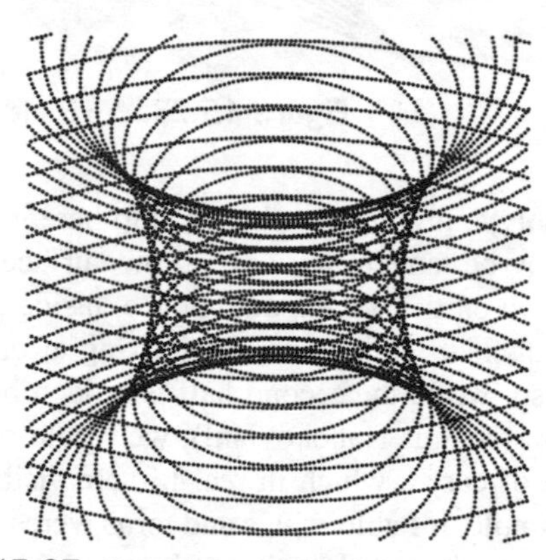

**Figure 15.27.** Central fragment of Figure 15.26, magnified

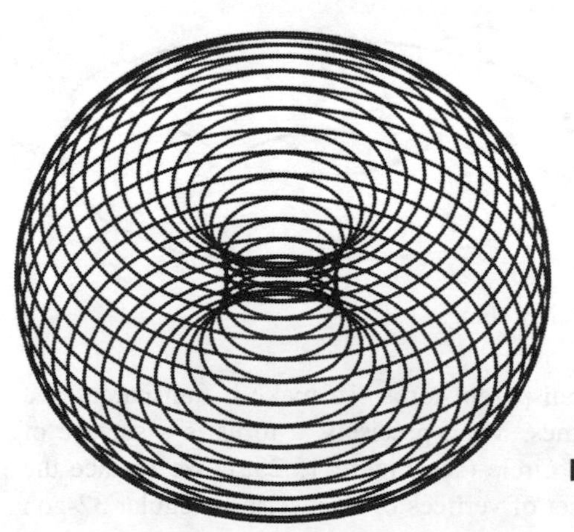

**Figure 15.28.** Another projection of a torus

**Figure 15.29.** Central fragment of Figure 15.28, magnified

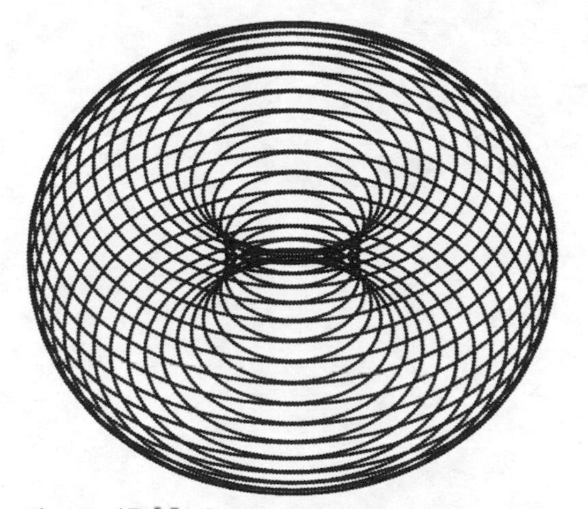

Figure 15.30. One more projection of a torus

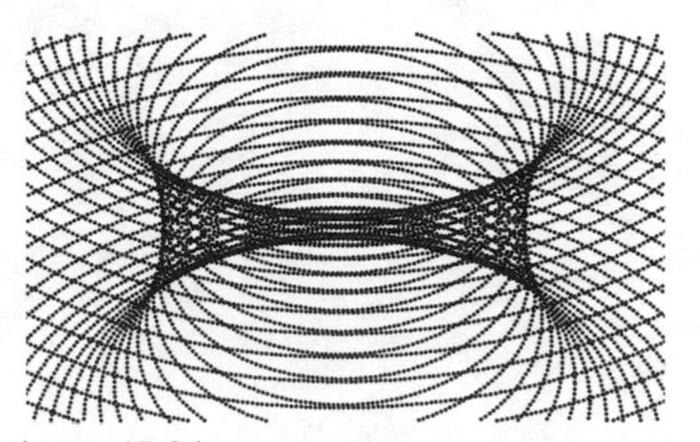

Figure 15.31. Central fragment of Figure 15.30, magnified

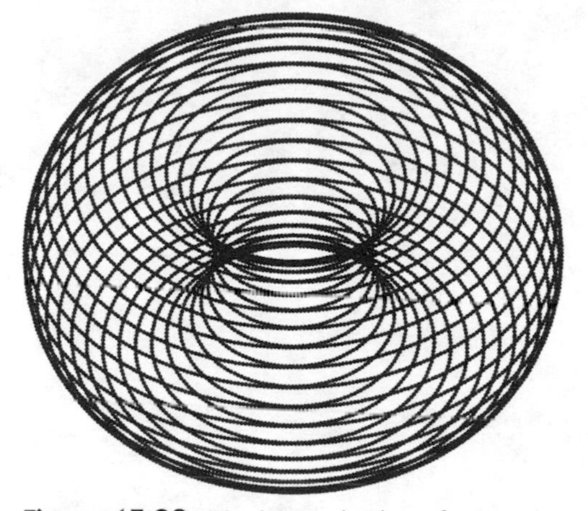

Figure 15.32. The last projection of a torus

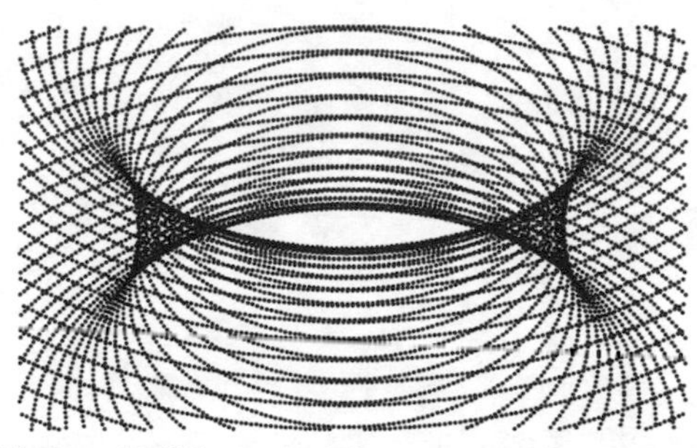

Figure 15.33. Central fragment of Figure 15.32, magnified

## References

[1] Arnold, V.I., *Catastrophe Theory,* 3rd revised and expanded edition, Berlin, NY, Springer Verlag, 1992.

[2] Hilbert, D., and S. Cohn-Vossen, *Geometry and Imagination,* Providence, RI, AMS Chelsea Publ., 1999.

# 16

# Triangles and Curvature

Richard Scott
*Santa Clara University*

## 1 Introduction

The study of a mathematical object often requires one to specify the transformations that preserve the relevant mathematical structure. For example, in studying a geometric object one would consider transformations called *isometries*, which preserve the distance between two points (such as a translation or rotation in the plane). If distance is not so important but preserving limits is the required property, then one might allow transformations called *homeomorphisms*, which permit twisting and stretching but not cutting or pinching. An even more general type of transformation is a *homotopy equivalence*, which allows such alterations as collapsing a disk to a single point or shrinking an annulus onto its central circle (Figure 16.1). Once one has decided on the relevant transformations, the study of the original object is then broadened to a study of all objects which can be obtained from the original by performing an allowed transformation. In particular, one is interested in attributes of the object, called *invariants*, that do not change when one of these transformations is performed.

The Gaussian curvature of a surface is a well-known example. Given a surface $S$, the Gaussian curvature is a function that assigns to each point on the surface a real number indicating the shape of the surface near that point. If $S$ has a "bump" at the point then the curvature is positive, if $S$ has a "saddle" at the point then the curvature is negative (Figure 16.2). This curvature function is

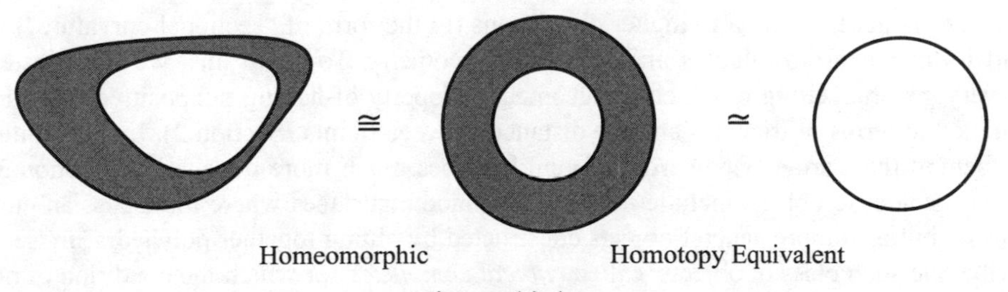


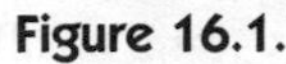
Figure 16.1.

**Figure 16.2.** Positive and negative curvature

a geometric invariant in the sense that any surface obtained from $S$ by an isometry will have the same curvature as $S$.

An example of a homotopy invariant (and hence also a homeomorphism invariant) is the *fundamental group*. Given a geometric object $M$, its fundamental group $\pi$ is obtained by considering all possible directed loops in $M$ which start and end at the same fixed point $p$. Two loops represent the same group element if one can be continuously deformed into the other. For example, suppose $M$ is the torus (the surface of a doughnut) in Figure 16.3. The two loops $A$ and $A'$ represent the same element of $\pi$, but $B$ represents a different element. Two loops can be combined to get a third loop by traveling the first loop and then traveling the second loop. This operation on loops makes the set $\pi$ into an algebraic object called a *group*, and it is a fact from topology that $\pi$ is a homotopy invariant of $M$. In other words, if two geometric objects are homotopy equivalent (or homeomorphic), then their fundamental groups are (algebraically) the same.

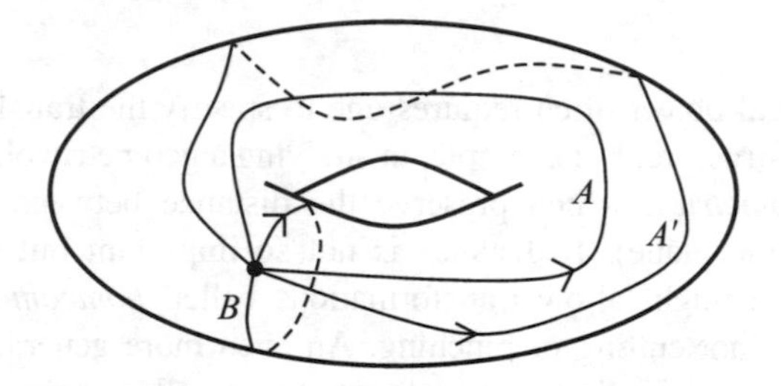


**Figure 16.3.** Loops on the torus

To distinguish homeomorphism and homotopy invariants from those attributes (such as Gaussian curvature) which are only invariant with respect to isometries, we shall refer to the former as "topological" and the latter as "geometric." Many important results in mathematics stem from the interplay between geometric and topological properties. The point of this article is to explore one instance of this—a famous theorem known as the Cartan-Hadamard Theorem. Roughly speaking, this theorem states that any surface whose Gaussian curvature is $\leq 0$ at every point can be "covered by" the plane (a more precise statement requires a discussion of *universal covers*, which will be treated in Section 3).

One of the drawbacks of attempting to describe results involving curvature is that Gaussian curvature and its generalization to higher dimensions (in the form of "sectional curvature") require ideas and formalism from calculus and differential geometry. To avoid this, we shall instead describe a very general setting in which the geometric property of having nonpositive curvature can be formulated in terms of triangles and the distance between points (Section 2). In this context, the intuition behind the Cartan-Hadamard Theorem becomes much more transparent (Section 3), and the relevant geometric objects include not only the smooth surfaces where the Gaussian curvature makes sense, but also more general objects constructed by gluing together polyhedra. In Section 4, we describe one such class of objects, called *cubical complexes* for which the condition of nonpositive curvature takes a particularly elementary form. Finally, in Section 5 we give an application of our Cartan-Hadamard Theorem which illustrates the close relationship between a geometric object and its fundamental group in the presence of nonpositive curvature.

## 2 Metric spaces

Any two points in a geometric object are separated by a well-defined distance, and the simplest mathematical object that captures this structure is a metric space.

**Definition 1** A *metric space* is a set $X$ with a prescribed distance between any two elements of $X$. The distance between points $x$, $y$ in $X$ will be denoted $d(x, y)$ and must satisfy the conditions:

- $d(x, y) \geq 0$ with equality holding if and only if $x = y$.
- $d(x, z) \leq d(x, y) + d(y, z)$ for all $x, y, z$ in $X$.

The notion of a straight line or line segment from classical geometry has a natural generalization in the setting of metric spaces. A *path* in a metric space $X$ is a set of points $\alpha(t)$ traced out in $X$. The variable $t$ (which we think of as the time at which we are at the point $\alpha(t)$) ranges over all numbers in some closed interval $[a, b]$. We assume further that this set of points is continuous, meaning there are no jumps or breaks as one travels the path. A path is *straight* (or *geodesic*) if it preserves distances, i.e., if $d(\alpha(t_1), \alpha(t_2)) = |t_2 - t_1|$ for all $t_1, t_2$ in the interval $[a, b]$. In this article, we shall only be interested in metric spaces with the property that any two points can be joined by a straight path. We call such a metric space a *geodesic space*. Informally, one can think of a geodesic space $X$ as a set of points such that for any two points $P$ and $Q$

(i) there is a distance $d(P, Q)$,

(ii) any path joining $P$ to $Q$ has length at least $d(P, Q)$, and

(iii) there is at least one path joining $P$ to $Q$ having length equal to $d(P, Q)$.

For example, any closed surface in 3-space is a geodesic space (here the distance between points is the length of a shortest path on the surface), and the plane minus the origin is an example of a metric space that is *not* a geodesic space (there is no path of length 2 joining the two points $P = (-1, 0)$ and $Q = (1, 0)$ on the $x$-axis). See Figure 16.4. A technical assumption we shall also impose on our geodesic spaces is that they be *complete*, meaning roughly that no limit points arc "missing" from the space. This assumption also rules out the example of the plane minus the origin, since the origin is a missing limit point.

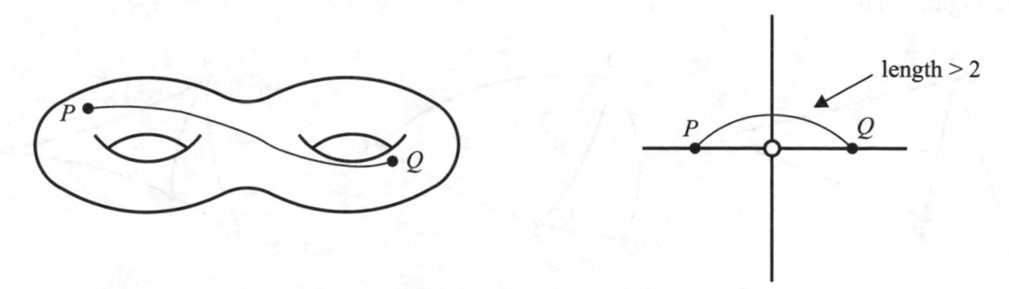


**Figure 16.4.** A geodesic space and a metric space that is *not* a geodesic space

There are three famous examples of geodesic spaces that deserve special attention.

1. The plane $\mathbb{R}^2$ with the usual Euclidean distance. Straight paths in $\mathbb{R}^2$ are just line segments (traveled at unit speed).

2. The unit sphere $\mathbb{S}$ with distance $d(A, B) = \angle AOB$ ($O$ is the center of the sphere). Note that the sphere is the *surface* of the ball, so although we use the center to define the distance between points, it is not actually a point on the sphere. Straight paths on $\mathbb{S}$ are arcs of great circles with length $\leq \pi$ (Figure 16.5).

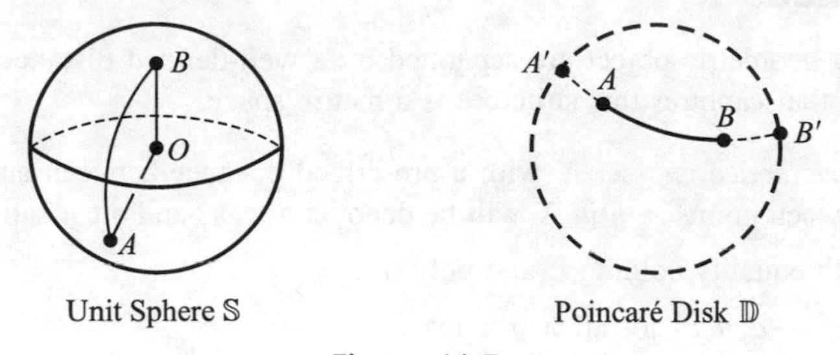


**Figure 16.5.**

3. The open (without boundary) unit disk $\mathbb{D}$ in $\mathbb{R}^2$ with distance between $A$ and $B$ given by the formula

$$d(A, B) = \frac{1}{2}\log\left(\frac{(AB')(BA')}{(AA')(BB')}\right)$$

where $A'$ and $B'$ are as in Figure 16.5 and $(PQ)$ denotes the Euclidean distance between the points $P$ and $Q$. Straight paths in $\mathbb{D}$ are segments of circles that are perpendicular to the boundary circle and straight lines through the center. The disk with this notion of distance and straight paths is called the Poincaré Disk.

Our formulation of "nonpositive curvature" for a metric space $X$ is obtained by comparing triangles in $X$ to regular Euclidean triangles. By definition, a *triangle* in a metric space $X$ is a set of three points joined by straight paths. Given any such triangle $\Delta$, we can compare it to a corresponding triangle $\Delta'$ in $\mathbb{R}^2$ with the same side lengths. Let $A$ be a vertex of $\Delta$ and let $P$ be a point on the side of $\Delta$ opposite $A$. There are corresponding points $A'$ and $P'$ in $\Delta'$, and we let $d'$ denote the Euclidean distance $d(A', P')$. We say that $\Delta$ is

- *flat* if $d(A, P) = d'$ for any choice of $A$ and $P$.
- *thick* if $d(A, P) > d'$ for any choice of $A$ and $P$.
- *thin* if $d(A, P) < d'$ for any choice of $A$ and $P$.

(See Figure 16.6.)

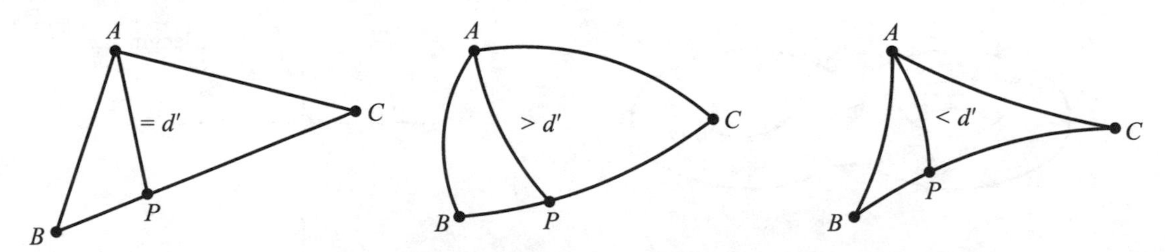


**Figure 16.6.** A flat, a thick, and a thin triangle

From the definition, any triangle in the Euclidean plane is flat. Intuitively, the pictures in Figure 16.7 suggest that triangles on the sphere are thick and triangles in the Poincaré disk (or on the saddle surface $z = x^2 - y^2$) are thin. This is true, however, the proofs require some fairly involved mathematics.

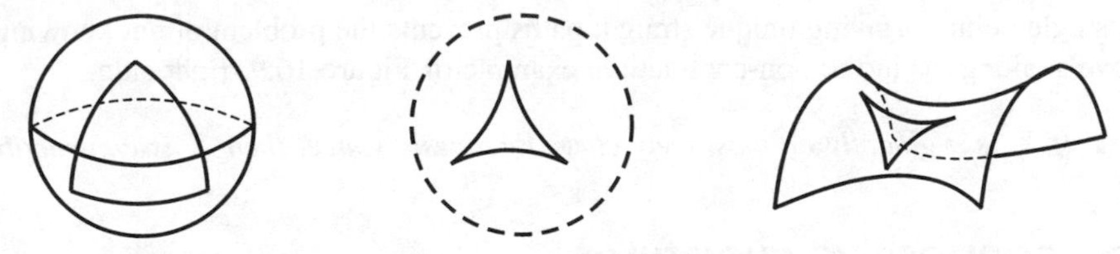

**Figure 16.7.** Triangles on the sphere, the Poincaré disk, and the saddle

From a topological standpoint, there is a significant difference between spaces with flat or thin triangles and spaces with thick triangles. Thus, we single out the former by defining a metric space $X$ to be *globally nonpositively curved* if every triangle in $X$ is either flat or thin. (We shall also refer to such an $X$ as having *global nonpositive curvature*, even though we have not actually defined a "curvature" quantity.) A simple but important consequence of having only flat or thin triangles is the following.

**Lemma 1** *Let $X$ be a globally nonpositively curved geodesic space. Then any two points in $X$ can be joined by a unique straight path.*

*Proof.* Suppose $B$ and $C$ are two points in $X$ that are joined by geodesic paths $\alpha$ and $\beta$ (Figure 16.8). For any point $A$ on the path $\alpha$, consider the resulting triangle $\Delta ABC$. Since

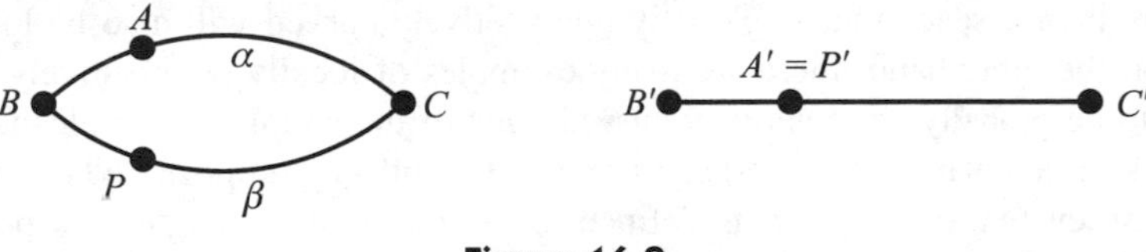


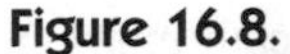

**Figure 16.8.**

$d(A,C) + d(A,B) = d(B,C)$, the corresponding Euclidean triangle $\Delta'$ will be a degenerate triangle with the point $A'$ lying on the segment $B'C'$. Letting $P$ be the point on $\beta$ corresponding to the point $P' = A'$, we have $d(A,P) \leq d(A',P') = 0$. This means $A = P$, so the point $A$ must lie on the path $\beta$. Since $A$ was arbitrary, every point on $\alpha$ must also lie on $\beta$, so $\alpha = \beta$. □

The property of having unique straight paths between any two points imposes a strict topological constraint on a space of global nonpositive curvature; namely, any such space is homotopy equivalent to a point, or *contractible*. To get a rough idea why this is true, fix a point $p$ in $X$ and imagine simultaneously pulling every other point $x$ toward $p$ along the unique straight path between $x$ and $p$ (Figure 16.9, left side). This procedure determines a continuous deformation of

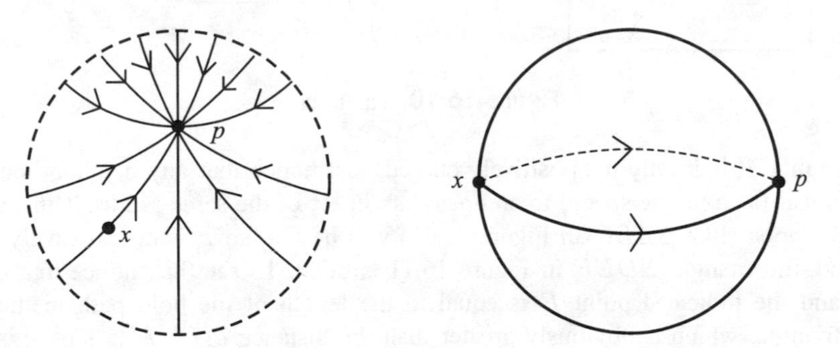


**Figure 16.9.** Contracting a globally nonpositively curved space (and why nonpositive curvature is needed)

$X$ into the single point $p$. Having unique straight paths prevents the problem of not knowing which path to move $x$ along (as in the non-contractible example of Figure 16.9, right side).

**Corollary 1** *If $X$ is a globally nonpositively curved geodesic space, then $X$ is contractible.*

## 3 Local nonpositive curvature

The traditional notion of curvature for a smooth surface in $\mathbb{R}^3$ assigns a curvature to every point of the surface. If the surface has a "bump" or a "dent" at the point then the curvature is positive there, whereas if the surface looks like a "saddle" at the point, then the curvature is negative there. Thus, the condition of having nonpositive curvature (in the Gauss sense) is a local condition, saying that in a small neighborhood of every point the surface is either flat or looks like a saddle. In contrast, our definition of nonpositive curvature in terms of triangles is global (since a single triangle can spread between points very far apart).

A local version of nonpositive curvature for a geodesic space $X$ can be defined as follows. Given a point $p$ in $X$, we define a *neighborhood* of $p$ to be any set in $X$ of the form $\{x \mid d(p,x) < r\}$ for some fixed positive number $r$ (called the *radius* of the neighborhood).

**Definition 2** A geodesic space $X$ is *locally nonpositively curved* (or simply *nonpositively curved*) if every point in $X$ has a neighborhood in which every triangle is thin or flat.

Obviously, any metric space that is globally nonpositively curved will also be locally nonpositively curved. On the other hand, there are many examples of locally nonpositively curved metric spaces that fail to be globally nonpositively curved. One of the simplest examples is the flat torus $T$, defined by taking a square with side-length one and identifying opposite edges (Figure 16.10). The distance between two points in $T$ is defined to be the minimal length of a path joining the two points (keep in mind that the shortest path might be one that crosses one of the sides of the square—as for points $A$ and $B$ in Figure 16.10 ). Topologically, $T$ is the same as (homeomorphic to) the usual torus sitting as a surface in $\mathbb{R}^3$, but there is no reason to expect them to be the same as metric spaces. In fact, they are not.

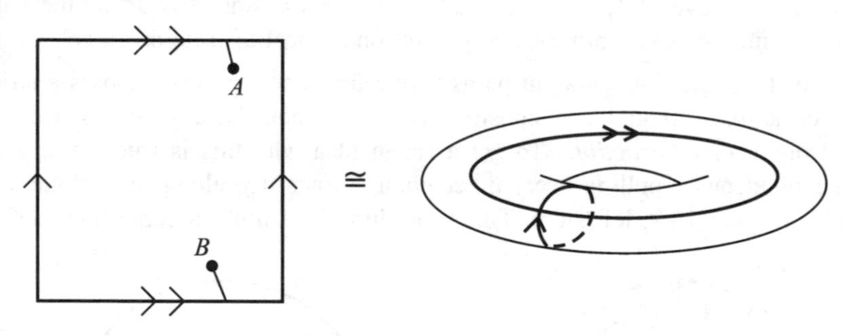


**Figure 16.10.** The torus

Now to see that $T$ is locally nonpositively curved, just notice that any neighborhood of radius $\leq 1/2$ is identical (as a metric space) to an open disk in $\mathbb{R}^2$ of the same radius. Thus, any triangle in this neighborhood (like $\Delta ABC$ in Figure 16.11) will be flat, so $T$ is nonpositively curved. On the other hand, the triangle $\Delta DEF$ in Figure 16.11 is thick. To see this, notice that the distance between $D$ and the indicated point $P$ is equal to the length of the bold path in the Euclidean comparison triangle, which is obviously greater than the distance $d(D', P')$. This shows that the flat torus $T$ fails to be globally nonpositively curved.

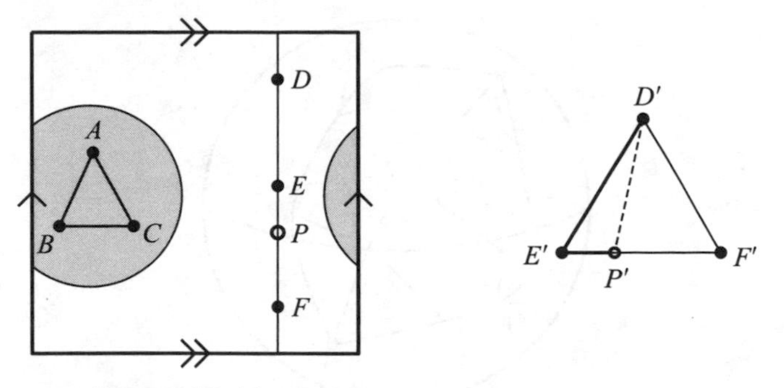


**Figure 16.11.** Flat and thick triangles on the torus

The example of the flat torus also illustrates that we should not expect the local condition of nonpositive curvature to impose the same topological constraint of contractibility as global nonpositive curvature does. The torus is not contractible. Nevertheless, there is still a restriction on the topology of a geodesic space that has local nonpositive curvature. To describe this restriction, we need a brief topological digression.

Given a geodesic space $X$ we can construct another geodesic space $\widetilde{X}$ whose points correspond to certain classes of paths in $X$. This space $\widetilde{X}$ is called the *universal cover of* $X$ and has the property that any loop in $\widetilde{X}$ can be continuously shrunk to a single point. To describe this universal cover, we first fix a point $p \in X$, and consider the set of all paths $\alpha$ that start at $p$. We then define two such paths $\alpha$ and $\alpha'$ to be equivalent if they have the same endpoints and they are homotopic (one can be continuously deformed into the other). For example, in Figure 16.12 $\alpha$ is equivalent to $\alpha'$ but not to $\beta$ or $\gamma$. The universal cover $\widetilde{X}$ is the set whose points correspond to families of equivalent paths based at $p$. Thus $\alpha$ and $\alpha'$ in Figure 16.12 represent the same point in $\widetilde{X}$, while $\alpha$, $\beta$, and $\gamma$ all represent different points.

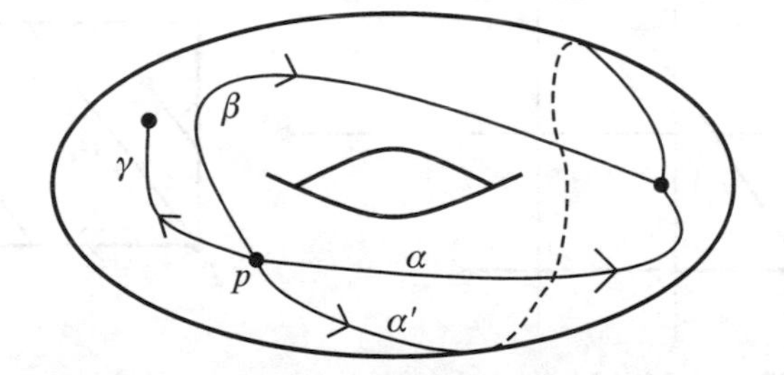


**Figure 16.12.** Paths based at $p$ on the torus

There is a natural mapping $\phi : \widetilde{X} \to X$, called the *covering projection*, which takes the class of any path $\alpha$ to its endpoint. Using this covering projection and the metric $d$ on $X$, one can define a unique metric $\tilde{d}$ on $\widetilde{X}$ with respect to which the covering projection is a local isometry. This means that sufficiently small neighborhoods in $\widetilde{X}$ are isometric with their images in $X$. In particular, *if* $X$ *is nonpositively curved, then so is* $\widetilde{X}$. To illustrate the construction of the universal cover, we give two examples: the universal cover of the unit sphere and of the flat torus.

1. *The unit sphere* $\mathbb{S}$ *in* $\mathbb{R}^3$ *is its own universal cover.* To see this, note that any two paths based at a point $p$ in $\mathbb{S}$ will be equivalent if and only if they have the same endpoints (Figure 16.13). Thus, there is precisely one point in $\widetilde{\mathbb{S}}$ for each point in $\mathbb{S}$, so the covering projection is one-to-one (and, in fact, an isometry).

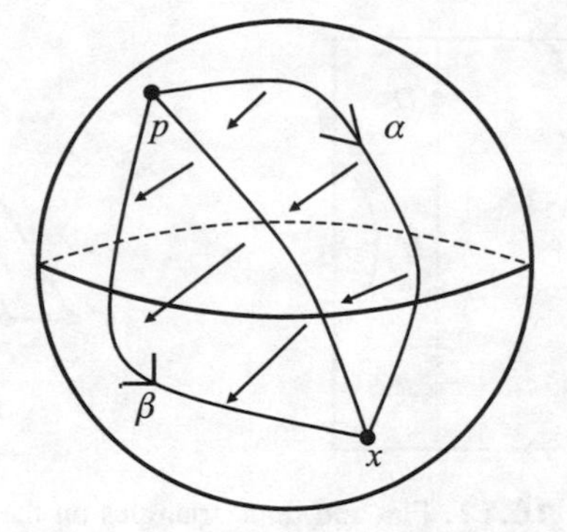


**Figure 16.13.** Equivalent paths in $\mathbb{S}$

2. *The universal cover of the flat torus $T$ is the Euclidean plane $\mathbb{R}^2$.* To describe the correspondence between points in $\mathbb{R}^2$ and paths in $T$, first note that every vertex of the square gets glued to a single point in the torus. We take this to be our fixed point $p$. Given a point $x$ in $\mathbb{R}^2$, we can form a path $\alpha_x$ in $T$, based at $p$, as follows. Tile the plane with identical squares as in Figure 16.14 and consider the line segment $L$ joining the origin to the point $x$. By translating all of the squares onto a single square and then gluing up the edges to obtain the torus, we see that the image of the original line $L$ is a path in $T$, based at $p$. This is $\alpha_x$. It can be shown that *any* path in $T$, based at $p$, is homotopic to exactly one of these "straight paths", hence $\widetilde{T}$ coincides with $\mathbb{R}^2$. The covering projection $\mathbb{R}^2 \to T$ is easily seen to be a local isometry, hence $\widetilde{T}$ and $\mathbb{R}^2$ also coincide as metric spaces.

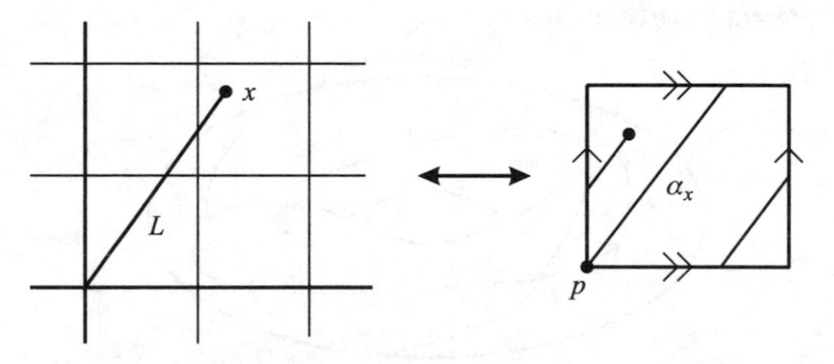


**Figure 16.14.** The correspondence between $\mathbb{R}^2$ and $\widetilde{T}$

Now, going back to our discussion of local versus global nonpositive curvature, notice that the thick triangle we found in the flat torus (Figure 16.11) corresponds to a loop in the torus which cannot be shrunk to a point. It turns out that in a space of nonpositive curvature, this is the only way a triangle can be constructed which is not flat or thin. Since passing to the universal cover removes any such loop (the path in the universal cover corresponding to such a loop will not close up), but retains nonpositive curvature, we obtain the following generalization of the Cartan-Hadamard Theorem from differential geometry.

**Theorem 1 (Cartan-Hadamard-Gromov)** *Let $X$ be a geodesic space of nonpositive curvature. Then the universal cover $\widetilde{X}$ has global nonpositive curvature, hence (by Corollary 1) is contractible.*

# 4 Cubical complexes

All of the examples of geodesic spaces we have seen so far are smooth surfaces, meaning that the classical Gaussian curvature (defined using calculus) makes perfect sense for them. In this section we shall describe some geodesic spaces that are not in general smooth. In this setting, Gaussian curvature does not apply, while our notion of nonpositive curvature does. These metric spaces are constructed by gluing together cubes along their faces.

Let $C_n$ denote the unit cube in $n$-dimensional Euclidean space. For example $C_1$ is the unit interval $[0, 1]$, $C_2$ is the unit square, $C_3$ is the ordinary unit cube, and so on. The Euclidean distance makes $C_n$ into a metric space. Each face of a cube can be identified with a unit cube in some lower dimension, and any two faces of the same dimension can be glued together in such a way that the distance between points on each face is preserved. One example of this is the construction of the flat torus, where opposite edges of the square are glued together preserving distances between points. More generally, one can take a finite collection of cubes (not exceeding some fixed dimension), and glue them together along faces of the same dimension (see Figure 16.15).

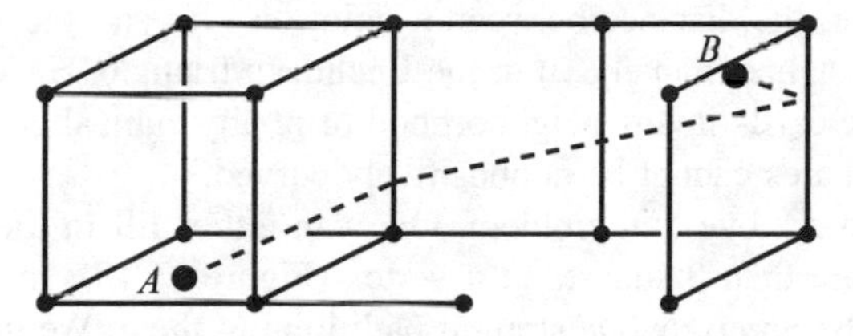


**Figure 16.15.** A straight path in a cubical complex

The resulting object $X$ is called a *cubical complex* and can be made into a metric space as follows. Given two points $A$ and $B$ in $X$, let $\alpha$ be a path joining $A$ to $B$ which restricts to a straight path (i.e., line segment) in each cube. Such a path has a well-defined length, namely, the sum of the lengths of the pieces in each cube, and we define the distance between $A$ and $B$ to be the minimum length of all such paths. It can be shown that any cubical complex is a geodesic space.

There is a nice combinatorial condition for when a cubical complex is nonpositively curved. To get a feel for this condition, consider the simple example in Figure 16.16 consisting of three unit squares glued to look like the corner of a cube. In any neighborhood of the corner vertex $p$, we can always find a triangle $\Delta ABC$ which is not thin or flat. This triangle is obtained by taking two straight paths of the same length joining $B$ and $C$ (one on each side of $p$) and then adding a point $A$ to one of these paths. We let $AB$, $BC$, and $AC$ denote the segments corresponding to the sides of the resulting triangle $\Delta ABC$. This triangle has the propery that the length of $BC$ is the sum of

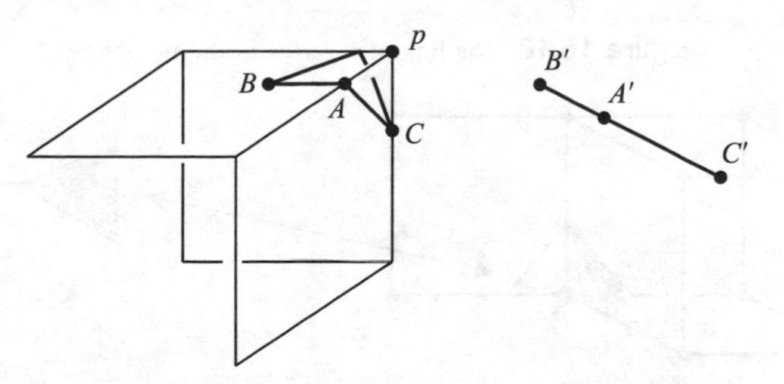


**Figure 16.16.** Three squares meeting at a vertex

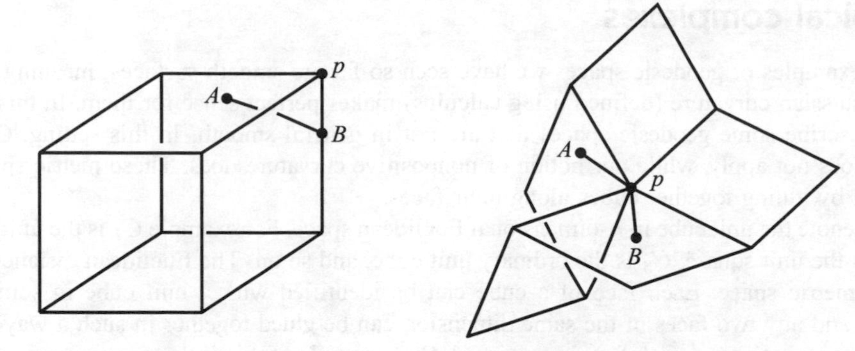


**Figure 16.17.** A 3-cube and a cubical complex with 5 squares

the lengths of $AB$ and $AC$; hence, the corresponding Euclidean triangle would be the degenerate triangle with point $A'$ lying on the line segment $B'C'$ (note the similarity with Lemma 1 and Figure 16.8). In particular, the distance between a point on $BC$ and the vertex $A$ will be greater than the corresponding distance (namely, 0) in the Euclidean triangle. Hence $\Delta ABC$ is not thin or flat. Since such a triangle exists in any neighborhood of *p*, any cubical complex containing such a configuration of three squares cannot be nonpositively curved.

There are two ways to remove this problem. One can either fill in the missing 3-dimensional cube or glue together more than 3 squares at a vertex (Figure 16.17). In either case, for any two points near *p*, there will be *precisely one* straight path joining them. We invite the reader to check this for the 5-square example in Figure 16.17.

To describe the general condition for a cubical complex to be nonpositively curved, we need to first describe the structure of how cubes come together at a vertex. Given an $n$-dimensional cube, if we slice off a corner vertex, we create a new face whose shape is called an $(n-1)$-*simplex.* For example, a 0-simplex is a point, a 1-simplex is a line segment, a 2-simplex is a triangle, and a 3-simplex is a tetrahedron (Figure 16.18). If we start with a vertex *p* in a cubical complex $X$, then all of the simplices surrounding *p* can be glued together along their faces to form a *simplicial complex*. This complex is called the *link of p*. See Figure 16.19.

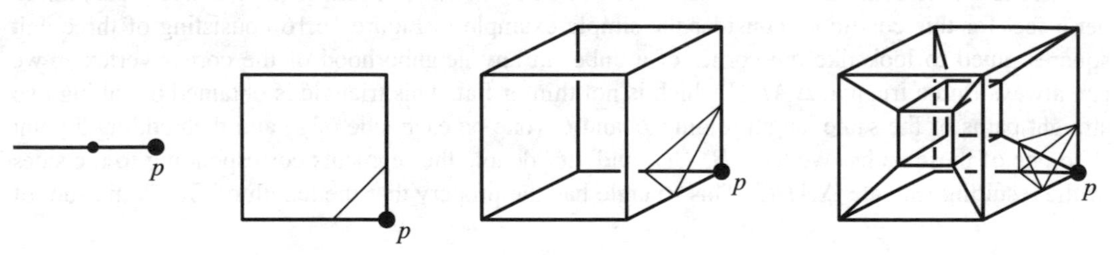


**Figure 16.18.** The link of a vertex in the $n$-cube

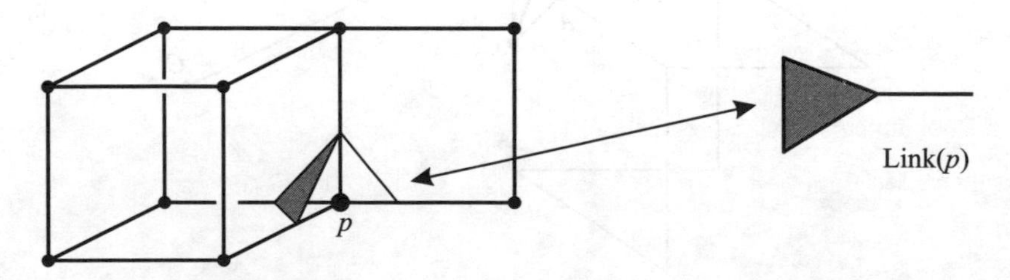


**Figure 16.19.** The link of a vertex in a cubical complex

We say that the link of $p$ has an *empty simplex* if all of the points and edges of a simplex occur in the link, but the simplex itself is missing. For example, in Figure 16.16, the link is an empty 2-simplex. On the other hand, neither of the links in Figure 16.17 have an empty simplex. The link on the left is a (filled in) 2-simplex, and the link on the right is an (empty) pentagon. See Figure 16.20.

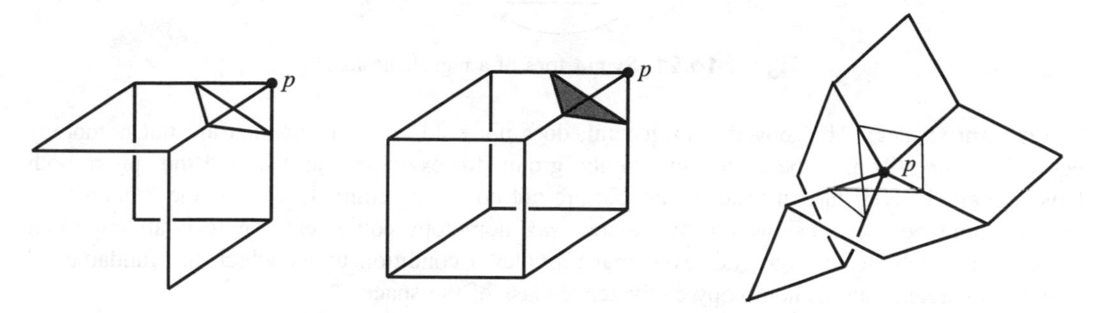


**Figure 16.20.** Vertex links from Figures 16.16 and 16.17

**Lemma 2 (Gromov)** *Let $X$ be a cubical complex. Then $X$ is nonpositively curved if and only if the link of every vertex contains no empty simplex.*

Combining this with our Cartan-Hadamard theorem, we obtain the following.

**Corollary 2** *Let $X$ be a cubical complex. If the link of every vertex contains no empty simplex, then the universal cover $\widetilde{X}$ is contractible.*

# 5 An application to group theory

By definition, a *group* is a set $G$ together with an operation that produces, for any two elements $g_1$ and $g_2$ in $G$, a third element which we denote by $g_1 g_2$. A group must satisfy the three axioms

(i) (Identity element) There exists an element $1 \in G$ such that $1g = g1 = g$ for all $g \in G$.

(ii) (Inverses) For every element $g \in G$ there exists an element $g^{-1} \in G$ such that $gg^{-1} = g^{-1}g = 1$.

(iii) (Associativity) For any three elements $g_1, g_2, g_3 \in G$, we have $(g_1 g_2) g_3 = g_1 (g_2 g_3)$.

Groups are essential to many different areas of mathematics because they form the basis of the mathematical formulation of symmetry. Colloquially, an object is symmetric if it looks the same from many different viewpoints. In mathematics, an object is symmetric if it has many different self-transformations which preserve the mathematical structure of the object. For example, a regular Euclidean polygon is symmetric since there are many rotations and reflections that map it back to itself (Figure 16.21). The set of self-transformations of a mathematical object (with composition as the operation) is always a group. Two groups $G_1$ and $G_2$ are considered the same, or *isomorphic*, if there is a correspondence between elements of $G_1$ and elements of $G_2$ that preserves the respective operations.

Recall from the introduction that to any (topological) space we can associate a group, the fundamental group, and that for any two homotopy equivalent spaces the resulting fundamental groups will be the same. This means that the fundamental group can be used to show that two spaces are *not* homotopy equivalent (if two spaces have different fundamental groups, they must

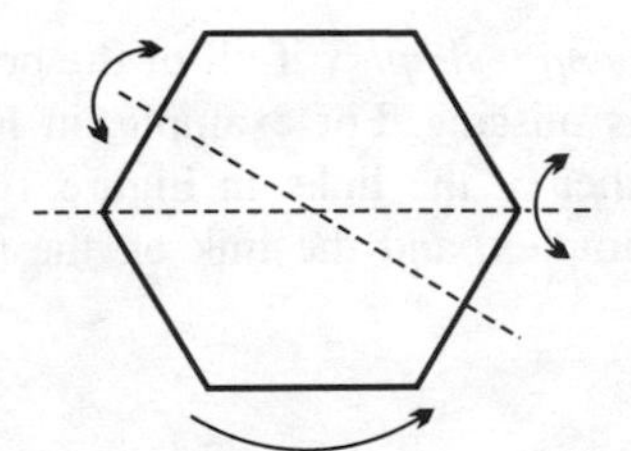

**Figure 16.21.** Symmetries of a regular hexagon

be different spaces). The converse, in general, does not hold. Two spaces that are not homotopy equivalent can still have the same fundamental group (for example, the disk and the sphere both have so-called trivial fundamental groups but are not homotopy equivalent), thus the fundamental group cannot be used to show that two spaces *are* homotopy equivalent. There is an important theorem in algebraic topology, however, that provides a condition under which the fundamental group *does* determine the homotopy equivalence class of the space.

**Theorem 2 (Eilenberg-MacLane)** *If $X_1$ and $X_2$ have the same fundamental group and their universal covers are contractible, then $X_1$ and $X_2$ are homotopy equivalent.*

Since our Cartan-Hadamard Theorem (Theorem 1) says that universal covers of nonpositively curved spaces are always contractible, an immediate consequence of this theorem is the following.

**Corollary 3** *Suppose $X_1$ and $X_2$ are nonpositively curved geodesic spaces. Then they are homotopy equivalent if and only if they have the same fundamental group.*

One implication of this corollary is that not only can fundamental groups be used to tell spaces apart, but in the presence of nonpositive curvature *the spaces can be used to tell fundamental groups apart*. For example, suppose we are given two groups $G_1$ and $G_2$ and we want to decide if they are isomorphic. If we can produce nonpositively curved spaces $X_1$ and $X_2$ with fundamental groups $G_1$ and $G_2$, respectively, and we can somehow show that $X_1$ and $X_2$ are not homotopy equivalent, then we know $G_1$ and $G_2$ cannot be isomorphic.

There is a standard method for describing groups which provides a natural setting for an application of this type, namely by giving a *group presentation*. Such a description for a group $G$ consists of:

(i) a list of elements in $G$, called *generators*, with the property that any element in $G$ can be expressed as a product of generators (or their inverses), and

(ii) a list of strings of generators (and their inverses), called *relations*, with the property that the product of elements in each string represents the identity in $G$, and such that any product of two elements in $G$ can be derived from the list of relations.

The common notation for a presentation is

$$G = \langle g_1, g_2, \dots \mid r_1, r_2, \dots \rangle$$

where the $g_i$'s are the generators and the $r_i$'s are the relations.

For example, the symmetry group of an equilateral triangle has a presentation

$$\langle a, b \mid a^2, b^2, (ab)^3 \rangle.$$

The generators $a$ and $b$ correspond to reflections across the two lines in Figure 16.22, the relations $a^2 = 1$ and $b^2 = 1$ correspond to the fact that performing a reflection twice returns the triangle to its original position, and the relation $(ab)^3 = 1$ corresponds to the fact that $ab$ is a rotation of 120° (so performing $ab$ three times returns the triangle to the original position). The six symmetries of the triangle can be written as strings of $a$'s and $b$'s as follows:

- the identity (do nothing) $= 1$
- the reflection fixing the bottom left vertex $= a$
- the reflection fixing the top vertex $= b$
- the reflection fixing the bottom right vertex $= aba$
- the 120° rotation clockwise $= ab$
- the 120° rotation counterclockwise $= ba$

To see how the relations can be used to determine products, consider the following example which computes the product of the rotation $ab$ with the flip $b$:

$$(b)(ab) = bab = (1)bab - (ab)^3bab = ababa(b^2)ab = ababa(1)ab$$
$$= abab(a^2)b = abab(1)b = aba(b^2) = aba(1) = aba.$$

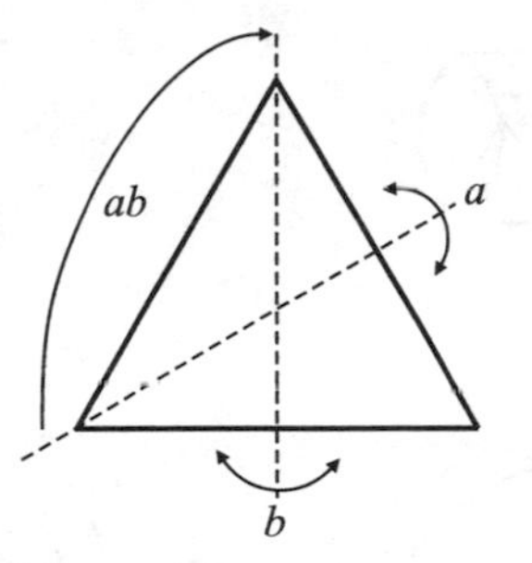


**Figure 16.22.** Symmetries of an equilateral triangle

There are two observations that make groups given by presentations particularly appropriate for the application we have in mind. First, a given group has many different presentations, and it is a very hard question in group theory to decide whether two presentations determine the same group. Second, there is a standard procedure for producing a geodesic metric space whose fundamental group coincides with a given group presentation. As an instructive (albeit contrived) illustration of how Corollary 3 can be applied to group presentations, we will address the question:

**Question.** Are the two groups $G_1 = \langle a, b \mid aba^{-1}b^{-1} \rangle$ and $G_2 = \langle a, b \mid abab^{-1} \rangle$ isomorphic?

To answer the question, we shall first describe the method for producing a geodesic space from a presentation. Suppose we are given a (finitely presented) group $G = \langle a_1, \ldots, a_n \mid r_1, \ldots r_m \rangle$. For each relation, one takes a regular polygon with the same number of sides as there are generators (or their inverses) appearing in the relation. Traveling the polygon counterclockwise, one then assigns a label to each directed edge corresponding to the relevant generator appearing in the relation. For example, the relation $aba^{-1}b^{-1}$ determines the labeled square on the left in Figure 16.23, while the relation $abab^{-1}$ determines the labeled square on the right. Having labeled the polygons for each relation in the presentation, one then constructs the space $X$ by gluing all of the vertices together and all of the edges with the same labels (if an edge label is the inverse of another, then the corresponding directed edges are glued with opposing directions). If in addition, all of the

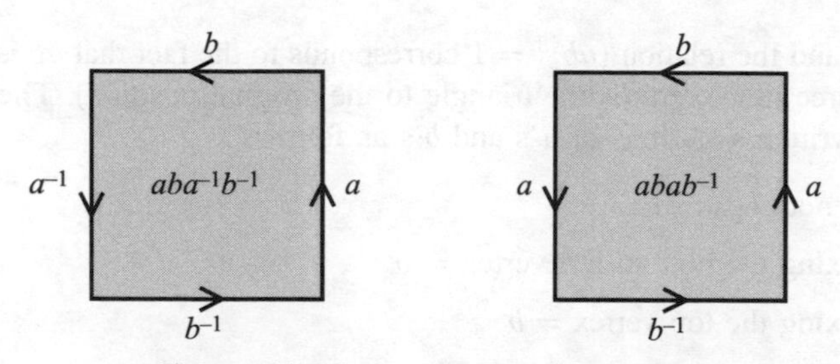


**Figure 16.23.** Relations for $G_1$ and $G_2$

polygons have unit side length, then the resulting space $X$ can be regarded as a geodesic space. Since each generator corresponds to a loop in $X$, and since each relation corresponds to a loop that can be continuously deformed to a point (just collapse the loop through the corresponding polygon), the fundamental group of $X$ coincides with the group given by our original presentation.

In the case of our two groups $G_1$ and $G_2$, the relevant spaces $X_1$ and $X_2$ are the flat torus and the so-called (flat) Klein bottle, respectively. Both of these can be viewed as cubical complexes constructed from a single unit square, and both have only a single vertex after gluing them (Figure 16.24). The link of this vertex in both cases is the boundary complex of a square (i.e.,

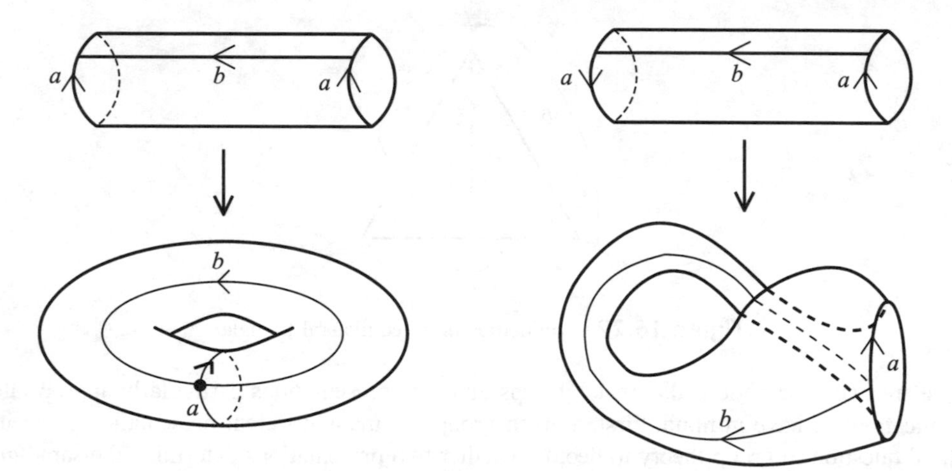


**Figure 16.24.** The torus and Klein bottle

four 1-simplices glued together in a circuit). Since this link has no empty simplices, Gromov's Lemma 2 implies that both $X_1$ and $X_2$ are nonpositively curved. By Corollary 3, then, $G_1$ and $G_2$ are isomorphic if and only if the torus and the Klein bottle are homotopy equivalent. But since the Klein bottle contains a Möbius strip (imagine a narrow strip with the loop $b$ as its center) and the torus does not, they cannot be homotopy equivalent. (That the property of containing a Möbius strip is a homotopy invariant for surfaces can be proved using homology groups, a topology topic beyond the scope of this paper.) Thus, $G_1$ and $G_2$ must be different groups.

# 6 Notes and further reading

The origins of our formulation of nonpositive curvature in the context of metric spaces are due to A. D. Aleksandrov, but many of the applications to topology and group theory were developed later by M. Gromov. A very readable exposition of these ideas, as well as further references and details for arguments we skipped over can be found in [BH]. There are several college-level textbooks, e.g., [A], that provide introductions to the topological ideas of cut-and-paste constructions, homeomorphisms, covering spaces, fundamental groups, and homology groups.

## References

[A] M.A. Armstrong. *Basic Topology.* Springer-Verlag, Berlin, Heidelberg, and New York, 1983.

[BH] M. Bridson and A. Haefliger. *Metric Spaces of Non-positive Curvature.* Springer-Verlag, Berlin, Heidelberg, and New York, 1999.

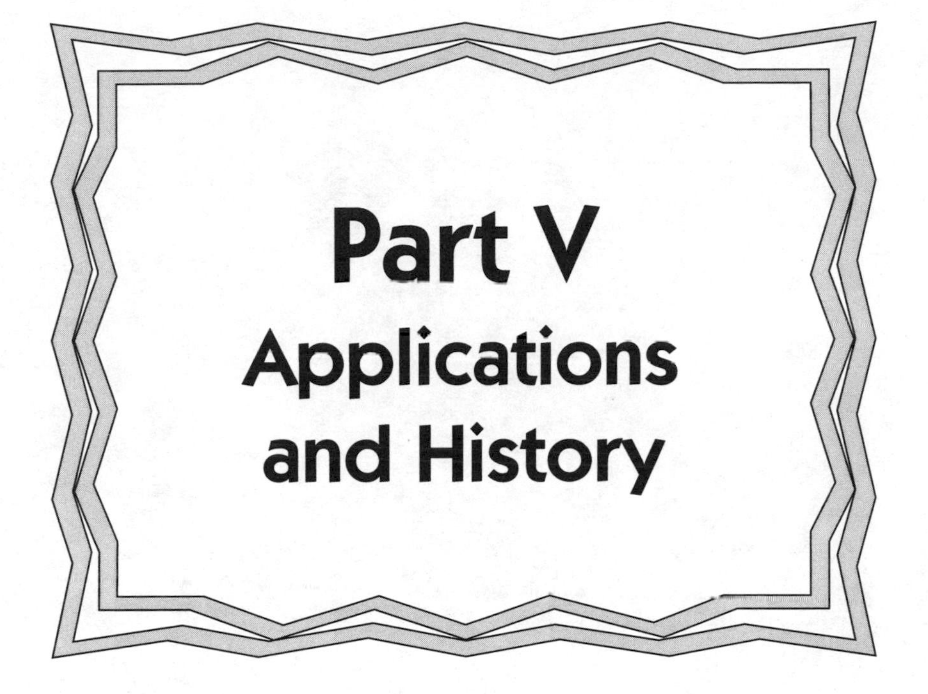

# Part V
# Applications and History

# 17

# Archimedes and his Floating Paraboloids

Sherman Stein
*University of California at Davis*

The most dazzling work of Archimedes is his analysis of the equilibrium of a floating body, specifically, a right section of a paraboloid of revolution. He pictures the paraboloid placed in a fluid with its base exposed, and answers the questions: Will it move to a vertical position? Will it assume a tilted position, with its base still exposed? Will it rest with the base just touching the surface of the water? Will it rest with the base partially submerged? He provides a complete analysis of all the possibilities in terms of the height of the section and the density of the material of which it is made, which we will take to be wood. We will also assume that the fluid is water and that its density is 1.

He also examines the same questions when the section is upside down, with its base initially submerged. The case of the exposed base is covered in Propositions 2, 4, 6, 8, and 10 in the Heath translation of Book 2, *On Floating Bodies* [2], and we will use this numbering as well as some of the notations there. Proposition 1 describes how much of a floating object is submerged. It asserts, in our terms, that if we take the density of water to be one, then the density of the object is equal to the volume of the submerged part divided by the total volume of the object. Propositions 3, 5, 7, and 9 concern the upside down case, where the base is submerged, and are the analogs of 2, 4, 6, and 8. Proposition 11 has not survived. We describe only the cases of the exposed base. We will follow the details of his reasoning as closely as possible, preserving the mathematical essence, which is invariant under translation. For other descriptions see [1] or [3].

We begin by collecting the properties of the parabola and paraboloid that he uses in his analysis. Most of these he treats as well known, probably in the lost works of Euclid and Aristaeus. The results on centers of gravity and volumes of sections of paraboloids are his own, developed in earlier works and described in [1, 2, 3].

## 1 Preliminaries

In Figure 17.1, $AM$ is the axis of the parabola and $QQ'$ is an arbitrary chord. $PV$, parallel to the axis of the parabola, is called the "axis" of the section. $V$ is the midpoint of $QQ'$. The tangent at $P$ is parallel to the chord.

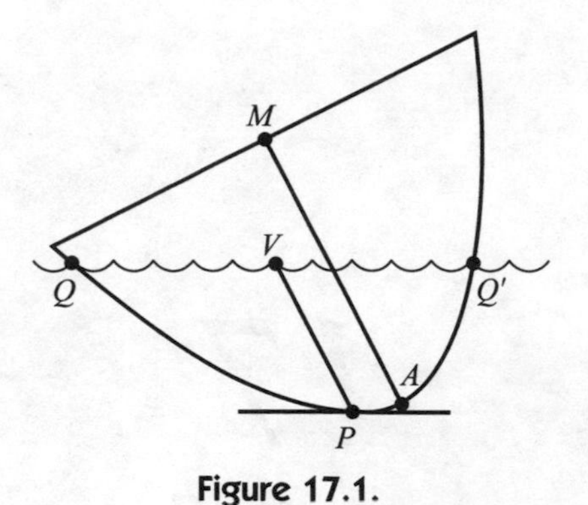


Figure 17.1.

In Figure 17.2, $A$ is the vertex of the parabola and $AM$ its axis. $P$ is any point on the parabola other than $A$. The tangent at $P$ meets the axis, extended, at $T$. $PR$ is a normal and $PN$ is perpendicular to the axis. Then, for any point $P$ on the parabola other than the vertex, $RN = 2f$, where $f$ is the distance from the focus to the vertex, and $AN = AT$. We shall call $RN$ the subnormal, its traditional name, now seldom used. Furthermore, $PN^2$ equals $2fNT$, which is closely related to our usual equation of a parabola.

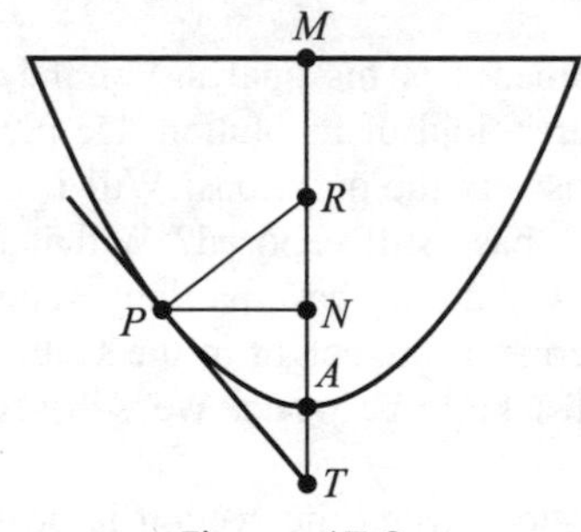


Figure 17.2.

In Figure 17.3, $QQ'$ is an arbitrary chord and $QM$ is perpendicular to the axis. $PV$ is the axis of the section determined by $QQ'$. $KO$ is perpendicular to the axis. Then

$$\frac{PV}{PK} \geq \frac{OM}{OA}.$$

Establishing this inequality by analytic geometry is an interesting exercise, which uses the fact that the square of a real number is nonnegative.

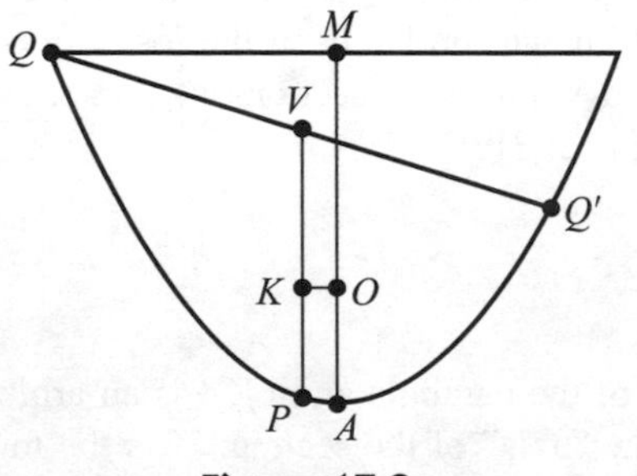


Figure 17.3.

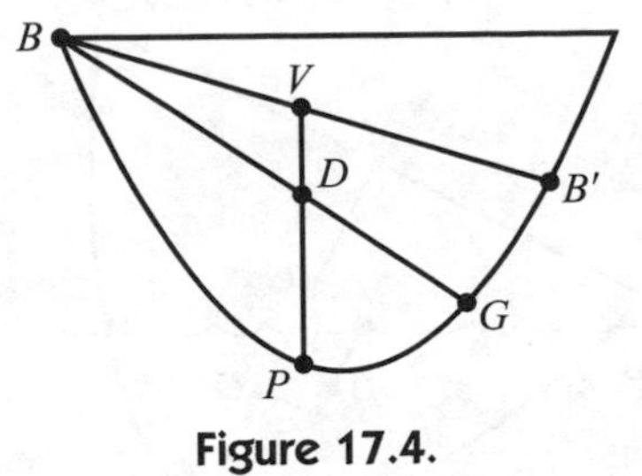


Figure 17.4.

In Figure 17.4 $BB'$ is an arbitrary chord of the parabola and $PV$ is the axis of the section cut off by this chord. The point $D$ on $PV$ has the property that $PD = 2VD$. The line through $B$ and $D$ meets the parabola also at $G$. Then $BD = \frac{3}{5}BG$. Note that the lowest $D$ occurs when $G$ is the vertex of the parabola.

Archimedes also uses properties of sections of paraboloids, which he established elsewhere. A paraboloid is formed by rotating a parabola about its axis. A plane perpendicular to the axis cuts off a right section; any other plane cuts off an oblique section.

Figure 17.5 shows the crossection of a right section and an oblique section through their axes. The tangent plane at $P$ is parallel to the plane cutting off the oblique section. $V$ is the center of the elliptical crossection, hence the midpoint of $QQ'$. $PV$ is parallel to the axis $AM$ and is called the "axis" of the section. The centroid of the oblique section is $D$, two-thirds of the way from $P$ to $V$. Similarly, the centroid of the right section is $C$, which is two-thirds of the way from $A$ to $M$. The volume of a section is proportional to $PV^2$. Hence the ratio between the volumes of the oblique and right sections is $PV^2/AM^2$. Throughout the proofs of the various propositions $D$ and $C$ will denote these centroids.

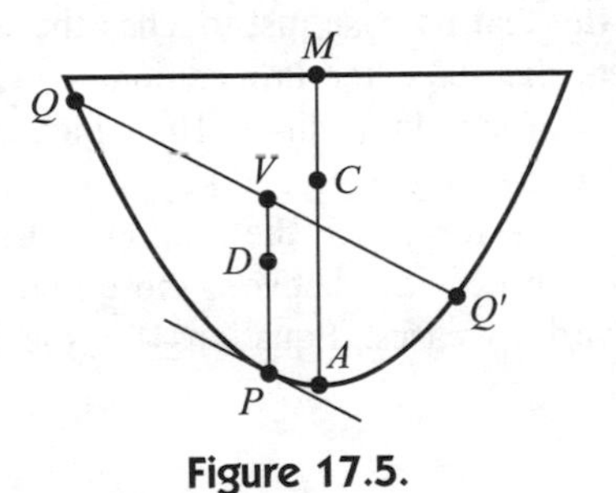


Figure 17.5.

In Archimedes' investigation, the right section corresponds to the paraboloid made of wood that has a density $s$, which is less than the density of water. The oblique section is the submerged part. Thus $QQ'$ represents the surface of the water, hence its wavy appearance. As a consequence of Proposition 1, Archimedes knows that $s = PV^2/AM^2$.

## 2 The key to Archimedes' technique

Figure 17.6 shows the typical situation that Archimedes faces. The tilted wood section, a right section, has axis $AM$ and centroid $C$, while the submerged section has axis $PV$ and centroid $D$. The point $O$ is a distance $2f$ below $C$ on the axis $AM$. The line perpendicular to $AM$ through $O$ meets $PV$ at a point $K$. Since the subnormal at $P$ has length $2f$, it follows by a little geometry that $CK$ is perpendicular to the water's surface and to the tangent at $P$. This line, $CK$, plays a key role in each proof.

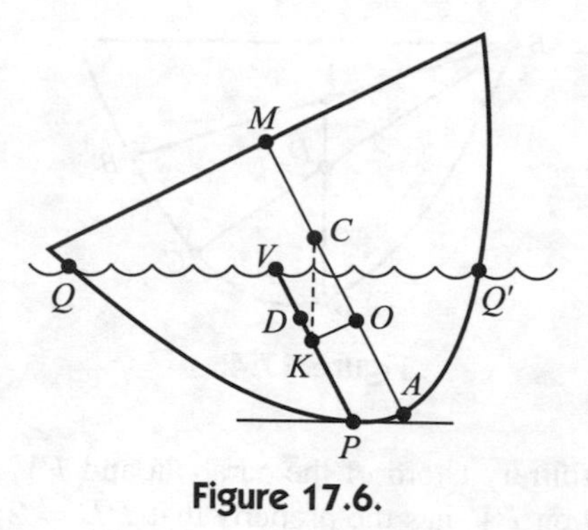


Figure 17.6.

Two forces operate on the paraboloid: gravity downward at $C$ and the water pressure upward at $D$. If $D$ is to the left of the vertical line through $C$, that is, above $K$, the paraboloid turns clockwise towards the vertical. If $D$ is to the right, that is, below $K$, the paraboloid tilts counterclockwise, further down. All of Archimedes' effort, then, is devoted to finding out which is higher, $D$ or $K$.

## 3 Summary of results

Proposition 2 shows that when the axis of the right section of a paraboloid is at most $3f$, the section is stable in the vertical position. Proposition 4 reaches the same conclusion if the wood is sufficiently dense, expressing the required density in terms of the length of the axis.

Proposition 6 treats a special case, showing that if the length of the axis is at most $\frac{15}{2}f$ and the section is placed in such a way that its base just touches the water's surface, the section will tend to move upwards. This sets the stage for Proposition 8, which determines the angle that the section then assumes in equilibrium. Proposition 10, which consists of several parts, treats sections whose axis has a length exceeding $\frac{15}{2}f$. The key to this proposition is first determining, for a given section, the two densities for which the section is in equilibrium with its base just touching the surface of the water. For densities between those two the base is partially submerged. Figure 17.7 summarizes these various cases. Equations for the two curves will be developed

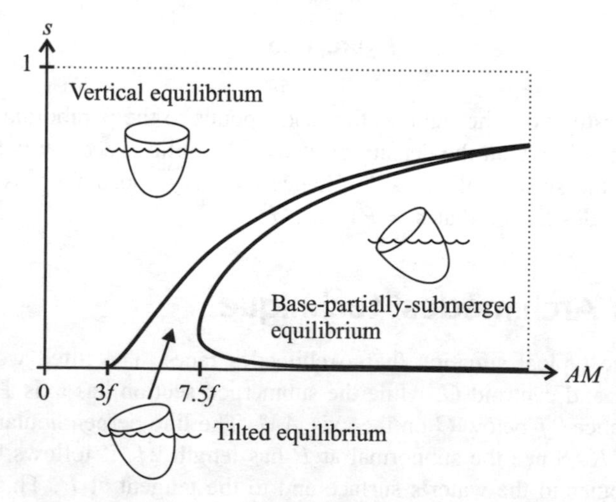


Figure 17.7.

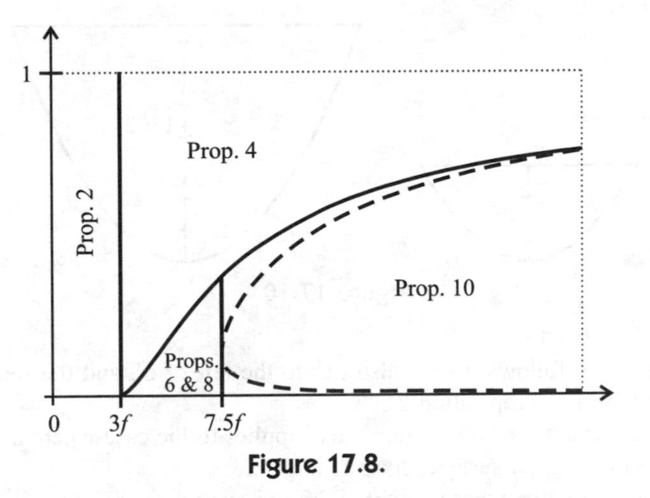


Figure 17.8.

later by translating Archimedes' geometrically described conditions into algebra. In terms of the Propositions, Archimedes' investigation is recorded by Figure 17.8.

## 4 Shallow sections

If the wood section is sufficiently shallow, it will be in stable equilibrium in a vertical position no matter how dense it is, as long as its density, $s$, is less than one. This is the substance of Archimedes' first case.

**Proposition 2.** *Assume that $AM < 3f$. Then if the section is placed in water with its base exposed, it will assume a vertical position.*

In Figure 17.9 $AM$ is less than $3f$, $C$ is the centroid of the entire section and $OC$ is $2f$. $D$ is the centroid of the submerged section.

Since $AC = \frac{2}{3}AM$ and $AM$ is less than $3f$, it follows that the length of $AC$ is less than $2f$, and $O$ is below $A$ on the axis $AM$ extended. If $P$ were to the right of the vertical line through $C$, then the subnormal at $P$ would lie completely between $A$ and $C$. But this is impossible since the subnormal has length $2f$, which is larger than the length of $AC$. Thus $P$ lies to the left of the

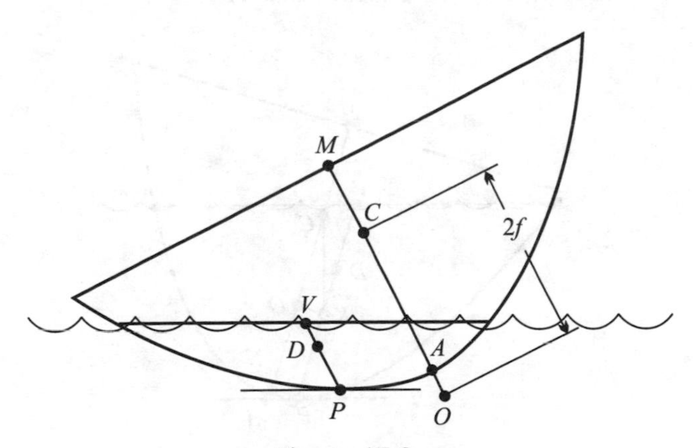


Figure 17.9.

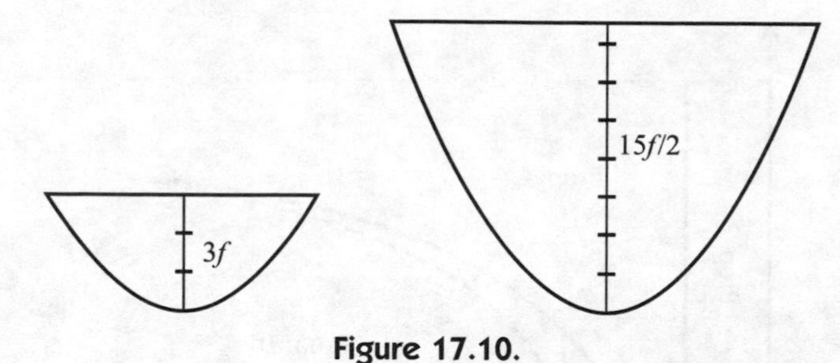


Figure 17.10.

vertical line through $C$. It follows that $D$ also lies to the left of $C$, and the section moves toward the vertical. This establishes Proposition 2.

Incidentally, this argument for Proposition 2 also applies to the case where there is no water and the section rests on a horizontal surface, such as a table.

Figure 17.10 shows a section whose height is $3f$ and one whose height is $\frac{15}{2}f$, which will play an important role in Propositions 6, 8, and 10. In all the remaining Propositions the height of the section, $AM$, will be assumed to be greater than $3f$.

## 5 Very dense sections

When the wood is very dense, the section sinks far into the water. In this case the force of the water may move the tilted section towards the vertical, making the vertical position stable. The next proposition shows how dense the wood must be for this to happen. Of course, the higher the section, the denser must the wood be.

**Proposition 4.** *Assume that $s$ is greater than $(AM - 3f)^2/AM^2$. Then if the section is placed in water with its base exposed, it will assume a vertical position.*

The pertinent diagram, Figure 17.11, is like Figure 17.9 except that now $O$ lies above $A$, since $AC$ is greater than $2f$.

Introduce the point $K$ on $PV$ such that $KO$ is perpendicular to $AM$ and the point $N$ on $AM$ such that $PN$ is also perpendicular to $AM$. Note that $CK$ is perpendicular to the surface of the water. Archimedes wishes to show that $D$ is above $K$, that is, $PD$ is greater than $PK$, which equals $NO$. To do this, he shows that $PD$ is greater than $AO$.

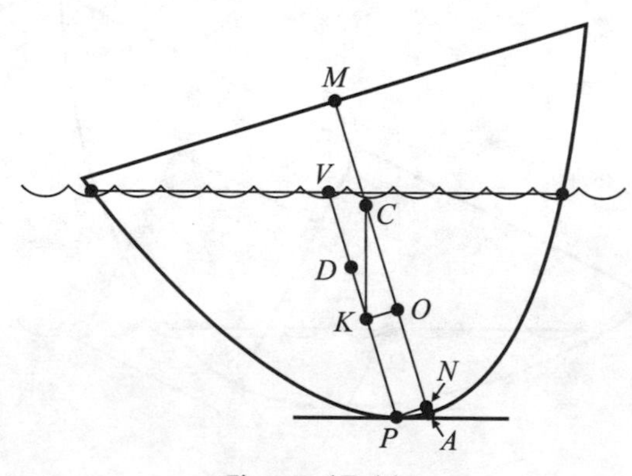


Figure 17.11.

Now, $PD = \frac{2}{3}PV$ and $AO = \frac{2}{3}AM - 2f$. Therefore he wants

$$\frac{2}{3}PV > \frac{2}{3}AM - 2f$$

or simply

$$PV > AM - 3f.$$

At this point he introduces the density $s = PV^2/AM^2$. To do this he divides the preceding inequality by $AM$ and squares, obtaining

$$\frac{PV^2}{AM^2} > \frac{(AM-3f)^2}{AM^2}.$$

Hence if the density satisfies the condition stated in the proposition, the section returns to the vertical.

There is a fine point here that did not disturb Archimedes, but does raise a question. All that is needed in the argument is that $PD$ is greater than $ON$. Could this happen with a density less than $(AM-3f)^2/AM^2$, or, what amounts to the same thing, $PV$ less than $AM - 3f$? In that case $PD$ is less than $AO$ and can be written $PD = AO - e$, where $e$ is a positive quantity independent of $P$. If $P$ is sufficiently near $A$, then $AN$ is less than $e$. It follows that for such $P$, $PD$ is less than $ON$. $D$ is then to the right of $C$ and the section veers away from the vertical. Therefore the paraboloid will not be in stable equilibrium in the vertical position.

Propositions 2 and 4 therefore completely cover all the cases where the section is stable in a vertical position: either it's shallow or its wood is suitably dense.

## 6 The base just touches the water

Propositions 6 and 8 are closely linked. It is here that sections with height $\frac{15}{2}f$ enter the scene. Proposition 6 shows that if a section whose height is less than $\frac{15}{2}f$ is placed with its exposed base touching the surface of the water at exactly one point, it will not remain in that position. Instead it will turn upwards and the base will remain exposed. Proposition 8 then determines the angle at which it is stable.

**Proposition 8.** *Assume that $AM < \frac{15}{2}f$. If the section is placed in the water with its exposed base touching the water surface at exactly one point, it will not stay in that position. Instead, it will rise.*

The surface of the water passes through the edge of the base, as shown in Figure 17.12.

Archimedes wishes to show that $D$ is to the left of $C$, or, equivalently, $D$ is above $K$, that is, $PD$ is greater than $PK$.

If $K$ is below $P$ on $PV$ extended, then $P$ is left of $C$; hence so is $D$. Consider therefore the case when $K$ is above $P$.

Now, the condition $PD > PK$ is equivalent to $PV/PK > PV/PD$, which is $\frac{3}{2}$. Since $PV/PK \geq OM/OA$, it suffices to show that

$$\frac{OM}{OA} > \frac{3}{2}.$$

Since $OM = \frac{1}{3}AM + 2f$ and $OA = \frac{2}{3}AM - 2f$, the preceding inequality is equivalent to

$$\frac{\frac{1}{3}AM + 2f}{\frac{2}{3}AM - 2f} > \frac{3}{2},$$

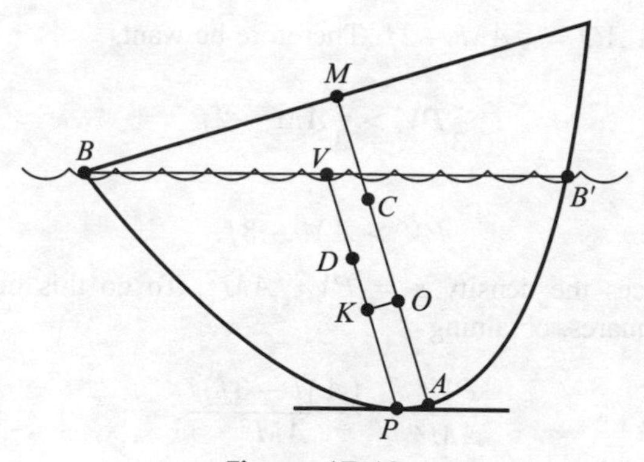


**Figure 17.12.**

which reduces to

$$AM < \frac{15}{2} f.$$

Since that inequality was assumed, the section rises from the position shown in Figure 17.12 toward the vertical.

This conclusion is the substance of Proposition 6.

**Proposition 6.** *Assume $AM < \frac{15}{2} f$. Then if the section is placed in the water with its exposed base touching the surface of the water at just one point, it will rise.*

## 7 The angle of tilt if the section is not too high

As we have seen, if $AM$ is less than $3f$, the section assumes a vertical position; if $3f < AM < \frac{15}{2} f$, it assumes a tilted position with its base exposed. In the latter case the next proposition shows at what angle the section is stable.

**Proposition 8.** *Assume that $3f < AM < \frac{15}{2} f$ and that $s < (AM - 3f)^2/AM^2$. If the section is placed in the water with its base exposed it will assume a position of equilibrium where the base is still exposed and at an angle to be described.*

We are seeking the angle $PTM$ in Figures 17.13a and 17.13b for which $PD = ON$. This angle will be described in terms of the ratio $PN/NT$.

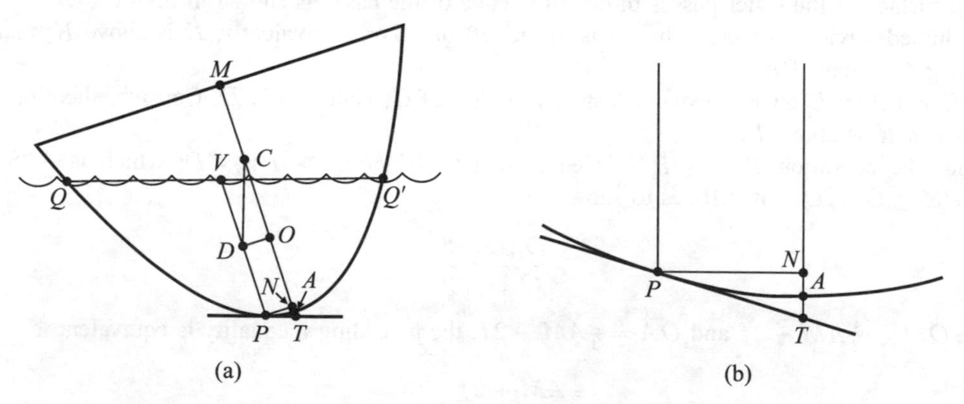


**Figure 17.13.**

It is not enough to determine when $D$ is directly below $C$. It is also necessary to show that in that situation the position is stable. So we must see what happens when $PD$ is less than $ON$ or greater than $ON$.

Consider first the inequality $PD < ON$, which Archimedes must translate into an inequality on $PN/NT$.

Since $PN^2 = 2fNT$, as mentioned in Section 1, we have $PN^2/NT^2 = 2f/NT = OC/NT$. The numerator, $OC$, is already expressed in terms of the geometry of the parabola, $OC = 2f$. It remains to express $NT$ in terms of the section and its density.

Since $PD < ON$, we have $AN < AO - PD$. Doubling both sides of this inequality yields $NT < 2(AO - PD)$. Thus

$$\frac{OC}{NT} > \frac{OC}{2(AO - PD)}$$

and therefore

$$\frac{PN^2}{NT^2} > \frac{OC}{2(AO - PD)}.$$

On the other hand, if we start with $PD > ON$, all the inequalities are reversed. Only when

$$\frac{PN^2}{NT^2} = \frac{OC}{2(AO - PD)}$$

is the section in equilibrium, and any deviation returns the section to that position.

All of the lengths $OC$, $AO$, and $PD$ are determined by the geometry of the section and its density. $AO$ is $\frac{2}{3}AM - 2f$ and $OC$ is $2f$. $PD$ is $\frac{2}{3}PV$, where $PV$ is determined by the equation

$$\frac{PV^2}{AM^2} = s.$$

In modern terms, the square of the tangent of the angle corresponding to equilibrium is

$$\frac{2f}{2(\frac{2}{3}AM - 2f - \frac{2}{3}AM\sqrt{s})},$$

which simplifies to

$$\frac{3f}{2(AM - 3f - AM\sqrt{s})}.$$

Archimedes describes this angle in terms of the sides of a certain right triangle. One leg, $ab$, has length $AC - OC - PD = AO - PD$. The other leg, $bd$, is a side of the square whose area is half that of the rectangle with sides $OC$ and $ab$. The angle in question is then the angle $dab$ in the right triangle shown in Figure 17.14.

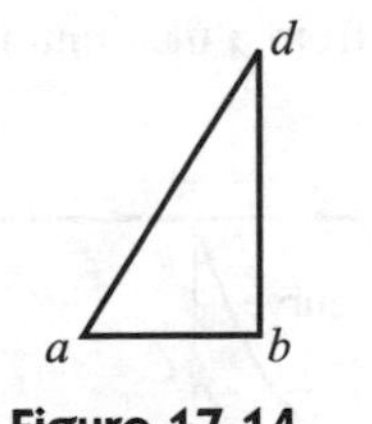


**Figure 17.14.**

As a check on this claim, we translate his geometric condition into algebra:

$$\frac{bd^2}{ab^2} = \frac{1}{2}\left(\frac{OCab}{ab^2}\right) = \frac{1}{2}\left(\frac{OC}{ab}\right) = \frac{OC}{2ab} = \frac{OC}{2(AO - PD)},$$

in agreement with our previous description of the angle.

## 8 Anything can happen when $AM$ is greater than $\frac{15}{2}f$

Archimedes finally considers the only remaining case, when the section is tall. He does this in Proposition 10, which has several parts. Now he assumes that $AM$ is fixed and is larger than $\frac{15}{2}f$. The behavior of the section then depends on its density, $s$.

Since the case where $s$ is greater than $(AM-3f)^2/AM^2$ has already been treated in Proposition 4, he assumes $s$ is less than $(AM-3f)^2/AM^2$.

The main part of his analysis is determining for which densities the section is in equilibrium with its base just touching the surface of the water. It turns out that there are two densities for which this happens, which we will call, for the time being, $s_1$ and $s_2$, with $s_1$ less than $s_2$. Proposition 10 then splits into three cases: $s < s_1$, $s_1 < s < s_2$, and $s > s_2$.

Archimedes determines these two densities and the angle the axis makes with the horizon in those cases.

Figure 17.15 shows the desired situation.

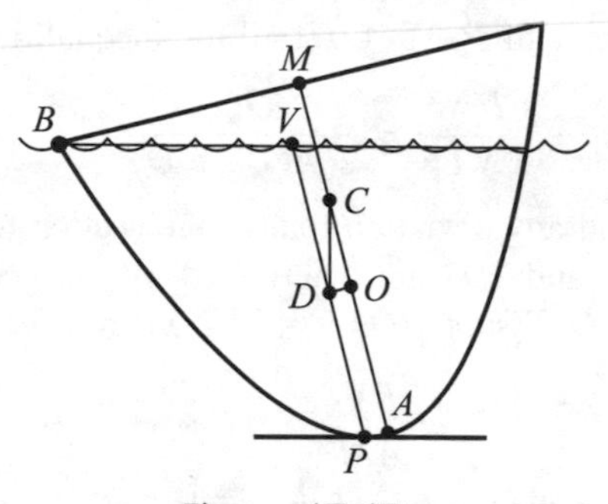


**Figure 17.15.**

Archimedes seeks an axis $PV$ such that the corresponding point $D$ is directly to the left of $O$. Each section made by a plane through $B$ corresponds to a choice of a chord $BB'$, or to its midpoint $V$, or to its centroid $D$, and ultimately to a density $s$. Archimedes introduces two curves that display the possible $V$'s and $D$'s. One is the set of all possible midpoints. This is a parabola half as large as the original one, going from $M$ to $B$. Then he introduces the curve swept out by all the corresponding possible centroids $D$. It stretches from $C$ to $B$. It also is a parabola, $\frac{3}{5}$ as large as the original parabola. Its lowest point occurs on the chord $BA$, $\frac{3}{5}$ of the way from $B$ to $A$. These parabolas are labeled the "$V$-curve" and the "$D$-curve" in Figure 17.16.

As $Q$ traces the original parabola from $B'$ to $B$, the volume of the corresponding sections diminishes. Hence $PV$ decreases, going from a maximum of $AC$ down to arbitrarily small values.

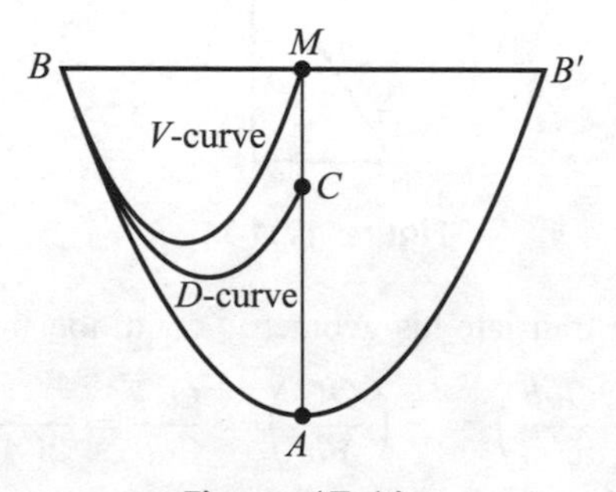


**Figure 17.16.**

Where a horizontal line through $O$ meets the $D$-curve determines the centroid of a section in which $D$ is directly below $C$. For this reason we should check that such a line does meet the $D$-curve.

The lowest point on the $D$-curve is at a distance $\frac{3}{5}AM$ below the line $BB'$. The point $O$ is at a distance $\frac{1}{3}AM + 2f$ below that line. The horizontal line through $O$ meets the $D$-curve twice if

$$\frac{3}{5}AM > \frac{1}{3}AM + 2f,$$

which is equivalent to

$$AM > \frac{15}{2}f.$$

Call the two intersection points $D_1$ and $D_2$, where $D_1$ is left of $D_2$. Call the corresponding points directly above them on the $V$-curve $V_1$ and $V_2$. Call the points directly below them on the original parabola $P_1$ and $P_2$. The tangent at $P_2$ meets the axis $AM$ extended at an angle $U$. The tangent at $P_1$ meets the same line at an angle $T$, which is less than $U$. Figure 17.17 displays this information as well as the submerged section that corresponds to the intersection $D_2$. The smaller section, whose surface passes through $V_1$ is not shown. Its surface is parallel to the tangent at $P_1$.

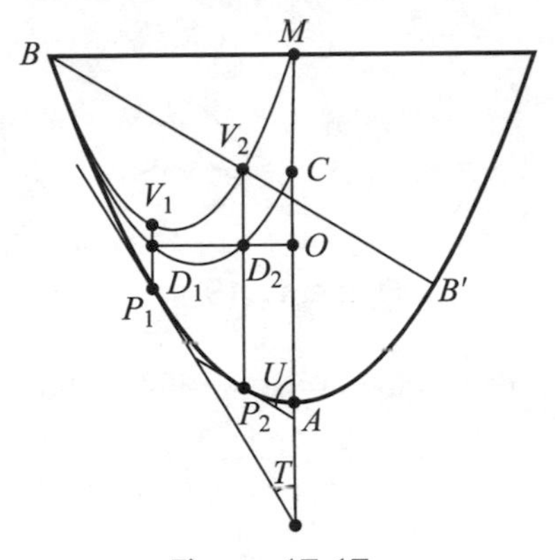


**Figure 17.17.**

If the density of the wood is $P_2V_2^2/AM^2$, the paraboloid is in equilibrium with its base just touching the surface of the water. It is tilted at an angle $U$ with the horizontal. Similarly, if the density is $P_1V_1^2/AM^2$, which is smaller, the paraboloid is in equilibrium and its axis makes the smaller angle $T$ with the horizontal. This settles the two critical cases of equilibrium (cases III a,b in [2]).

The remaining cases can be analyzed easily with the aid of Figure 17.18.

Take the case when $s$ is greater than $P_2V_2^2/AM^2$. For the corresponding section, its axis $PV$ is larger than $P_2V_2$. There is exactly one vertical chord of that length that joins the $V$-curve to

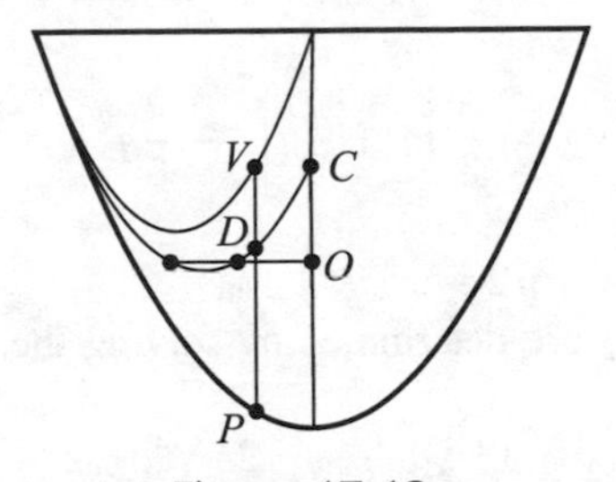


**Figure 17.18.**

the original parabola. The corresponding point $D$ on the $D$-curve lies above the horizontal line through $O$, as shown in Figure 17.18. Therefore the paraboloid moves upward toward the vertical and will assume an angle of equilibrium greater than $U$. As shown earlier, this angle is not a right angle.

When $s$ is less than $P_1V_1^2/AM^2$ similar reasoning shows that the paraboloid assumes a position of equilibrium, where its angle with the horizon is less than $T$. In both cases the base is exposed.

If $s$ lies between the two critical densities, the corresponding point on the $D$-curve lies below the horizontal line through $O$. Thus the paraboloid moves downward and its base is partly submerged.

The behavior of a section with axis greater than $\frac{15}{2}f$ is summarized in Figure 17.19. When the wood is very dense, the section is in stable equilibrium in the vertical position. If the density is a bit less, it is in equilibrium in a tilted position, but with the base still exposed. As the density decreases to a certain value, the section is in equilibrium with its base touching the water at one point. As the density decreases further, the section tilts further and is in equilibrium with its base partly submerged. When the density shrinks further to a certain value, near $\frac{1}{625}$, it is again in equilibrium with the base just touching the surface of the water. Finally, for smaller densities it is in equilibrium with the base exposed. As the density decreases in these steps, less of the section is submerged.

**Figure 17.19.**

This completes Archimedes' investigation of the paraboloid floating with its base exposed.

# 9 Comments

There seems to have been no immediate application of this work of Archimedes. Certainly the upside down case was hardly of practical significance. Using his theory he could easily have chosen a closer approximation to a ship's hull, namely a parabolic prism. After all, he had shown elsewhere that the centroid of a parabolic section lies $\frac{3}{5}$ of the way from its vertex to the midpoint of its base. The only differences between the two cases is the replacement of $\frac{2}{3}$ by $\frac{3}{5}$ and the fact that the area of the parabolic section is proportional to $PV^{3/2}$.

In order to translate Proposition 10 into modern terminology, we begin by taking the paraboloid to be obtained by rotating the parabola $y = x^2$. The two points at the ends of its base are $B = (-q, q^2)$ and $B' = (q, q^2)$. Hence $A = (0,0)$, $M = (0, q^2)$, and $f = \frac{1}{4}$. The $V$-curve in Proposition 10 has the equation $y = 2(x + q/2)^2 + q^2/2$. The $D$-curve is

$$y = \frac{5}{3}\left[x + \frac{2}{5}q\right]^2 + \frac{2}{5}q^2.$$

The distance $PV$ is $2(x + q/2)^2 + q^2/2 - x^2$.

Therefore the points $D_1$ and $D_2$ are determined by solving the equation

$$\frac{2}{3}q^2 - \frac{1}{2} = \frac{5}{3}\left(x + \frac{2q}{5}\right)^2 + \frac{2}{5}q^2.$$

Hence

$$x = -\frac{2}{5}q \pm \sqrt{\frac{4}{25}q^2 - \frac{3}{10}}.$$

For these values of $x$,

$$PV = \frac{1}{50}\left(26q^2 - 15 \pm 6q\sqrt{16q^2 - 30}\right).$$

Since $AM = q^2$ and $s = PV^2/AM^2$ it follows that

$$s = \frac{1}{2500}\left(26 - \frac{15}{AM} \pm 6\sqrt{16 - 30/AM}\right)^2.$$

Note that $s$ is defined when $AM$ is greater than $\frac{30}{16}$, which is $\frac{15}{2}f$. As $AM$ approaches infinity the two values of $s$ approach 1 and $\frac{1}{625}$. The graph of $s$ in terms of $AM$ is the right-hand curve in Figure 17.7. It has two asymptotes, the lines $y = 1$ and $y = \frac{1}{625}$. The second asymptote implies that when the section is very tall and its density is a certain number near $\frac{1}{625}$, it is in equilibrium when its base just touches the water.

The curve that plays a role in the first two propositions is $s = (AM - 3f)^2/AM^2$. Since we are taking $f$ to be $\frac{1}{4}$, this is $s = \left(1 - \frac{3}{4}/AM\right)^2$, the left curve in Figure 17.7, which is defined only when $AM$ is at least $3f$.

As we look back at his various proofs we may well marvel at Archimedes' ingenuity and his mastery of the parabola and paraboloid. If we also recall that he developed the theory of centroids single-handedly, without the aid of integral calculus, but axiomatically, we are even further impressed. As a sustained mathematical tour de force, this analysis of the equilibrium of a floating body far exceeds that which led to his determination of the volume and surface area of a ball or his upper and lower bounds on $\pi$. It is an impressive piece of pure mathematics inspired by a question about the real world.

In conclusion I wish to thank Anthony Barcellos, who prepared the illustrations using CoHort graphics and Don Chakerian, Jerry Alexanderson, and Tatiana Shubin for suggestions that substantially clarified the exposition.

## References

[1] Dijksterhuis, E. J., *Archimedes,* Princeton University Press, Princeton, 1987.

[2] Heath, T. L., *The Works of Archimedes,* Dover, New York, 1953.

[3] Stein, S., *Archimedes: What did he do besides cry eureka?,* Mathematical Association of America, Washington, DC, 1999.

# 18

# Mathematical Mapping from Mercator to the Millennium

Robert Osserman
*Mathematical Sciences Research Institute*

**Note:** The material presented here ranges from elementary descriptive material all the way to recently developed ideas in complex analysis. It is written throughout in a manner designed to convey the intuitive and geometric ideas behind the mathematics, so that readers may be able to get something out of even the parts that they cannot follow in detail. By way of specific background, section 1 uses only algebra and trigonometry plus the definition of a derivative. Section 2 adds elementary facts about linear algebra and $2 \times 2$ matrices. Section 3 deals with maps of the plane, while the remaining sections focus on functions of a complex variable. Readers should not feel discouraged if they find that later sections require more mathematical experience than they currently possess.

It came as something of a revelation, after years of working in and around the subject, to discover that the single, simple, intuitive notion of the *scale* of a map underlies an astonishingly wide swath of basic mathematics—from differential calculus and linear algebra to conformal and quasiconformal mapping and functions of a complex variable. Furthermore, the process of constructing maps with given properties based on scale leads directly to the integral calculus. In the particular case of the Mercator map, finding an explicit formula for its construction led to the formulation and solution of a problem in calculus, as so beautifully told in the article by Rickey and Tuchinsky on "An Application of Geography to Mathematics" [1980].

In view of these multiple connections, one might rightly suspect that the notion of the scale of a map, although indeed intuitive, is not actually all that simple. Our goals, then, will be

first: define exactly what is meant by the scale of a map, both in its simplest form and in its more refined senses,

second: describe a number of historically important geographical maps, many of which are defined in terms of certain scaling properties,

third: explain how a number of purely mathematical notions are related to the concept of scaling, and

fourth: review some of the major mathematical developments of the past 400 years where these mathematical notions are involved.

The term "mathematical mapping" in the title will be used in two ways. First, among geographical maps, some are of the freeform variety, giving the general "lay of the land" but not purporting to convey precise information about shapes and sizes. The maps that concern us have the feature that they are based on some specific mathematical principle. Oddly enough, the use of the word "map" in mathematics itself is extrapolated from the former, less "mathematical" kind of maps, which allow any method at all of assigning to each point of the original—say an area in the countryside—a point in our image: the "map" of that countryside, allowing some parts that interest us to be enlarged and others of less interest to be contracted or even shrunk to a point. The particular mathematical maps that we will deal with here will all be connected in some way to our central theme: scale.

## 1 Scale

We start with the simplest example of scaling: an architect's drawings showing, say, the floor plan or the front view of a house. The plans will always be drawn to a certain *scale*: the ratio of the size of any object shown in the drawing to the actual size of the object it represents.

Exactly the same idea is used for making maps of cities, states, or other geographical entities. There are three ways commonly used to indicate the scale of a map: arithmetical, geometrical, and verbal.

**Arithmetical:** The scale is indicated in the form 1:6,000,000, meaning that the ratio of distances on the map to distance on the ground is 1/6,000,000.

**Geometrical:**

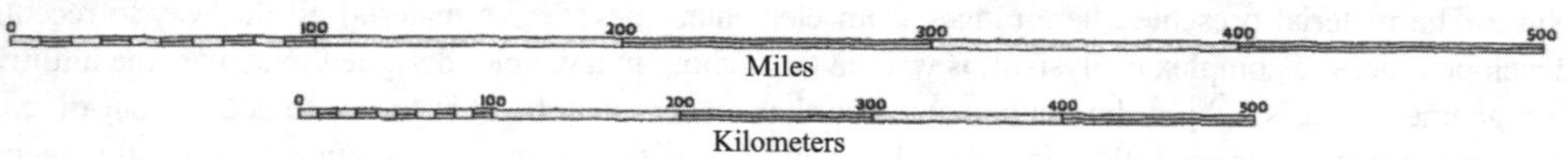


**Verbal:** 94.7 miles to the inch.

Of course, this last is an approximation. The exact scale is 6 million inches to an inch, but since a mile is 5,280 feet times 12 inches to the foot, we get immediately the rough estimate that 6 million inches is somewhat under a hundred miles. It goes without saying that the same scale is far more easily expressed verbally as 60 kilometers to the centimeter.

Whatever notation one uses, the meaning is the same: saying that the scale $s$ of a map is $1:n$ or $1/n$ means that if any two points on the map are a distance $d$ apart, then the corresponding points on the region being mapped are a distance $nd$ apart.

There is only one difficulty with this beautifully simple concept; a mathematical theorem states that, for the surface of the earth, no such map exists. In its simplest form, where we take the surface of the earth to be a sphere, the theorem goes back to Euler.

**Theorem 1 (Euler [1775])** *It is impossible to make an exact scale map of any part of a spherical surface.*

To risk stating the obvious, when we speak of a map in this context, we refer to a map drawn on a flat sheet of paper. One can always make an exact scale "map" of the earth in the form of a globe. What Euler's Theorem says is that if we draw on a flat piece of paper a map of some region on a spherical surface, there is bound to be some distortion. One sometimes sees the statement that "one can preserve the size or shape, but not both." In fact, Johann Lambert [1772 ] gave the first general mathematical treatment of cartography, defining precise versions of preserving "size" and "shape." He also gave many new constructions, including some of the most widely used maps since that time. However, the intuitive notions of preserving size or shape had long been known, together with the empirical fact that one could not do both; one could not construct an exact scale map.

Mercator, in constructing his famous map in 1569, opted for very good reasons to abandon size in favor of shape. We shall come back later to give exact definitions, but we start by explaining the intuition behind Mercator's map. The key idea is that we cannot have a fixed scale for the map, but we *can* have a fixed scale in certain directions.

To make these ideas more concrete, let us examine a particular class of maps known in cartography as "cylindrical projections." In order to define them, we first recall some standard terminology.

The **equator** is the great circle equidistant from the North and South Poles.

The **meridians** are the circular arcs joining the North and South Poles. They are perpendicular to the equator, and are the curves one traverses when traveling due north or south from any point.

The **parallels of latitude**, or **parallels** for short, are the circles perpendicular to the meridians. They are also the circles at fixed distance from the North or South Pole, and they are the curves one traverses when traveling due east or west from any point (other than the poles.)

A **cylindrical projection** is a map constructed as follows: The equator is represented by a horizontal line segment. The length of the segment determines *the scale of the map along the equator*; that is, if $L$ is the length of the equator, and $w$ the width of the map—the length of the horizontal segment representing the equator—then all distances along the equator are represented by the fixed factor $w/L$. The meridians are represented by vertical lines whose length may be either finite or infinite, and the parallels of latitude by horizontal line segments of the same fixed length $w$ as the equator. As a result, all cylindrical projections have the following properties:

**1.** The map is in the form of a rectangle or infinite vertical strip representing all of the earth except for the poles, with the two vertical sides corresponding to a single meridian, and every other meridian corresponding to a unique vertical line. (In the infinite case, this is of course the theoretical map, with the actual finite map cut off to represent the portion of the earth between two fixed parallels of latitude.)

**2.** The map has a fixed scale along each parallel of latitude. Namely, the quarter circle along a meridian from the equator to the North or South Pole is divided into ninety degrees, and the *latitude* of any point is the number of degrees along the meridian north or south of the equator. It follows that at latitude $\varphi$ north or south, the parallel is a circle of radius $R\cos\varphi$ where $R$ is the radius of the earth and the length of the equator is $L = 2\pi R$. Hence, at latitude $\varphi$, the parallel is mapped with fixed scale

$$\frac{w}{2\pi R\cos\varphi} = \frac{w}{L\cos\varphi} = s\sec\varphi,$$

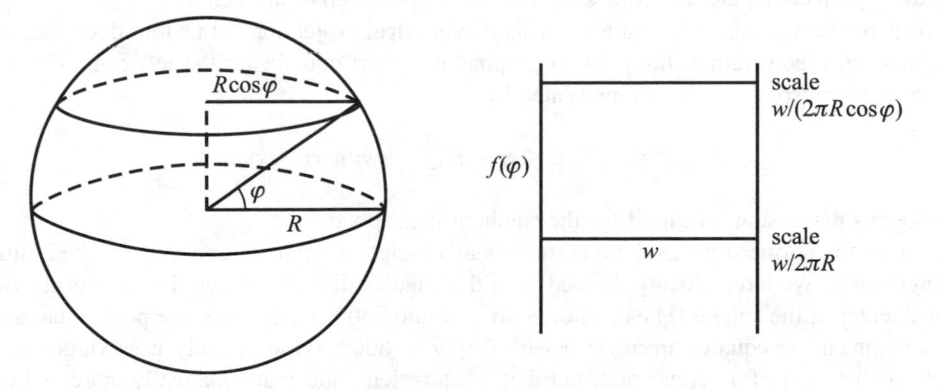


**Figure 18.1.** Cylindrical map

where $s = w/L$ is the scale of the map along the equator.

These two properties illustrate Euler's Theorem in action. The whole familiar class of maps in which "north" is the vertical direction on the map and east-west is horizontal, and that have a fixed scale along some east-west line, are of necessity a portion of a cylindrical projection and cannot have a fixed scale for the whole map. The reason such maps are able to indicate a "scale" is that the factor $\sec\varphi$ in the expression $s\sec\varphi$ for the scale will not vary enough over a small portion of the earth's surface to make any practical difference. (Other inaccuracies in the map are bound to be far greater.)

Among the so-called "cylindrical projections" is one that is a true "projection" in the mathematical sense. It can be defined geometrically as follows. Let $S$ be a globe: a sphere depicting the surface of the earth, and $C$ a circular cylinder tangent to $S$ along the equator. Project $S$ onto $C$ along rays from the center $O$ of $S$.

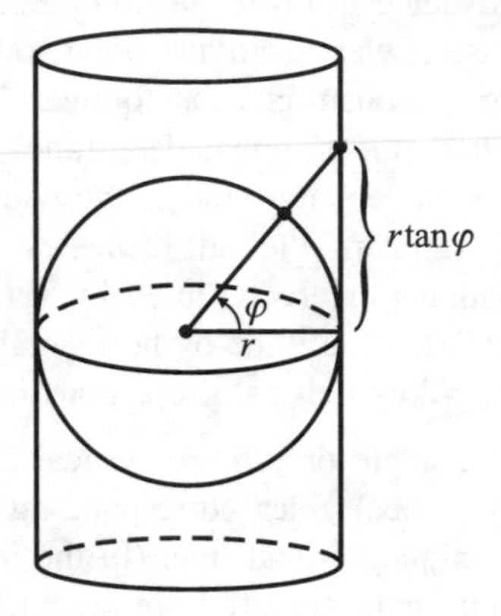


**Figure 18.2.** Central cylindrical projection

Each meridian on the sphere will clearly map onto a vertical line, and each parallel of latitude will map onto a circle on the cylinder parallel to the equator.

Cut the cylinder along a vertical line and unroll it onto a vertical strip in the plane. The result is a particular case of cylindrical projection known as a "true cylindrical projection" or "central cylindrical projection." When reading about map-making one must be careful to distinguish the way the word "projection" is used there, to mean any systematic form of representation, from its more narrow use in mathematics, or for that matter in art and in everyday parlance, where one thinks of a "projector" such as a slide projector or of a shadow projected on a wall. In cartography, these "true" projections are sometimes referred to as "perspective projections."

In order to give a precise formula for a general cylindrical projection, let us introduce rectangular coordinates with the origin at the point corresponding to the equator on the left edge of the map. The northern hemisphere will be represented by

$$0 \leq x \leq w, \quad 0 \leq y \leq H, \qquad \text{with } H \leq \infty.$$

An analogous discussion will hold for the southern hemisphere.

A point in the northern hemisphere is traditionally assigned a pair of coordinates: the latitude $\varphi$ and longitude $\theta$. We have already defined $\varphi$ as the angular distance above the equator as viewed from the center of the sphere. The equator is divided into 360° starting at some point. That assigns to each point on the equator an angle $\theta$ with $0 \leq \theta < 360°$. (One actually uses values of $\theta$ up to 180° east or west of a given point, but that is equivalent, and mathematically more awkward.) The *longitude* of any point is the value of $\theta$ where the meridian through the point hits the equator.

Thus, every cylindrical projection is given explicitly by the equations

$$x = \frac{w\theta}{360}, \quad y = f(\varphi) \tag{1}$$

for some monotonically increasing function $f$, with $f(0) = 0$, $f(90) = H$. For example, it follows immediately from the geometric definition of a true cylindrical projection that if we want a map of width $w$, we choose a globe of radius $r = w/2\pi$, and the equations become

$$x = \frac{w\theta}{360}, \quad y = \frac{w}{2\pi}\tan\varphi. \tag{2}$$

As we saw earlier, the scale of every such map along the parallel at latitude $\varphi$ is

$$s_\varphi = s\sec\varphi \tag{3}$$

where

$$s = \frac{w}{L} \tag{4}$$

is the scale along the equator.

We can now describe exactly what it was that Mercator was trying to do, and how he went about doing it.

Mercator wanted his map to have two properties: first, it should be a cylindrical projection so that at any point of the map, the vertical direction represents north, and second, the map should not distort shapes. Now it is clear intuitively that if a map has different scales in the horizontal and vertical directions, then shapes will be distorted. (Think of looking at yourself in a fun-house mirror.) Since Mercator knew the horizontal scaling factor at each latitude, either given trigonometrically as in equations (3), or else geometrically via the ratio of the two horizontal segments in Figure 18.3, he simply had to adjust the vertical scale accordingly.

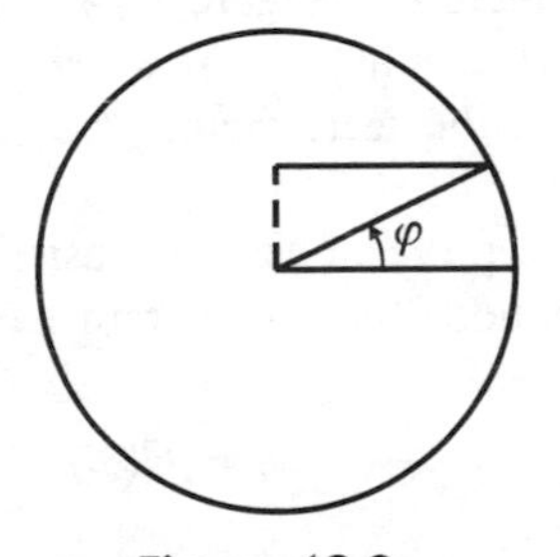


**Figure 18.3.**

Mercator did not divulge the exact procedure he used to construct his map, but it seems most likely that he proceeded somewhat as follows: divide up the region of the earth between two fixed latitudes into thin strips, each bounded by a pair of nearby parallels of latitude. The horizontal scale would be approximately constant in each strip, and one could use, for example, the exact value at the center parallel of the strip. Map that strip onto a horizontal strip in the plane using the same scale in the vertical direction. Stacking these strips one on top of another would give Mercator the result he was seeking.

Needless to say, this gives only an approximation to a "true" Mercator map, where the vertical scale exactly equals the horizontal scale at each point. However, all actual cartographic maps are only approximate. The real problem was that anyone else wanting to make a "Mercator map" of a part of the earth's surface would have to either copy the original or else repeat the whole tedious procedure. What was wanted was the actual function $f(\varphi)$ in the equations (1) that resulted in

Mercator's map; that is, the function $f(\varphi)$ for which horizontal and vertical scaling are everywhere equal. In order to find such a function, we have to make precise what we mean by the "scale" of a map in the vertical direction at a given point when that scale is constantly changing.

In the case of a cylindrical projection given by equations (1), an arc of a meridian is defined by an interval of latitude, $a < \varphi < b$, while the image of this arc on the map will be a vertical line segment $f(a) < y < f(b)$. Assuming the earth is a perfect sphere, the length of the arc will be $L(b-a)/360$, and the image on the map will have length $f(b) - f(a)$. The overall scale factor for this arc of the meridian is therefore

$$\frac{360}{L}\frac{f(b)-f(a)}{b-a}.$$

The value of this scale factor over smaller and smaller intervals of arc will be closer and closer to the exact scale factor at a point, leading us inevitably to the

**Definition.** The *vertical scale* of a cylindrical projection given by equation (1) at a point at latitude $\varphi = a$ is equal to

$$\lim_{b\to a}\frac{360}{L}\frac{f(b)-f(a)}{b-a}.$$

In other words, the notion of the *scale at a point* when the scale is continually changing is precisely the notion of a derivative:

$$f'(a)\cdot\frac{360}{L}.$$

The extra factor $360/L$ arises because we have measured distance along the meridian in degrees, rather than arc length, with one degree of latitude having length $L/360$.

In fact, given any monotonically increasing function $y = f(x)$, we may picture it either via a graph or as a map of an interval $I$ of the $x$-axis onto an interval $J$ of the $y$- axis. The two interpretations are connected via the picture in Figure 18.4.

*The scale factor of the map $I \to J$ at any point $p$ is exactly equal to the derivative $f'(p)$.* In particular, the map *shrinks* distances if the scale factor $f'(p)$ is less than 1, and *stretches* them if $f'(p) > 1$.

Returning to our case of cylindrical projections, it is easier to work with them mathematically if we express the latitude $\varphi$ and longitude $\theta$ in radians rather than degrees. The equations (1) then take the form

$$x = \frac{w\theta}{2\pi}, \quad y = F(\varphi), \tag{5}$$

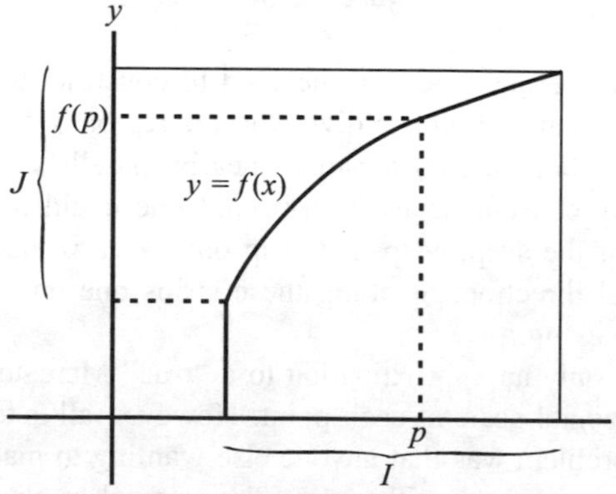

y
f(p)
J
y = f(x)
x
p
I

Figure 18.4.

where again, $w$ is the width of the map, and the scale along the equator is $s = w/L$, with $L$ the length of the equator. Since arc length along the meridian is given by $\varphi L/2\pi$, the vertical scale at latitude $\varphi$ is

$$s_v(\varphi) = F'(\varphi) \cdot \frac{2\pi}{L}. \tag{6}$$

As we saw earlier in equations (3) and (4), the horizontal scale at latitude $\varphi$ is

$$s_h(\varphi) = s \sec\varphi, \quad s = \frac{w}{L}. \tag{7}$$

Mercator's goal was to have

$$s_v(\varphi) = s_h(\varphi) \qquad \text{for all } \varphi,$$

which reduces to

$$F'(\varphi) = \frac{w}{2\pi} \sec\varphi. \tag{8}$$

The procedure we have outlined that Mercator presumably used in constructing his map was precisely a numerical integration of this equation. That was made explicit by English mathematician Edward Wright who used the method to construct a set of tables [1610] that would allow anyone to draw a much more accurate "Mercator" map than Mercator himself was able to do.

We now know the exact solution to equation (8). With $F(0) = 0$, it is

$$F(\varphi) = \frac{w}{2\pi} \log(\sec\varphi + \tan\varphi)$$

which Mercator could not possibly have known, since logarithms had yet to be invented, to say nothing of the derivatives and integrals used to derive the equation. Again we refer to the article by Rickey and Tuchinsky [1980] for the many steps leading to this result.

In general, constructing a map in which the pointwise vertical scaling is given in advance amounts precisely, according to equation (6), to carrying out an integration. In other words, *the two fundamental operations, differentiation and integration, correspond precisely to determining the scale of a variable scale map and constructing a map when given the (variable) scale.*

As another example, if our goal was to preserve size rather than shape, then instead of having the horizontal and vertical scaling factors be equal, we would make them reciprocal, so that the stretching in one direction would match the shrinking in the other. Going back to equations (6) and (7), we see that we need to make

$$F'(\varphi) \sec\varphi \equiv c, \quad \text{a constant}$$

or

$$F'(\varphi) = c \cos\varphi$$

so that

$$F(\varphi) = c \sin\varphi.$$

We can choose the constant $c$ to make the horizontal and vertical scales equal at any given latitude and hence have a very good map near that latitude. For example, if we let $c = w/2\pi$, then by equations (6) and (7), the vertical and horizontal scales will be equal at the equator, and we get one of Lambert's maps: the cylindrical equal-area map (see Figure 18.5).

As a final example of a cylindrical projection, we note that one of the most simple-minded of all goes back to antiquity and was commonly used in the 16th century under the name "plate carrée." It uses a fixed scale along each meridian—the same scale as that along the equator. It has the advantage that the distance between any pair of points on the same meridian can be read off directly from the map. Also, even though it preserves neither shape nor size, it does not have some

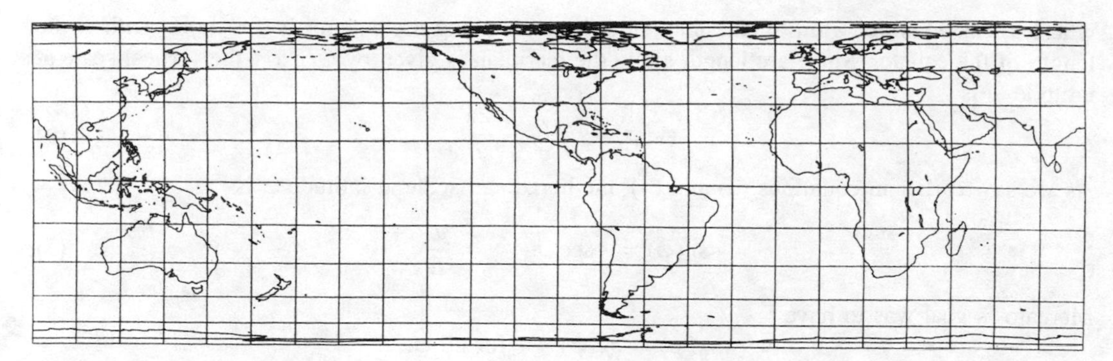

**Figure 18.5.** Lambert Cylindrical Equal-Area projection with shorelines, 15° graticule. Standard parallel 0°. Central meridian 90° W.

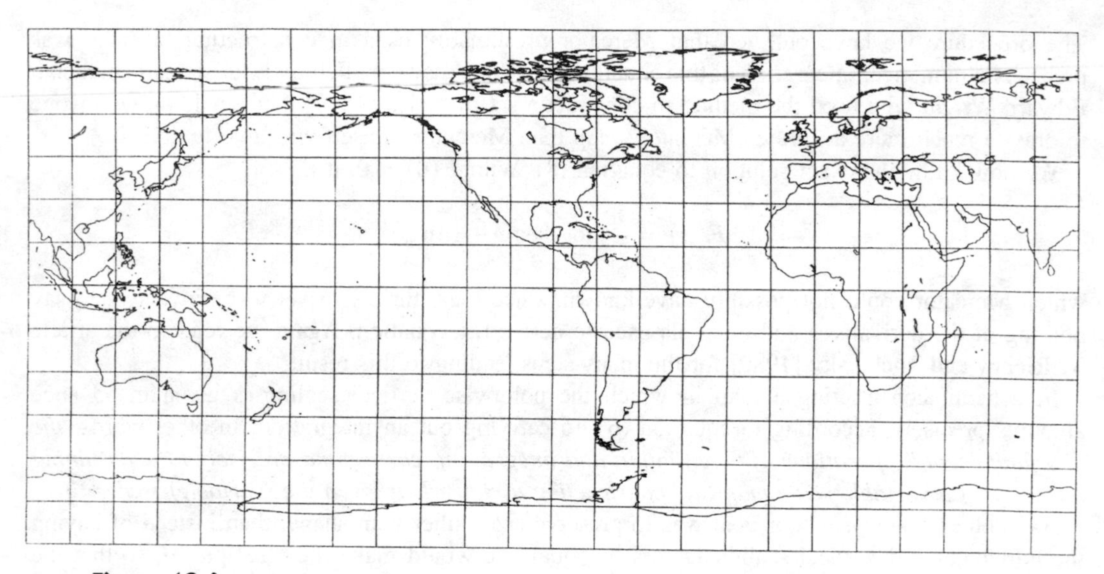

**Figure 18.6.** Plate Carrée projection with shorelines, 15° graticule. Central meridian 90° W.

of the extreme distortions of the Mercator or Lambert equal-area maps, but serves as a kind of compromise between the two (see Figure 18.6).

The equations for the plate carrée could not be simpler:

$$x = c\theta, \quad y = c\varphi.$$

## 2 Maps of the plane

Suppose we use any of the cylindrical projections described above to make a map of Australia. If we compose that map with a map of the plane into the plane, we then get a new map of Australia. Conversely, if we make any two different maps of Australia, then they are related to each other by a map of the plane into the plane. Our goal, then, will be to look more closely at maps of the plane into the plane, with our focus again on questions related to scale. That will also allow us to make precise the intuitive notions of preserving shape and preserving size.

We start with the simultaneously simplest and most important case: that of linear maps. We let $T$ be a linear transformation of the $x, y$-plane into the $u, v$-plane, which we may write in the form

$$u = ax + by \qquad (9)$$
$$v = cx + dy$$

where $a$, $b$, $c$, $d$ are fixed real numbers. An arbitrary line through the origin in the $x, y$-plane may be given parametrically by expressing $x$ and $y$ as constant multiples of a parameter $t$. Substituting these expressions in equations (9) gives $u$ and $v$ as constant multiples of $t$, hence defines a line through the origin in the $u, v$-plane, the image of the original line under the transformation $T$. This mapping of a line into a line will have a fixed scale $s$, which represents the ratio of the distances between any two points on the image line and the distances of their preimages. In general we will have $s > 0$, but we may have the degenerate case $s = 0$ when the entire first line maps onto the origin. As we rotate the original line around the origin, the scale $s$ will attain a maximum $s_1$ and a minimum $s_2$ with $0 \leq s_2 \leq s_1$. A basic result of linear algebra is the following.

**Lemma 1** *One can choose new axes $X$, $Y$ in the $x, y$-plane and $U$, $V$ in the $u, v$-plane such that the transformation $T$ takes the form*

$$U = s_1 X, \quad V = s_2 Y. \qquad (10)$$

Said differently, the matrix

$$A = \begin{pmatrix} a & b \\ c & d \end{pmatrix}$$

of the linear transformation $T$ can be reduced to a diagonal matrix by pre- and post-multiplication by orthogonal matrices. We also have

$$|\det A| = |ad - bc| = s_1 s_2. \qquad (11)$$

In particular, the scale factor is always positive whenever $A$ is nonsingular.

To be a bit more specific, we can always choose new axes in the $x, y$-plane by a rotation of the axes, and if $\det A > 0$, then we may do the same in the $u, v$-plane. If $\det A < 0$, then we must make an additional reflection in order to put the transformation in the form (10), with the scale factors both positive.

One immediate consequence of equations (10) is that distances get scaled by the factors $s_1$ and $s_2$ in two orthogonal directions, giving us the result:

**Corollary 1.1** *Under the transformation $T$, all areas are multiplied by the factor $s_1 s_2$. In particular, areas are preserved under $T$ if and only if $s_1 s_2 = 1$.*

A second consequence follows by noting that a line making an angle $\alpha$ with the $X$-axis is given parametrically by $X = t\cos\alpha$, $Y = t\sin\alpha$, while its image under $T$ has the equations $U = ts_1\cos\alpha$, $V = ts_2\sin\alpha$ and makes an angle $\beta$ with the $U$-axis, where $\tan\beta = (s_2/s_1)\tan\alpha$. We therefore conclude:

**Corollary 1.2** *Angles are preserved under $T$ if and only if $s_1 = s_2$.*

The unit circle $C$ is given parametrically by $X = \cos t$, $Y = \sin t$, and its image under $T$ is given by $U = s_1\cos t$, $V = s_2\sin t$, which is a circle if $s_1 = s_2$ and an ellipse with major and minor axes along the $U$ and $V$ axes and area $\pi s_1 s_2$ if $s_1 > s_2$.

That leads us to look at two separate cases.

**Case 1:** $s_1 > s_2$. The image of the unit circle under the transformation $T$ is an ellipse whose major and minor axes determine the $U$ and $V$ axes respectively. The pre-images of the $U$ and $V$ axes are the $X$ and $Y$ axes: the directions of maximum and minimum scaling. The key properties of the mapping are therefore made graphically clear by drawing the unit circle in the $x, y$-plane with the directions of the $X$ and $Y$ axes indicated, and the image ellipse in the $u, v$-plane. *The size of the ellipse shows the area distortion and the eccentricity of the ellipse indicates the shape distortion.*

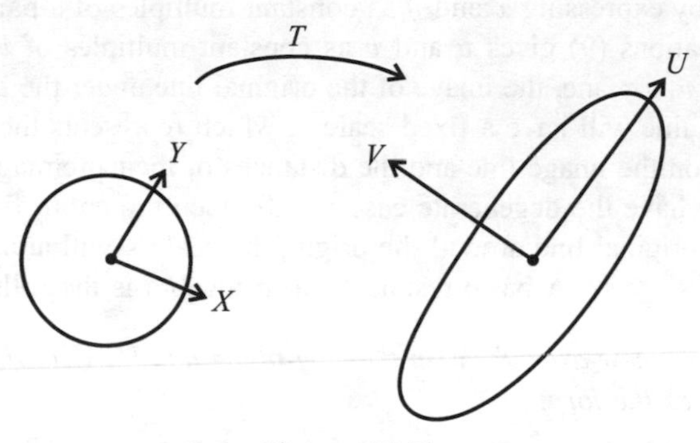


**Figure 18.7.**

Precisely this device is used by cartographers to give map-viewers an instantaneous overview of the nature of the distortion for a given map. For example, Figures 18.8 and 18.9 are the pictures for the plate carrée and the Lambert equal-area map shown above.

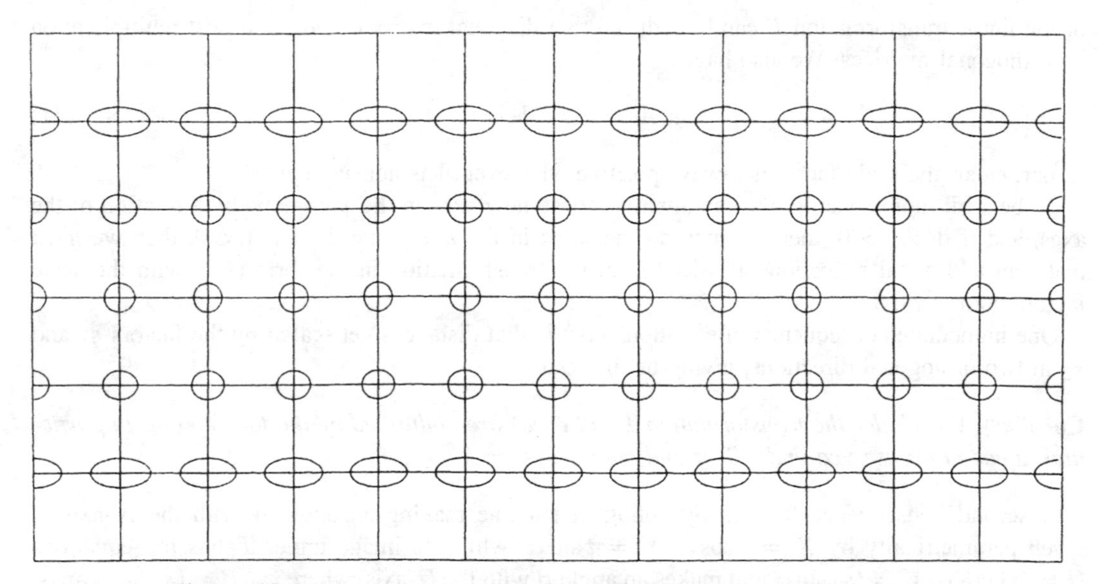

**Figure 18.8.** Plate Carrée projection with Tissot indicatrices, 30° graticule.

Notice that in both cases there are circles along the equator, indicating no shape distortion —that is, equal scaling in the horizontal and vertical directions. Also in both cases, at all points off the equator the ellipses have major axes in the horizontal direction, which is therefore the direction of maximum stretching, but in the case of the plate carrée the ellipses grow larger and larger toward the poles, indicating area distortion also, whereas in the equal-area map, the ellipses all have the same area as the circles along the equator. The ellipses used in these distortion diagrams are called

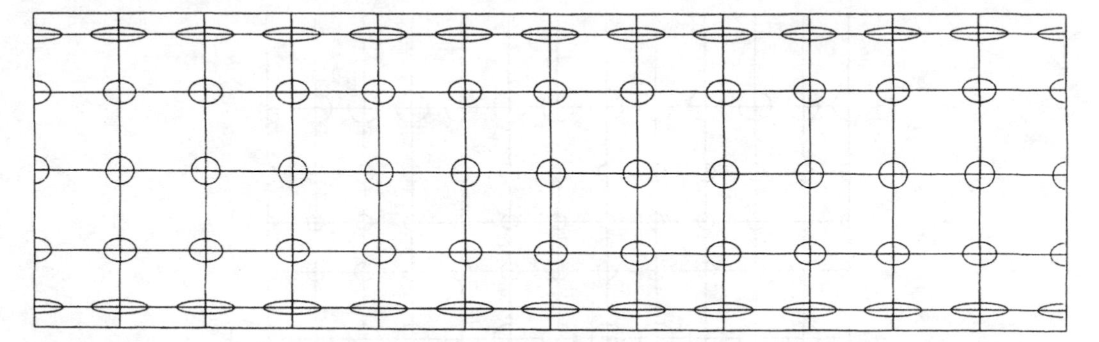

**Figure 18.9.** Lambert Cylindrical Equal-Area projection with Tissot indicatrices. 30° graticule. Standard parallel 0°. All ellipses have the same area, but shapes vary.

"Tissot indicatrices." At any point of a map, the Tissot indicatrix shows the directions and relative amounts of maximum and minimum scaling.

In contrast to the figures above, the Tissot indicatrices show that for the central cylindrical projection (Figure 18.10) there is distortion in the vertical direction, whereas for the Mercator projection (Figure 18.11), there is less size distortion and no distortion of shape: each Tissot indicatrix is a circle.

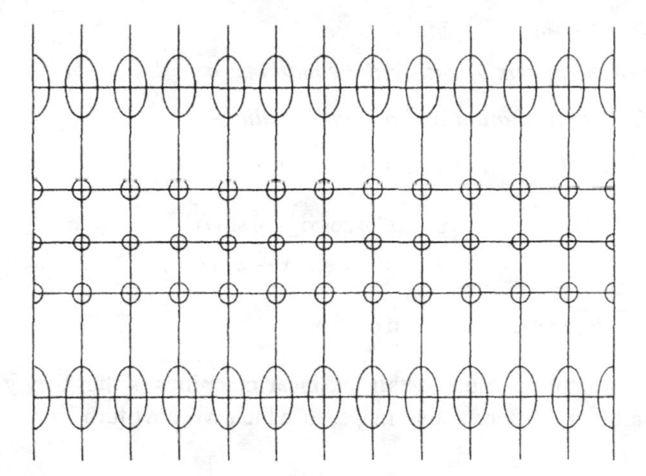

**Figure 18.10.** Central Cylindrical projection with Tissot indicatrices, 30° graticule.

**Case 2:** $s_1 = s_2$.

**Definition.** A linear transformation is called a *simple scaling* or a *homothety* if it is of the form $u = sx$, $v = sy$.

A linear transformation is said to "preserve shape" if it is a *similarity transformation*; that is, it maps every triangle onto a similar triangle.

From the above discussion, together with a bit more argumentation, we can give an extended characterization of these transformations.

**Proposition 1** *Let $T$ be a linear transformation (9) with matrix $A$, and assume that $T$ is orientation-preserving; that is,* $\det A > 0$. *(If $T$ is orientation reversing:* $\det A < 0$, *then we may apply the statements below to the transformation consisting of $T$ followed by a reflection—for example replacing $v$ by $-v$.) The following are equivalent:*

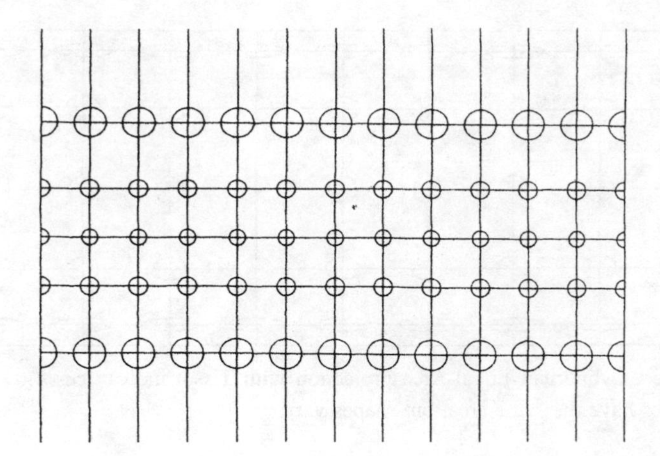

**Figure 18.11.** Mercator projection with Tissot indicatrices, 30° graticule. All indicatrices are circular (indicating conformality), but areas vary.

(a) *$T$ is a similarity transformation,*

(b) *all distances are multiplied by a fixed factor $s$,*

(c) *all angles are preserved,*

(d) *the coefficients of the matrix $A$ satisfy the equations: $a = d$, $b = -c$,*

(e) *$T$ is a composition of a rotation and a simple scaling,*

(f) *the equations* (9) *for $T$ can be written in the form*

$$u = s(x\cos\alpha + y\sin\alpha)$$
$$v = s(-x\sin\alpha + y\cos\alpha)$$

*where $s = \sqrt{a^2+b^2}$, $\cos\alpha = a/s$, $\sin\alpha = b/s$.*

The reason for going into this much detail on linear mappings is that they govern the behavior near each point for arbitrary differentiable maps, to which we now turn.

## 3 Smooth maps

Let $F$ be a continuously differentiable map of the $x, y$-plane into the $u, v$-plane. The differential $dF$ of $F$ at a point $P$ is the linear transformation $T$ whose matrix consists of the partial derivatives of $F$ at $P$; that is,

$$a = u_x(P),\ b = u_y(P),\ c = v_x(P),\ d = v_y(P). \tag{12}$$

There are two common interpretations of the differential. One is that it is the best linear approximation to the map $F$ in a neighborhood of $P$. The other is that it is the *tangent map* to $F$; that is, if $C$ is any smooth curve through $P$, then the tangent vector to the image of $C$ under $F$ at the point $F(P)$ depends only on the tangent vector to $C$ at $P$, and this induces a linear map of tangent vectors at $P$ to tangent vectors at $F(P)$, which is precisely the differential $dF$ at $P$. Since the angle between a pair of smooth curves intersecting at $P$ is by definition the angle between their tangent vectors, it follows that the map $F$ preserves angles at $P$ if and only if $dF$ is a similarity transformation.

**Definition.** A map $F$ is a *conformal map* if it is a diffeomorphism that preserves angles at every point; that is, $F$ is a one-to-one continuously differentiable map with a differentiable inverse, and $dF$ is a similarity transformation at every point.

We shall come back in the next section to a more detailed look at conformal maps in the plane. We note here that the cylindrical projections described in section 1 are all maps of the sphere into the plane, and the differential of any such map can be defined exactly as in the case of plane maps as maps of tangent vectors of curves on the sphere at a point to tangent vectors of their image curves. From Proposition 1 it follows that the angle between any two curves at a point on the sphere equals the angles between their image curves if and only if the maximum and minimum scaling factors are equal; that is, the scaling factor is the same in all directions at the point. But the way we constructed Mercator's map was precisely from that property. We therefore conclude:

**Proposition 2** *Mercator's map is the unique cylindrical projection that preserves angles.*

It follows that Mercator's map is the unique map with the two key properties

(i) the vertical direction on the map corresponds to the north/south direction,

(ii) given any two points on the map corresponding to a pair of locations on the earth, if the straight line joining them on the map makes a given angle with the vertical, then starting at the first location and following the fixed compass direction determined by that angle will lead to the second location.

It was this second property that made Mercator's map indispensable for navigation for a very long time.

It is the combination of angle-preserving and the fixed vertical direction for north that gives the second property above. If one does not specify vertical for north, then there are many other possibilities for angle-preserving maps. Indeed, one of them goes back to antiquity. It is called *stereographic projection.* It is a true projection, in which the sphere is projected from some point on it—often chosen to be the North Pole—onto either the plane tangent to the sphere at the antipodal point—say the South Pole—or else the plane parallel to that one through the center of the sphere; which of those two planes one projects onto is immaterial, since the resulting maps will differ by a similarity transformation between the two planes.

For stereographic projections from the North or South Pole, one has the property that the parallels of latitude map onto concentric circles about the origin, and the meridians map onto rays extending outward from the origin. The fact that stereographic projection preserves angles appears to have first been pointed out and proved by Edmund Halley, of comet fame, in 1695.

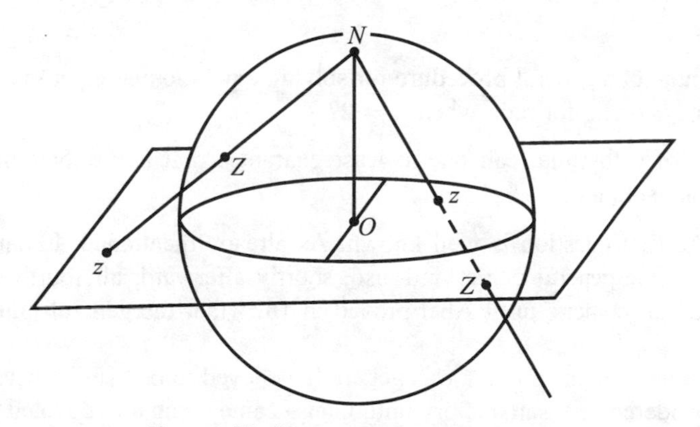


**Figure 18.12.** Stereographic projection

Our principal interest will be in conformal maps of the plane into the plane. By virtue of condition (d) in Proposition 1 and equations (12), we have the following elementary but fundamental result.

**Proposition 3** *A smooth map $F$ defined in a domain $D$ in the $x, y$-plane and mapping $D$ into the $u, v$-plane preserves angles if and only if $dF$ is nonsingular everywhere and satisfies the equations*

$$u_x = v_y, \quad u_y = -v_x. \tag{13}$$

These equations are called the *Cauchy-Riemann equations.* They are simply the statement that the tangent map to $F$ at each point is a similarity transformation. In view of condition (e) of Proposition 1, the geometric content of the Cauchy-Riemann equations is that near each point, the map $F$ behaves like a rotation composed with a simple scaling. This is an important picture to keep in mind in the following sections.

## 4 The complex plane

Solving algebraic equations is one of the oldest and most fundamental problems in mathematics. A particular case of importance is that of polynomial equations, such as

$$x^2 = 1, \quad x^2 + 1 = 0, \quad x^3 = 15x + 4, \quad x^5 = 1.$$

The first of these has the two obvious solutions, $x = 1$ and $x = -1$. The second has no real solutions, but has two solutions if one introduces the imaginary number $i = \sqrt{-1}$, namely, $x = i$ and $x = -i$. The third is of interest because it has three real solutions, but if one uses the general formula for solving cubics developed by a series of Italian mathematicians in the 16th century (see chapter 1 of Nahin [1998] for an excellent review of this subject) then one finds that the expressions for two of these real roots involve imaginary numbers too. This led to the introduction of complex numbers: expressions of the form $a + bi$ where $a$ and $b$ are real numbers. We denote $|a + bi| = \sqrt{a^2 + b^2}$. The last equation above has the one real root $x = 1$ and four complex roots.

A number of facts and a number of questions soon emerged regarding general polynomial equations—that is, an equation with a finite number of terms, each consisting of a constant times a power of the unknown $x$. The *degree* of the equation is the highest power of the unknown that occurs. The facts were

1. An equation of degree $n$ can have at most $n$ solutions.
2. Solutions may be real or complex or some of both.

The questions were

1. Is there a formula, or a general procedure for solving a polynomial equation of degree $n$, such as the familiar quadratic formula when $n = 2$?
2. In the absence of a formula, can one at least guarantee that any polynomial equation does have at least one solution?

The answer to the first question is well known. As already mentioned, Italian mathematicians found the solution of the general cubic, and also, shortly afterward, all fourth degree equations. Degree five stumped all comers, until Abel proved in 1824 that the general quintic equation had no such solution.

As to the second question, the answer was generally believed to be "yes" but various attempts at proofs were not considered very satisfactory until Gauss came along and devoted his PhD thesis of 1799 to the subject. According to Gauss' own description in a letter, about a third of the thesis is devoted to a proof of the theorem that every polynomial can be written as a product of linear and quadratic factors, while the rest is devoted to history and criticisms of previous "proofs" including those of d'Alembert, Euler, and Lagrange.

Gauss returned repeatedly throughout his life to this question, providing a number of different proofs. The result became known as "the fundamental theorem of algebra." In fact, he returned to it in 1849, giving a variant of his first proof, but making more explicit use of the complex plane.

The idea of representing complex numbers by points in the plane appears to have occurred to several people independently towards the end of the 18th century. One gets a geometric picture of the purely algebraic (and abstract) entities—complex numbers—by associating to each number $a+bi$ the point $(a,b)$ in the plane. Or viewed the other way around, the "complex plane" is simply the ordinary Euclidean plane, where to each point $(a,b)$ one assigns the complex "coordinate" $a+ib$. Finding real roots of a real polynomial equation—that is, real values of $x$ such that $P(x)=0$, where $P(x)$ is a polynomial with real numbers as coefficients—can be pictured geometrically as finding a point where the graph of the equation $y=P(x)$ crosses the $x$-axis. How does one picture geometrically a complex root of the equation? The answer, not surprisingly, is by means of mappings.

Let $x$ and $y$ be real variables, and let $z=x+iy$. Then a polynomial $P$ assigns to every complex number $z$ another complex number $w=P(z)$, and so $P$ defines a map from the complex $z$-plane to the complex $w$-plane. Equivalently, letting $w=u+iv$, $P$ defines a smooth map from the $x,y$-plane to the $u,v$-plane. There is always the trivial case to consider—a polynomial of degree zero, which has only a constant term, and does not actually depend on $z$. Considered as a map, such a polynomial maps the whole $z$-plane onto a single point—the value of the constant term. What the fundamental theorem of algebra states is that *for every polynomial $P$ of degree $>0$, $P(z)$ maps the $z$-plane onto the entire $w$-plane.* This says that for any complex number $c$, the equation $P(z)=c$ has a (complex) solution. It is obvious, but worth stating, that the solvability for every polynomial of degree $n>0$ for every value of $c$ is completely equivalent to solving $P(z)=0$ for every such polynomial, since we can just transfer the value $c$ to the left side of the equation.

This picture of a polynomial mapping the $z$-plane onto the $w$-plane became the impetus for 200 years of further developments, some of them spectacular, that will be the subject of the remainder of our discussion.

## 5 Analytic functions

An obvious next step up from polynomials, which have a finite number of terms, is to a kind of "infinite polynomial": $\sum_{j=0}^{\infty} c_j z^j$. Such an infinite sum will define a unique complex number providing it converges, which is equivalent to saying that the infinite sums of the real and imaginary parts of each term converge. In general, any such power series will converge for all $z$ satisfying $|z|<R$ for some $R>0$, and fail to converge for $|z|>R$, in which case $R$ is called the *radius of convergence* of the series. There are also two extreme cases: the series may not converge for any $z\neq 0$, in which case one says $R=0$, or it may converge for every value of $z$, so that $R=\infty$. In this last case, the infinite series will define a function $w=F(z)$ that can again be pictured as a map $F$ of the $z$-plane into the $w$-plane. Such functions are called *entire functions.* We shall return to them later for a closer look. The fundamental result at the heart of the subject is the following:

**Proposition 4** *Let $F$ be a smooth map of a domain $D$ in the $x,y$-plane into the $u,v$-plane. Let $w=f(z)$ be the complex function defined by the map $F$, where $z=x+iy$ and $w=u+iv$. Then the following are equivalent:*

(a) *$u$ and $v$ satisfy the Cauchy-Riemann equations* (13) *at every point of $D$,*

(b) *$f(z)$ has a complex derivative at every point of $D$,*

(c) *for every point c of D, the function f can be written as a power series in* $z - c$ *with a radius of convergence* $R > 0$.

**Definition.** The complex function $f(z)$ is said to be *analytic* in a domain $D$ if any (hence all) of conditions (a), (b), or (c) holds.

The equivalence of three so diverse-appearing conditions has no analog in the theory of real functions, and is a signal that one can expect complex analytic functions to enjoy many special and sometimes surprising properties. The first of those is that by Proposition 3 complex analytic functions define angle-preserving maps at all points where the differential is not zero. But the complex derivative of a complex function $f(z)$ is given in terms of the real and imaginary parts by $f'(z) = u_x + iv_x$, so that combined with the Cauchy-Riemann equations, $f' = 0$ at a point $z$ if and only if all partial derivatives of $u$ and $v$ with respect to $x$ and $y$ are zero at that point. One consequence of condition (c) above is that if an analytic function is not constant, then its derivative can vanish only at isolated points; everywhere else, it defines a conformal mapping.

One of the most important analytic functions is the *exponential function* define by

$$\exp z = 1 + z + \frac{z^2}{2} + \frac{z^3}{3!} + \cdots + \frac{z^k}{k!} + \cdots$$

which has the following properties:

1) the series converges for all $z$, so that $\exp z$ is an entire function
2) $\exp 1 = e$
3) $\exp x = e^x$ for $x$ real
4) $\exp iy = \cos y + i \sin y$ for $y$ real
5) $\exp(a + b) = \exp a \exp b$
6) $\exp(z + 2\pi ni) = \exp z$ for every integer $n$
7) $|\exp z| = e^x \neq 0$ for all $z$.

One form of the fundamental theorem of algebra is that if $P(z)$ is a polynomial of degree $n$, and $c$ any complex number, then the polynomial $P(z) - c$ is also of degree $n$ and can be written as the product of $n$ linear factors, each of which contributes one solution to the equation $P(z) = c$, and some of which may be equal. Hence, there are at most $n$ solutions to the equation $P(z) = c$ for any value of $c$. Furthermore, there are "in general" exactly $n$ distinct solutions, the exceptions being those (at most $n - 1$) values $c$ of the form $P(a)$ where $P'(a) = 0$. Looked at in terms of mappings, a complex polynomial $P$ of degree $n$ defines a mapping of the complex $z$-plane to the complex $w$-plane with the property that for every point in the $w$-plane, its inverse image consists of exactly $n$ points, with at most a finite number (in fact, at most $n - 1$) exceptions where the inverse image consists of fewer than $n$ points.

If we think of entire functions as "polynomials of infinite degree" we might expect something analogous to be true. We shall examine that question more closely in the following section, but it is instructive to examine the case of the exponential function in more detail. We see immediately one important difference. By virtue of property 7 above, the equation $\exp z = 0$ has *no* solutions. For any complex number $c \neq 0$, to find all solutions of the equation $\exp z = c$, we write $c$ in polar form as $c = r(\cos\theta + i \sin\theta)$, where $r = |c|$. Combining properties 3, 4, and 5 of the exponential function, we find that $\exp z = \exp(x + iy) = e^x(\cos y + i \sin y)$ so that

$$\begin{aligned} \exp z = c &\iff e^x = r \text{ and } y = \theta + 2\pi ni \text{ for an arbitrary integer } n \\ &\iff x = \log r \text{ and } y = \theta + 2\pi ni \text{ for an arbitrary integer } n. \end{aligned} \tag{14}$$

In other words, *the equation* $\exp z = c$ *has an infinite number of solutions for every* $c \neq 0$, *and no solutions for* $c = 0$.

It is clear from equation (14) that the way to visualize the function $w = \exp z$ as a map is to use rectangular coordinates $x, y$ in the $z$-plane and polar coordinates $r, \theta$ in the $w$-plane. We note first that the image of the entire $x$-axis is the positive $u$-axis or the ray $\theta = 0$, while the mapping itself is given by the ordinary exponential function $u = e^x$. Each horizontal line $y = b$ maps in the same way, by $r = e^x$ onto the ray $\theta = b$. We should picture the effect dynamically, as the horizontal line in the $z$-plane moves upward from $y = 0$ to $y = 2\pi$, the image ray rotates once around from $\theta = 0$ to $\theta = 2\pi$. The effect is that the infinite horizontal strip $0 \leq y < 2\pi$ maps onto the whole $w$-plane minus the origin. The same process is repeated for $2\pi \leq y < 4\pi$ and so on, with each horizontal strip of width $2\pi$ mapping onto the whole plane minus the origin. Said differently, as the horizontal lines in the $z$-plane sweep out the plane, the image rays rotate around and around infinitely often.

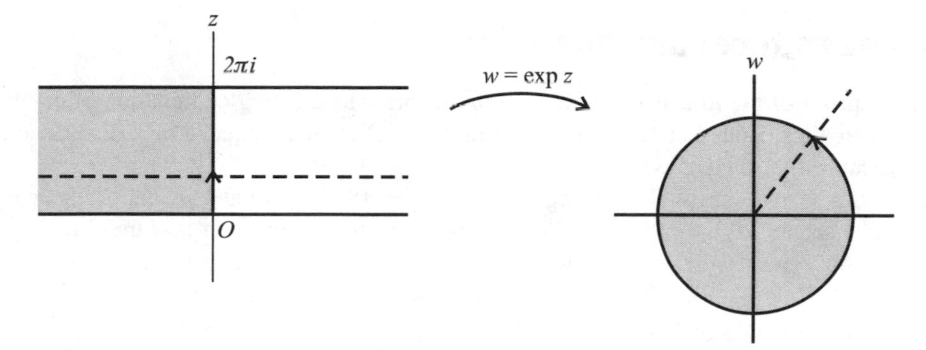

**Figure 18.13.**

If we restrict to a single point moving vertically, tracing out the line $x = a$, say, the image will trace out the circle of radius $e^a$; in particular, the $y$-axis maps onto the unit circle, the left half-plane maps onto the interior of the unit circle except for the origin, and the right half-plane maps onto the exterior of the unit circle.

Next, we may view equations (14) in the reverse direction, as defining a map of the whole $w$-plane minus the origin onto the horizontal strip $0 \leq y < 2\pi$ in the $z$-plane. The map is given explicitly in terms of polar coordinates in the $w$-plane, by

$$x = \log r, \quad y = \theta \tag{15}$$

and is called the complex logarithm, written $z = \log w$. It maps rays emanating from the origin onto horizontal lines, and circles centered at the origin onto vertical line segments.

The application of all this to geographic maps is gradually becoming better known, but not nearly as well-known as it should be. Namely, if we map the globe by stereographic projection from the North Pole onto the plane, the South Pole maps onto the origin, the meridians map onto rays emanating from the origin, and the parallels of latitude onto circles centered at the origin. If we compose this map with the complex logarithm defined by (15), then the meridians map onto horizontal lines and the parallels onto vertical line segments. Furthermore, both stereographic projection and the complex logarithm preserve angles, hence so does the composition. If we now follow by a rotation of 90° in the positive direction, then meridians map onto vertical lines, and parallels onto horizontal line segments, so that we have an example of a cylindrical projection. But it is also a conformal map, and as we have seen, that determines it uniquely—it must be our good old standby, Mercator's projection. Summing up, *to give explicit equations for the Mercator map, we simply make a stereographic projection and follow it by the complex logarithm, (15), and rotation through* 90°.

One final note on the complex logarithm. What we have defined is actually just a branch of the logarithm, defined by restricting $\theta$ to the interval $0 \leq \theta < 2\pi$. However, if we allow all possible values of $\theta$ in the polar representation of a complex number $w$, then equations (15) define the general complex logarithm as a multiple-valued function whose different values all differ by integer multiples of $2\pi i$, just as the inverse trigonometric functions are multiple-valued, with values differing by multiples of $\pi$ or $2\pi$. However, that is a whole other story that we need not go into here.

For the geometric view of the complex exponential and logarithm, and of complex functions in general, with many beautiful and instructive pictures, we recommend the book *Visual Complex Analysis* [1997] by Tristan Needham.

## 6 Nineteenth century highlights

Gauss' 1799 proof of the fundamental theorem of algebra was a fitting culmination of eighteenth-century mathematics. Both Gauss and his theorem went off in unanticipated new directions as the nineteenth century got underway.

To start with Gauss, he published two papers in the 1820's devoted to questions regarding mappings. The second of the two, from 1827, is the more fundamental. It lays the foundation for much of the subsequent work in the field of differential geometry. The most famous result in the paper is known as Gauss' *Theorema Egregium* or "most excellent theorem." One corollary of the theorem is a far-reaching generalization of Euler's theorem that there is no exact scale map of any region of a sphere onto a plane. What Gauss proved was that the sphere in this theorem is the rule rather than the exception. Namely, if $S$ is *any* surface, with a very few exceptions, then *there is no exact scale map of any region of S onto the plane.* The exceptions are the so-called "developable" surfaces, which may be obtained by simply rolling up a sheet of paper in various forms; for instance, cylinders and cones and a class of surfaces known as "tangent developables."

The other of Gauss' papers, from 1822, contains a positive result about mapping. It says that *any sufficiently smooth surface has a mapping into the plane that is locally conformal.* Said differently, all small regions on the surface can be represented by a plane map that preserves angles. Of course, if the surface is a sphere, then Mercator's map and stereographic projection are examples, but Gauss' theorem states that conformal maps exist "in general."

Gauss' disciple and successor Bernhard Riemann also made two major contributions of relevance to us. The first is known as the *Riemann Mapping Theorem.* It was a complete departure from earlier work, in that instead of asserting that there was *some* conformal mapping in a given situation, it said that you could actually prescribe the shape of the image. So for example, if you took any two plane domains each bounded by a simple closed curve—one might be an ellipse, and the other a rectangle—then Riemann's theorem states that there exists a one-to-one angle-preserving map between the two domains. The way the theorem is usually stated, one of the two domains is a circular disk, which is sufficient, since if you can map each of the two domains conformally onto a circular disk, then you can by composition map them conformally onto each other.

Riemann stated the theorem in his PhD thesis of 1851 and gave what he thought was a proof, but there turned out to be a gap in his reasoning. It took much of the remainder of the century to provide a complete proof and also to find ways to construct explicit mappings for simple cases, such as the ellipse and a rectangle. A key figure in that work was H. A. Schwarz [1869–70].

We will come back to the second of Riemann's contributions shortly, but first jump ahead to one of the biggest surprises of the century in the theory of complex functions. It was proved in

1879 by the young French mathematician, Émile Picard. It may be viewed as the direct successor, 80 years on, to the fundamental theorem of algebra.

**Theorem 2 (Picard's Theorem [1879])** *Let $f(z)$ be a nonconstant entire function. Then the equation $f(z) = c$ has a solution for every complex number $c$ with at most one exception.*

It is probably safe to say that up to the time that Picard proved his theorem, there was no evidence at all that such a result would be true. Looked at from the point of view of a mapping, it said that every "infinite polynomial" maps the whole plane either onto the whole plane or onto the plane minus a single point. The fact that there may be an exceptional point was of course well known from the example of the exponential function, where $\exp z = 0$ has no solution.

Again thinking in terms of the fundamental theorem of algebra, as the degree of the polynomial goes up, so does the number of solutions to the equation $P(z) = c$, and so one might expect that an "infinite polynomial" would have an infinite number of solutions. Picard indeed went on to show that a much stronger version of his first theorem was true.

**Theorem 3 (Picard's "big" Theorem [1879])** *Let $f(z)$ be an entire function that is not a polynomial. Let $R$ be an arbitrary positive number. Then for every complex number $c$ with at most one exception, the equation $f(z) = c$ has a solution with $|z| > R$.*

**Corollary 3.1** *For an entire function $f(z)$, not a polynomial, the equation $f(z) = c$ has an infinite number of solutions for all values of $c$ with at most one exception.*

*Proof.* Case 1. For every $R > 0$ the equation $f(z) = c$ has a solution with $|z| > R$ for every $c$, with no exceptions. Let $z_1$ be any solution. Choose $R > |z_1|$ and choose a corresponding solution $z_2$ with $|z_2| > R$. Proceeding in the same way gives an infinite number of solutions.

Case 2. For some $R > 0$ there is a value of $c$ such that the equation $f(z) = c$ does not have a solution with $|z| > R$. Then apply the same reasoning above to any complex number $\neq c$.

These theorems of Picard set the stage for much of the research in the theory of functions of a complex variable for the next hundred years and beyond. That will be the subject of our next section. Let us mention here just one immediate corollary of Picard's Theorem that requires the introduction of a new concept.

**Definition.** A function $f$ is called *meromorphic* in a domain $D$ if, for every point $a$ in $D$, there is a neighborhood of $a$ in which $f(z)$ can be represented by a power series in $(z - a)$ plus a polynomial in $1/(z - a)$. A point $a$ at which positive powers of $1/(z - a)$ occur is called a *pole* of $f$.

Examples of meromorphic functions in the whole plane are *rational functions*: quotients $f(z) = P(z)/Q(z)$ of two polynomials. If $P$ and $Q$ have no factors in common, then the poles of $f$ are simply the zeros of the denominator.

For a rational function $f$, the equation $f(z) = c$ has a solution for every value of $c$, since one can multiply through by the denominator and apply the fundamental theorem of algebra. However, one cannot view a rational function, or a meromorphic function, as a map into the complex plane, since it has no finite value at a pole. One traditionally talks about a map into the *extended plane* consisting of the ordinary complex plane plus a single point at infinity. Then if $a$ is a pole of $f$, we write $f(a) = \infty$.

Riemann's second contribution, referred to earlier, was to give a beautiful geometric interpretation of the extended plane. He simply imported stereographic projection from cartography to map the ordinary plane onto a sphere minus a point, and then the point at infinity fills in the missing point on the sphere. Using the standard stereographic projection from the North Pole, it is the North

Pole on the sphere that corresponds to the point at infinity. It is easy to show that if $a$ is a pole of $f$, then as $z \to a$, $f(z)$ composed with the inverse of stereographic projection tends to the North Pole, so that a meromorphic function in a plane domain $D$ can be considered as a continuous map of $D$ into the sphere. A closer look at this map shows that at the North Pole the map is not only continuous but has exactly the same behavior as at any other point on the sphere. In particular, at a *simple pole,* where the term $1/(z-a)$ occurs, but no higher powers, the map into the sphere is conformal at $a$. If higher powers of $1/(z-a)$ occur, then the map into the sphere behaves exactly the same as at ordinary points where $f$ is analytic, but $f' = 0$.

The unit sphere viewed this way as the extended complex plane via stereographic projection is called the *Riemann sphere.* It has the effect of taming the "point at infinity" in the complex plane and making it essentially no different from any other point. Another way of thinking about it is that if $f(z)$ is a meromorphic function with a pole at $a$, then one can compose $f$ with stereographic projection taking $f(a)$ to the North Pole, and follow with a stereographic projection from any other point on the sphere taking the North Pole onto a finite point. Then the composed map of the plane into the plane will be an ordinary analytic function in a neighborhood of the point $a$.

Picard's two theorems have an immediate extension to meromorphic functions:

**Theorem 4** *Let $f(z)$ be a nonconstant meromorphic function in the entire plane. Then viewed as a map into the Riemann sphere, the image of $f$ covers the entire sphere with at most two exceptions. Furthermore, if $f$ is not a rational function, then the same is true for $f(z)$ with $|z| > R$ for any* $R > 0$.

*Proof.* If the image omits three points on the sphere, and if $c \neq \infty$ is one of them, then the function $g(z) = 1/\big(f(z)-c\big)$ will be an entire function omitting two points, contradicting Picard's Theorem.

Another way to think of it is that if the image omits three points, then make a stereographic projection from one of them onto the plane; that will again give an entire function omitting two points.

When these results were first announced, they probably appeared to be the culmination of a century's work in the subject, and in fact, Picard's proofs used some of the most sophisticated developments of the previous years. However, as is so often the case, Picard's theorems turned out to be just the starting point for a whole array of further investigations. They will be the subject of our final section.

## 7 The twentieth century

The years following the publication of Picard's theorems saw many new proofs, generalizations, and refinements, but nothing to compare with the sweep and depth of a paper by a young Finnish mathematician, Rolf Nevanlinna, in 1925. That paper inaugurated a whole branch of complex function theory called "value distribution theory" or simply "Nevanlinna theory." What Nevanlinna did was to look at the range of values taken on by an analytic or meromorphic function, and introduce highly refined measures of the relative frequency with which the function took on those values. Now for entire functions or meromorphic functions in the whole plane other than rational functions, all values are assumed infinitely often, with one or two possible exceptions, so that what one is comparing is not simply the size of these sets, but rather a kind of measure of their density. Roughly speaking, one compares the number of solutions of the equation $f(z) = c$ inside a circle of radius $R$ for different values of $c$, and then sees what happens as $R$ tends to infinity. What Nevanlinna proves is

(i) for "almost all" values of $c$, in a very strong sense, the solutions of $f(z) = c$ have "the same order of magnitude" or "the same density" in a very precise sense.

(ii) for the exceptional values of $c$ where the equation has "fewer solutions" there is a precise measure of the size of the solution set, called the *defect* of c, and the sum of all the defects is at most 2.

The second of these results is Nevanlinna's far-reaching generalization of Picard's theorem. It follows immediately from the definition of the defect that if a meromorphic function in the plane omits a value altogether, then the defect of that value is equal to 1. Hence there can be at most two such values. For an entire function, the point at infinity is omitted, hence has defect equal to 1. It follows that the sum of all the other defects can be at most equal to 1, and in particular, at most one finite value can be omitted altogether.

Among the many other results proved by Nevanlinna in this groundbreaking paper, we note one of particular interest. For every value c, Nevanlinna introduces in addition to the defect, a "ramification index" of c, and obtains a result similar to (ii) above for the sum of the ramification indices. He also makes the following definition.

**Definition.** Let $f(z)$ be an analytic or meromorphic function in a domain $D$. A value $c$ of $f(z)$ is *totally ramified* if whenever $f(b) = c$, $f'(b) = 0$.

In other words, if $f$ is viewed as a mapping from the $z$-plane to the $w$-plane, then at none of the points that maps onto $c$ does $f$ define locally a one-to-one conformal map, as would be the case if the derivative were different from zero. Rather, at each of the pre-images of c, $f$ behaves like the function $z^n$ in a neighborhood of the origin, for some $n > 1$. The image is then said to be "branched" or "ramified" in a neighborhood of c.

**Theorem 5 (Nevanlinna [1925])** *An entire function can have at most two totally ramified values. A meromophic function in the plane can have at most four totally ramified values.*

Both numbers "two" and "four" in this theorem are sharp. For example, the complex sine and cosine are entire functions defined by

$$2\cos z = \exp iz + \exp(-iz), \quad 2i\sin z = \exp iz - \exp(-iz).$$

It follows that just as for real values of $z$, $\sin^2 z + \cos^2 z \equiv 1$ and $\cos z$ is the derivative of $\sin z$. Hence $\sin z = \pm 1 \Leftrightarrow \cos z = 0$, which means that the two values 1 and $-1$ are totally ramified for $f(z) = \sin z$. Similarly, Nevanlinna points out that examples of meromorphic functions with exactly four totally ramified values include classical elliptic functions such as the Weierstrass $\wp$-function. We give a more geometric description of such a function that will be of particular interest in the sequel.

Let a regular tetrahedron be inscribed in the unit sphere, with one vertex at the North Pole. The four vertices of the tetrahedron will form four equally-spaced points on the sphere. Project the tetrahedron onto the sphere from the center of the sphere. The result will be a tiling of the sphere by four congruent spherical triangles. We want to construct a meromorphic function in the plane with the property that each of the four vertices will be totally ramified, and everywhere else the map will be an unramified conformal map. To do so, we make a stereographic projection from the North Pole, under which the four spherical triangles map onto one domain $D$ bounded by three circular arcs meeting at 120°, together with three unbounded domains each having one side in common with a side of $D$, and the other two sides consisting of rays from the endpoints of that side out to infinity (see Figure 18.14).

We assume that configuration to lie in the $w$-plane, and we use the Riemann mapping theorem to define a conformal map $f(z)$ of the interior of an equilateral triangle in the $z$-plane onto the domain $D$. That can be done in a way that takes the vertices into the vertices and the center

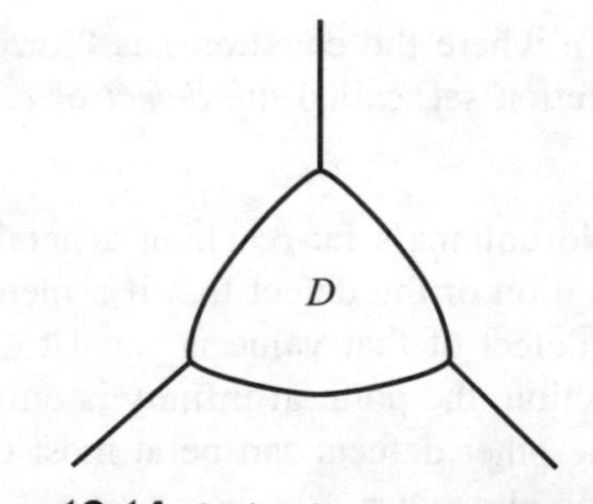


**Figure 18.14.** 3 circular arcs meeting at 120°

into the center. A fundamental result of H. A. Schwarz that he obtained while investigating the Riemann mapping theorem tells us that the map $f(z)$ can be extended to a meromorphic function in the whole $z$-plane by a process known as the "Schwarz reflection principle" [1869–70 (a)]. If one pictures the $z$-plane tiled by the equilateral triangles obtained by successive reflection across the sides of the original triangles and their images, then the extended function $f(z)$ will satisfy $f'(z) \neq 0$ everywhere except at the vertices of the tiling, where the mapping will behave like $z^2$.

We turn next to a 1926 paper of André Bloch. Bloch's goal was two-fold. First, to give an elementary proof of Picard's Theorem, whose original proof and subsequent refinements tended to be anything but elementary. Second, to implement in this case a general policy that has come to be known as "Bloch's Principle." The idea is that if one has a theorem that applies to functions of a certain class defined in the whole plane, then one should seek a finite version—say in a disk of radius $R$—that yields the original theorem in the limit as $R \to \infty$.

**Theorem 6 (Bloch [1926])** *There exists a positive constant $B$ with the following property. Let $w = f(z)$ be analytic in the unit disk and be normalized so that $f'(0) = 1$. Then for every $r < B$ there exists a disk of radius $r$ in the $w$-plane that is the one-to-one conformal image under $f$ of a domain inside the unit disk of the $z$-plane.*

The largest value of $B$ for which Bloch's Theorem holds is known as *Bloch's constant.* Its precise value is not known.

**Corollary 6.1** *The mapping of the $z$-plane into the $w$-plane defined by an arbitrary nonconstant entire function $w = f(z)$ has the property that for every $R > 0$, there is a disk of radius $R$ in the $w$-plane that is the one-to-one image under $f$ of some domain in the $z$-plane.*

*Proof.* We can first normalize $f$ so that $f'(0) = 1$. Choose $\lambda > R/B$, where $B$ is Bloch's constant. Then apply Bloch's Theorem to the function $g(z) = f(\lambda z)/\lambda$ in the unit disk. Let $r = R/\lambda$ . Since $r < B$, we conclude that the image of the disk of radius $\lambda$ under $f$ includes a disk of radius $\lambda r = R$.

*Note.* Corollary 6.1 had been proved earlier by Valiron [1926].

**Corollary 6.2** *Picard's Theorem.*

To derive Picard's theorem from Corollary 6.1, suppose that the image of a nonconstant entire function $f$ omits two points $a$, $b$. Then the function $g(z) = (f(z) - a)/(f(z) - b)$ is an entire function that omits the values 0, 1. We can then compose $g$ with the complex logarithm to get an entire function that omits the points $2\pi ni$ for all integers $n$. Another slightly more complicated, but still elementary composition yields an entire function that omits a rectangular lattice-type array of points with the property that for some $R$ sufficiently large, every disk of radius $R$ contains one of those points. This gives an entire function that violates the conclusion of Corollary 6.1. Hence the assumption that the original function could omit two distinct values is false.

The 1930's saw a series of dazzlingly original papers on Nevanlinna theory by another Finnish mathematician, Lars V. Ahlfors, for which he received one of the first two Fields Medals awarded in 1936. In those papers, Ahlfors re-frames, re-formulates, and re-proves the main results of Nevanlinna theory in far more geometric fashion than in the original papers. One of those papers in particular, "On the theory of covering surfaces" from 1935 was singled out by the committee choosing the Fields medalists and was later described by Ahlfors himself as a "much more radical departure from Nevanlinna's own methods" which is indeed the case. We cite just one of the most striking results from that paper.

**Theorem 7 (Ahlfors [1935])** *Let $w = f(z)$ be a nonconstant complex function defined in the whole z-plane.*

1) *If $f$ is entire, then given any two disjoint disks in the w-plane, the interior of at least one of them is the image under $f$ of some domain in the z-plane;*
2) *If $f$ is entire, then given any three disjoint disks in the w-plane, the interior of at least one of them is the one-to-one conformal image under $f$ of some domain in the z-plane;*
3) *If $f$ is meromorphic in the whole plane, then the same holds for any five disjoint disks on the Riemann sphere.*

The third statement here is known as Ahlfors' "five islands" theorem.

This three-part theorem of Ahlfors has some of the same aspects of astonishing simplicity of statement and totally unanticipated result that characterizes Picard's original theorem. Part 1 of the theorem is of course a far-reaching generalization of Picard, since if a nonconstant entire function were to omit two values, one could choose a disk about each in contradiction to Ahlfors' result. In the other direction, if one starts with the two disks, one knows from Picard's Theorem that one of them at least must be completely covered by the image of the function, but it might well be in many bits and pieces, whereas Ahlfors' theorem says it is the image of a single connected domain (an "island").

Similarly, part 2 of Ahlfors' theorem implies Corollary 6.1 to Bloch's theorem above, since one can make all three of the given disks as large as one wants. But nothing in Bloch's theorem and its corollaries implies that one can pick specific disks in advance, only that *somewhere* in the image is a disk with the specified property.

Finally, part 3 is an equally surprising generalization of Nevanlinna's Theorem about the maximum number of totally ramified values for a meromorphic function in the plane, since if there were five totally ramified values one could choose five disjoint disks about them and obtain a contradiction to Ahlfors' result.

If Ahlfors had stopped there he probably would still have been awarded the Fields Medal. But in answer to "can you top this?", he did. To fully understand the icing on this cake, one must take note of a particular property of analytic and meromorphic functions that was generally understood to account for the vast majority of special properties that they enjoy. That is the property known as "rigidity." What that means in this context, is that if you change an analytic function in some neighborhood—no matter how small—of a point, then it changes everywhere. Said differently, the values of an analytic or meromorphic function are determined over its whole domain of definition by its values in an arbitrarily small neighborhood of any point in that domain. That property is not shared by even infinitely differentiable real functions, which may be pushed and pulled locally without affecting them elsewhere and still kept infinitely differentiable. The aspect of Ahlfors' paper that must have been the most counterintuitive based on all that came before—where all the elaborate machinery developed specifically for analytic and meromorphic functions had been invoked—was that his methods showed that none of the special properties of those classes of functions were needed, but rather that the results and all their corollaries actually remain true for an enormously broader class of mappings, with no rigidity properties whatever.

**Definition.** A smooth map between plane domains, or more generally between surfaces (such as the plane and the sphere) is *quasiconformal* if there is a uniform bound to the ratio of maximum to minimum scaling factors at each point.

To understand the significance of this condition, recall from sections 2 and 3 that the differential of a smooth map at a point is a linear transformation that maps a circle onto an ellipse, and the ratio of major to minor axes of that ellipse is precisely the ratio of maximum to minimum scaling factor at the point. For a conformal map, the ellipse reduces to a circle at each point. In general, as for example, in Lambert's equal area map, or the plate carrée, the ellipses get more and more distorted toward the poles, so that there is no uniform bound on the ratio of maximum to minimum scale factor. Those maps are therefore not quasiconformal. However, the class of quasiconformal maps is far larger than that of conformal maps. What Ahlfors proved was

**Theorem 8 (Ahlfors [1935])** *The conclusions of Theorem 7 are all valid for arbitrary quasiconformal maps of the plane into the sphere.*

After many detours, this whole circle of ideas reached its culmination in the final year of the twentieth century with a theorem of Mario Bonk and Alexandre Eremenko [2000]. In order to state the theorem, we recall the example we gave of a meromorphic function that is totally ramified over the four vertices of an equilateral tetrahedron inscribed in the unit sphere. Let $C'$ be a circle passing through three of those vertices, and let $D'$ be the circular disk on the sphere bounded by $C'$.

**Theorem 9 (Bonk and Eremenko [2000])** *Let $D$ be any circular disk on the Riemann sphere smaller than the disk $D'$ described above. Then for any (nonconstant) meromorphic function $f(z)$ in the plane, there is a domain in the $z$-plane mapped one-to-one conformally by $f$ onto a disk of size $D$.*

One of the many remarkable features of this theorem is that unlike Bloch's theorem and other similar ones, there is no normalization required. The bound given on the size of the image holds for *all* meromorphic functions in the plane. Furthermore, also unlike the original Bloch's Theorem, where the precise value of Bloch's constant remains unknown, the bound given here is best possible, since any disk larger than $D'$ would include one of the vertices of the tetrahedron in its interior, and the example we have constructed would contradict the conclusion.

But the most remarkable feature of this theorem is that, as the authors show, it implies all the previous theorems of Nevanlinna, Bloch, and Ahlfors described above. The proof employs an eclectic array of tools, from classical spherical geometry to quasiconformal mappings. And it may be worth a passing comment that in order to get a sharp result, the authors resort to the 2,000 year-old device of using stereographic projection from the plane onto the sphere.

**Acknowledgement** The maps and Tissot indicatrices for maps in Figures 18.5, 18.6, 18.8, 18.9, 18.10, and 18.11 were all taken from "An Album of Map Projections" by John P. Snyder and Philip M. Voxland (U.S. Geological Survey Professional Paper 1453, 1989).

## References

1569. G. Mercator, *Nova et aucta orbis terrae descriptio ad usum navigantium emendate accomodata* (A new and enlarged description of the earth with corrections for use in navigation).

1610. E. Wright, *Certain Errors in Navigation,* London.

1695. E. Halley, An easy Demonstration of the Analogy of the Logarithmic Tangents, to the Meridian Line, or sum of the secants: with various Methods for computing the same to the utmost Exactness, *Philosophical Transactions* 19, 202–214.

1772. J. H. Lambert, Anmerkungen und Zusätze zur Entwerfung der Land- und Himmelskarten (English translation: Notes and Comments on the Composition of Terrestial and Celestial Maps, Ann Arbor, University of Michigan 1972.)

1775. L. Euler, On representations of a spherical surface on the plane, in *Collected Works,* Series 1, Vol. 28, 248–275. (See pp. 251–253.)

1799. C. F. Gauss, *Demonstratio nova theorematis omnem functionem algebraicam rationalem integram unius variabilis in factores reales primi vel secundi gradus resolvi posse,* in *Collected Works,* Vol. III, 3–30.

1822. C. F. Gauss, Allgemeine Auflösung der Aufgabe die Theile einer gegebenen Fläche auf einer andern gegebnen Fläche so abzubilden, dass die Abbildung dem Abgebildeten in den kleinsten Theilen ähnlich wird, in *Collected Works,* Vol. IV, 189–216.

1824. N. H. Abel, *Mémoire sur les équations algébriques,* Christiana.

1827. C. F. Gauss, *Disquisitiones Generales Circa Superficies Curvas,* in *Collected Works,* Vol. IV, 219–258. (Original and a translation in P. Dombrowski, *Astérisque* 62, Soc. Math. de France, Paris 1979.)

1851. B. Riemann, Grundlagen für eine allgemeine Theorie der Functionen einen veränderlichen complexen Grösse. (Ph D thesis, Göttingen), in *Collected Works,* 3–45.

1869–70. H. A. Schwarz, (a) Ueber einige Abbildungsaufgaben, J. reine u. angewandte Math. 70, 105–120; *Collected Works II,* 65–83.
(b) Zur Theorie der Abbildung; *Collected Works II,* 108–132.
(c) Ueber eine Grenzübergang durch alternirendes Verfahren, Vierteljahrschrift der Naturforschungen Gesellschaft in Zürich, 15. Jahrgang, 272–286, *Collected Works II,* 133–143.

1879. E. Picard, Sur une propriétée des fonctions entiéres, *C. R. Acad. Sci. Paris* 88, 1024–1027.

1925B. A. Bloch, Les Théorèmes de M. Valiron sur les fonctions entières et la théorie de l'uniformisation, *Ann. Fac. Sci. Univ. Toulouse III,* 17, 1–22.

1925N. R. Nevanlinna, Zur Theorie der meromorphen Funktionen, *Acta Math.* 46, 1–99.

1926. G. Valiron, Sur les Théorèmes des MM. Bloch, Landau, Montel et Schottky, *C.R. Acad. Sci. Paris* 183, 728–730.

1935. L. V. Ahlfors, Zur Theorie der Überlagerungsflächen, *Acta Math.* 65, 157–194.

1980. F. V. Rickey and P. M. Tuchinsky, An Application of Geography to Mathematics, *Math. Magazine* 53, 162–166.

1997. T. Needham, *Visual Complex Analysis,* Clarendon Press, Oxford.

1998. P. J. Nahin, *An Imaginary Tale: The Story of* $\sqrt{-1}$, Princeton University Press.

2000. M. Bonk and A. Eremenko, Covering properties of meromorphic functions, negative curvature and spherical geometry, *Annals of Math.* (2) 152, 551–592.

# 19

# Alice in Numberland

## an informal dramatic presentation in 8 fits

Robin Wilson
*The Open University, UK*

This dramatic presentation has been performed several times, with casts ranging from six to twenty-five performers. In particular, performances were given to capacity audiences at the joint AMS-MAA Winter Meetings in San Antonio (1999) and New Orleans (2001); these were given in costume, and were illustrated by overhead transparencies and coloured slides. The presentation lasts about one hour, and notes and production suggestions are given at the end of the script.

**Dramatis Personae**

***Oxford Characters***

Charles Lutwidge Dodgson (Lewis Carroll)
Alice Pleasance Liddell
Mrs Liddell
Revd. Robinson Duckworth
Louisa, Elizabeth, and Mary Dodgson
Edith Rix
Bartholomew Price
Queen Victoria
Finals examiner
Senior Censor
Sir Michael Dummett
Three college scouts
Three Oxford students

***Fictional Characters***

Mock Turtle and Gryphon
White Queen and Red Queen
Humpty Dumpty
Hatter
Butcher
Mein Herr, The Earl, and Lady Muriel
Poet
Euclid and Minos
Tweedledum and Tweedledee

## Fit the First: Letters to my Child-Friends

**Dodgson** Letter to my child-friend, Margaret Cunnynghame, Christ Church, Oxford; January 30th 1868.

*[reading]* Dear Maggie,
*No carte has yet been done of me*
*that does real justice to my smile;*
*and so I hardly like, you see,*
*to send you one. Meanwhile,*
*I send you a little thing*
*to give you an idea of what*
*I look like when I'm lecturing.*
*The merest sketch, you will allow*
*—yet I still think there's something grand*
*in the expression of the brow*
*and in the action of the hand.*
*Your affectionate friend, C. L. Dodgson*
*P.S. My best love to yourself—to your Mother*
*my kindest regards—to your small,*
*fat, impertinent, ignorant brother*
*my hatred. I think that is all.*

**Mrs Liddell** Charles Dodgson, mathematics lecturer at Christ Church, the beautiful Oxford college where my husband is Dean, was an inveterate letter writer. During the last thirty-seven years of his life he wrote and received over 98,000 letters and kept a detailed register of all of them.

**Dodgson** I find I write about 20 words a minute, and a page contains about 150 words—that's about seven-and-a-half minutes to a page. So the copying of 12 pages takes about one-and-a-half hours, and the original writing two-and-a-half hours or more.

Sometimes I hardly know which is me and which the inkstand. The confusion in one's *mind* doesn't so much matter—but when it comes to putting bread-and-butter and orange marmalade into the *inkstand*; and then dipping pens into *oneself* and filling *oneself* up with ink, you know, it's horrid!

**Mrs Liddell** Although many letters were written to his brothers and sisters or to distinguished figures of the time, the most interesting ones were to his child-friends, often containing poems, puzzles, and word-games.

Because of a stammer and partial deafness, he was often ill-at-ease in adult company, and frequently sought the company of young children instead. He had a deep understanding of their minds and an appreciation of their interests, qualities that stemmed from his own happy childhood experiences.

Most of his friendships were with young girls, such as with my own dear daughters Alice, Edith, and Lorina—

**Dodgson** I am fond of children (except boys).

**Mrs Liddell** —and here's part of a letter to his young friend Isabella Bowman:

**Dodgson** *[reading] It's all very well for you and Nellie and Emsie to unite in millions of hugs and kisses, but please consider the time it would occupy your poor old very busy Uncle!*

*Try hugging and kissing Emsie for a minute by the watch, and I don't think you'll manage it more than 20 times a minute. "Millions" must mean 2 millions at least.*

*Now, 20 into 2 million hugs and kisses is 60 into a hundred thousand minutes, which is 12 into 1666 hours, or 6 into 138 days (at twelve hours a day), giving 23 weeks.*

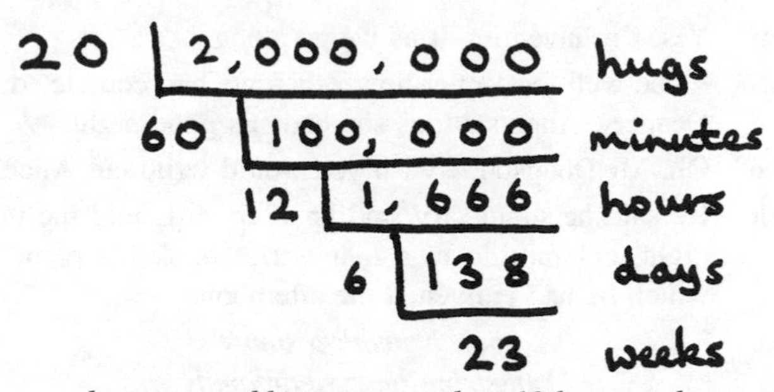


*I couldn't go on hugging and kissing more than 12 hours a day, and I wouldn't like to spend Sundays that way. So you see, it would take 23 weeks of hard work. Really, my dear child, I cannot spare the time.*

*Please give my kindest regards to your mother, and $\frac{1}{2}$ of a kiss to Nellie and $\frac{1}{200}$ of a kiss to Emsie, and one two-millionth of a kiss to yourself,*

*Your loving Uncle, C. L. Dodgson.*

**Mrs Liddell** But here's a rare letter to a young lad of fourteen, Wilton Rix.

**Dodgson** *[reading] Honoured Sir,*

*Understanding you to be a distinguished algebraist (that is, distinguished from other algebraists by different face, different height, etc.), I beg to submit to you a difficulty which distresses me much.*

*If $x$ and $y$ are each equal to 1, it is plain that*
$2 \times (x^2 - y^2) = 0$,
*and also that* $5 \times (x - y) = 0$.
*Hence* $2 \times (x^2 - y^2) = 5 \times (x - y)$.
*Now divide each side of this equation by* $(x - y)$.
*Then* $2 \times (x + y) = 5$.
*But* $(x + y) = (1 + 1)$, *i.e.,* $= 2$.
*So that* $2 \times 2 = 5$.

*Ever since this painful fact has been forced upon me, I have not slept more than 8 hours a night, and have not been able to eat more than 3 meals a day.*

*I trust you will pity me and will kindly explain the difficulty to*

*Your obliged, Lewis Carroll.*

## Fit the Second: A Visit to Wonderland

**Mrs Liddell** His pen-name Lewis Carroll derived from his real name: Carroll (or Carolus) is the Latin for Charles, and Lewis is a form of Lutwidge, his middle name and mother's maiden name. He used it when writing for children, and in particular for his Alice books—*Alice's adventures in Wonderland* and *Through the looking-glass and what Alice found there*—that featured my daughter Alice.

Let's recall that famous day, the fourth of July 1862, when my daughters rowed to Godstow with Dodgson and his friend the Reverend Robinson Duckworth.

**Duckworth** I rowed stroke and he rowed bow in the famous Long Vacation voyage to Godstow, when the three Miss Liddells were our passengers, and the story was actually composed and spoken over my shoulder for the benefit of Alice Liddell, who was acting as cox of our gig. I remember turning round and saying 'Dodgson, is this an extempore romance of yours?' And he replied:

**Dodgson** Yes, I'm inventing it as we go along.

**Duckworth** I also well remember how, when we had conducted the three children back to the Deanery, Alice said, as she bade us good-night:

**Alice** Oh, Mr Dodgson, I wish you would write out Alice's adventures for me.

**Duckworth** He said he should try, and he afterwards told me that he sat up nearly the whole night, committing to a manuscript book his recollections of the drolleries with which he had enlivened the afternoon.

**Alice**

**A** *boat, beneath a sunny sky*
**L***ingering onward dreamily*
**I***n an evening of July—*
**C***hildren three that nestle near,*
**E***ager eye and willing ear,*
**P***leased a simple tale to hear—*

**L***ong has paled that sunny sky,*
**E***choes fade and memories die;*
**A***utumn frosts have slain July.*

**S***till she haunts me, phantomwise,*
**A***lice moving under skies*
**N***ever seen by waking eyes.*

**C***hildren yet, the tale to hear,*
**E***ager eye and willing ear,*
**L***ovingly shall nestle near.*

**I***n a Wonderland they lie,*
**D***reaming as the days go by,*
**D***reaming as the summers die:*

**E***ver drifting down the stream—*
**L***ingering in the golden gleam—*
**L***ife, what is it but a dream?*

**Duckworth** The Alice books contain many allusions to Oxford: I, Duckworth, am the duck, the lory is Alice's sister Lorina, and with his stammer, the dodo is—

**Dodgson** Do-do-dodgson.

**Duckworth** The other animals are ones that Alice saw in her visits with us to the new Oxford University Museum.

As you might expect from a mathematics lecturer, the Alice books are brimming with arithmetical allusions.

**Mock turtle** We went to school in the sea. The master was an old turtle—we used to call him Tortoise—

**Alice** Why did you call him tortoise if he wasn't one?

**Mock turtle** We called him tortoise because he taught us. Really you are very dull!

**Gryphon** You ought to be ashamed of yourself for asking such a simple question.

**Mock turtle** I only took the regular course.

**Alice** What was that?

**Mock turtle** Reeling and writhing, of course, to begin with. And then the different branches of arithmetic—ambition, distraction, uglification and derision.

**Alice** I never heard of 'uglification.' What is it?

**Gryphon** Never heard of uglifying! You know what to beautify is, I suppose?

**Alice** Yes. It means—to—make—anything—prettier.

**Gryphon** Well then—if you don't know what to uglify is, you are a simpleton.

**Alice** And how many hours a day did you do lessons?

**Mock turtle** Ten hours the first day, nine hours the next, and so on.

**Alice** What a curious plan!

**Gryphon** That's the reason they're called lessons—because they lessen from day to day.

**Alice** Then the eleventh day must have been a holiday.

**Mock turtle** Of course it was.

**Alice** And how did you manage on the twelfth?

**Gryphon** That's enough about lessons. You're about to be examined by the Red Queen and the White Queen.

**Red queen** You can't be a queen, you know, until you've passed the proper examination. And the sooner we begin it the better.

**White queen** Can you do addition? *What's one and one and one and one and one and one and one and one and one and one?*

**Alice** I don't know. I lost count.

**Red queen** She can't do addition. Can you do subtraction? *Take nine from eight.*

**Alice** Nine from eight I can't, you know; but—

**White queen** She can't do subtraction. Can you do division? *Divide a loaf by a knife.* What's the answer to that?

**Alice** I suppose—

| | |
|---|---|
| **Red queen** | Bread-and-butter, of course. Try another subtraction sum. *Take a bone from a dog*: what remains? |
| **Alice** | The *bone* wouldn't remain, of course, if I took it—and the *dog* wouldn't remain: it would come to bite me—and I'm sure *I* shouldn't remain! |
| **Red queen** | Then you think nothing would remain? |
| **Alice** | I think that's the answer. |
| **Red queen** | Wrong as usual. The dog's temper would remain. |
| **Alice** | But I don't see how— |
| **Red queen** | Why look here! The dog would lose its temper, wouldn't it? |
| **Alice** | Perhaps it would. |
| **Red queen** | Then if the dog went away, its temper would remain! |
| **Both queens** | She can't do sums a bit! |

| | |
|---|---|
| **Duckworth** | Another character who couldn't do sums was Humpty Dumpty. |
| **Alice** | What a beautiful belt you've got on! |
| **Humpty** | It's a cravat, child, and a beautiful one, as you say. It's a present from the White King and Queen. They gave it me—for an un-birthday present. |
| **Alice** | I beg your pardon? |
| **Humpty** | I'm not offended. |
| **Alice** | I mean, what is an un-birthday present? |
| **Humpty** | A present given when it isn't your birthday, of course. |
| **Alice** | I like birthday presents best. |
| **Humpty** | You don't know what you're talking about! How many days are there in a year? |

**Alice** Three hundred and sixty-five.

**Humpty** And how many birthdays have you?

**Alice** One.

**Humpty** And if you take one from three hundred and sixty-five, what remains?

**Alice** Three hundred and sixty-four, of course.

**Humpty** *[doubtful]* I'd rather see that done on paper.

$$\begin{array}{r} 365 \\ -\quad 1 \\ \hline 364 \\ \hline \end{array}$$

**Alice** *[slowly]* Three hundred and sixty-five . . . minus one . . . is three hundred and sixty-four.

**Humpty** That seems to be done right—though I haven't time to look over it thoroughly right now—and that shows that there are three hundred and sixty-four days when you might get un-birthday presents.

**Alice** Certainly.

**Humpty** And only one for birthday presents, you know. There's glory for you!

## Fit the Third: Snarks and Twists

**Mrs Liddell** Mathematical ideas also appear in his other children's books. In Fit 5 of *The Hunting of the Snark,* the Butcher tries to convince the Beaver that 2 plus 1 is 3:

**Butcher** *Two added to one—if that could be done,*
*It said, with one's fingers and thumbs!*
*Recollecting with tears how, in earlier years,*
*It had taken no pains with its sums.*
*Taking Three as the subject to reason about—*
*A convenient number to state—*
*We add Seven, and Ten, and then multiply out*
*By One Thousand diminished by Eight.*
*The result we proceed to divide, as you see,*
*By Nine Hundred and Ninety and Two:*
*Then subtract Seventeen, and the answer must be*
*Exactly and perfectly true.*

$$\frac{(\underline{3} + 7 + 10)(1000 - 8)}{992} - 17 = \underline{3}$$

**Mrs Liddell** And in his last major novel, *Sylvie and Bruno concluded,* Dodgson's ability to illustrate mathematical ideas in a painless and picturesque way is used in the construction of *Fortunatus's purse* from three hankerchiefs. This purse has the form of a projective plane, with no inside or outside, and so contains all the fortune of the world. Since it cannot exist in three dimensions, he ceases just before the task becomes impossible.

We are sitting in a shady nook. Lady Muriel is sewing while her father, the Earl, looks on. Up comes Mein Herr, a distinguished German professor.

**Mein Herr** Hemming pocket handkerchiefs? So that is what the English miladies occupy themselves with, is it?

**The Earl** It is the one accomplishment in which Man has never yet rivalled Woman!

**Mein Herr** You have heard of Fortunatus's purse, Miladi? Would you be surprised to hear that, with three of these little handkerchiefs, you shall make the purse of Fortunatus quite easily?

**Lady Muriel** Shall I indeed? Please tell me how, Mein Herr! I'll make one before I touch another drop of tea!

**Mein Herr** You shall first join together these upper corners, the right to the right, the left to the left—and the opening between them shall be the *mouth* of the Purse.

**Lady Muriel** And if I sew the other three edges together, the bag is complete?

**Mein Herr** Not so, Miladi. The *lower* edges shall *first* be joined—ah, not so!—turn one of them over, and join the *right* lower corner of the one to the *left* lower corner of the other, and sew the lower edges together in what you would call *the wrong way.*

**Lady Muriel** I see! And a very twisted, uncomfortable, uncanny-looking bag it makes! But the moral is a lovely one. Unlimited wealth can only be obtained by doing things the wrong way! And how are we to join up these mysterious—no, I mean *this* mysterious opening? Yes, it is one opening—I thought it was *two,* at first.

**Mein Herr** You have seen the puzzle of the Paper Ring? Where you take a slip of paper, and join its ends together, *first twisting one,* so as to join the upper corner of one end to the lower corner of the other?

**The Earl** *[holding up Möbius strip]* I saw one made only yesterday—Muriel, my child, were you not making one, to amuse those children you had to tea?

**Lady Muriel** Yes, I know that puzzle. The Ring has *only one surface* and *only one edge.* It's very mysterious!

**The Earl** The *bag* is just like that, isn't it? Is not the *outer surface* of one side of it continuous with the *inner surface* of the other side?

**Lady Muriel** So it is! Only it *isn't* a bag just yet. How shall we fill up this opening, Mein Herr?

**Mein Herr** Thus!! The edge of the opening consists of *four* handkerchief edges, and you trace it continuously, round and round the opening—*down the right edge* of one handkerchief, *up the left edge* of the other—and then *down the left edge* of the one, and *up the right edge* of the other!

**Lady Muriel** So you can! And that proves it to be *only* one opening!

**Mein Herr** Now this *third* handkerchief also has four edges, which you can trace continuously round and round. All you need to do is join its four edges to the four edges of the opening. The Purse is then complete, and its outer surface—

**Lady Muriel** I see!—its *outer surface* will be continuous with its *inner surface*. But it will take time. I'll sew it up after tea.
But why do you call it *Fortunatus's purse,* Mein Herr?

**Mein Herr** Don't you see, my child? Whatever is *inside* that purse is *outside* it, and whatever is *outside* it is *inside* it. So you have *all the wealth of the world* in that little purse.

**Lady Muriel** *[clapping her hands in unrestrained delight]* I see! I'll certainly sew the third handkerchief in—*some other* time.

**Mrs Liddell** This formidable trio then got on to topics of a more geographical nature.

**Mein Herr** There's another thing we've learned from your Nation—*map-making*. But we've carried it much further than you. What do you consider the *largest* map that would be really useful?

**The Earl** About six inches to the mile.

**Mein Herr** Only *six inches*! We very soon got to six *yards* to the mile. Then we tried a *hundred* yards to the mile. And then came the grandest idea of all! We actually made a map of the country, on the scale of *a mile to the mile*!

**Lady Muriel** Have you used it much?

**Mein Herr** It has never been spread out, yet. The farmers objected: they said it would cover the whole country and shut out the sunlight! So we now use the country itself, as its own map, and I assure you it does nearly as well.

**Mrs Liddell** And talking of maps, here's a little map-colouring game for two people that Mr Dodgson used to play with his young child-friends. I believe it later became quite a celebrated puzzle.

**Dodgson** *A is to draw a fictitious map divided into counties.*
*B is to colour it—or rather mark the counties with names of colours—using as few colours as possible.*
*Two adjacent counties must have different colours.*
*A's object is to force B to use as many colours as possible. How many can he force B to use?*

## Fit the Fourth: Early Days

**Mrs Liddell** But I've been talking all this time as if you already know Mr Dodgson. My first contact with him was when my husband became Dean of Christ Church in 1855. This was the year when Mr Dodgson, already ensconced there, was appointed the mathematics lecturer. But what about his earlier years?

**Dodgson** I was born in 1832 into a 'good English church family' in Daresbury in Cheshire, where my father, the Reverend Charles Dodgson, was the incumbent until 1843, when we all moved to Croft Rectory in Yorkshire. There I and my seven sisters and three brothers enjoyed a very happy childhood, with lots of games to play and delightful walks in the Yorkshire countryside. I used to write and paint and build mechanical things, such as a miniature railway in our garden. When I was fourteen I was sent to Rugby School, where I delighted in mathematics and the classics, but I was never happy with all the rough-and-tumble.

In 1850 I was accepted at Oxford, and went up in January 1851 to Christ Church, the largest college, where I was to spend the rest of my life. Regrettably, I had to return home after only two days, as my mother died suddenly and unexpectedly.

My University course consisted mainly of mathematics and the classics, and involved three main examinations, starting in summer 1851 with my Responsions exams.

**Louisa** *[reading]* Letter written in old English from Christ Church, June 1851:
*My beloved and thrice-respected sister,*

*Onne moone his daye nexte we goe yn forre Responsions, and I amme uppe toe mine eyes yn worke.*

*Thine truly, Charles.*

**Dodgson** In early July I went with Aunt Charlotte to the Great Exhibition. I think the first impression produced on you when you get inside is of bewilderment. It looks like a sort of fairyland.

The next year I took my second Oxford examination—Moderations.

**Elizabeth** *[reading] December 9th 1852*
*Dearest Elizabeth,*

*You shall have the announcement of the last piece of good fortune this wonderful term has had in store for me, that is, a 1st class in Mathematics. Whether I shall add to this any honours at Collections I can't at present say, but I should think it very unlikely, as I have only today to get up the work in The Acts of the Apostles, 2 Greek Plays, and the Satires of Horace and I feel myself almost totally unable to read at all: I am beginning to suffer from the reaction of reading for Moderations.*

*I am getting quite tired of being congratulated on various subjects: there seems to be no end of it. If I had shot the Dean, I could hardly have had more said about it.*

**Dodgson** In the Summer of 1854, shortly before my Finals examinations I went on a reading party to Whitby in Yorkshire with my tutor, Bartholomew Price, the Sedleian Professor of Natural Philosophy *[Price stands up and bows]*—everyone called him 'Bat' Price, because his lectures were way above the audience.

I remembered him later when writing the Hatter's song:

**Hatter** *Twinkle, twinkle, little bat,*
*How I wonder what you're at . . .*
*[to Alice]* You know the song, perhaps?

**Alice** I've heard something like it.

**Hatter** It goes on, you know, in this way:
*Up above the world you fly,*
*Like a tea-tray in the sky,*
*Twinkle, twinkle, twinkle, . . .*

**Dodgson** My Finals examinations took place in December 1854, and ranged over all areas of pure and applied mathematics. Here's the first paper I sat, on Geometry and Algebra:

**Examiner** *Question 1.* Compare the advantages of a decimal and of a duodecimal system of notation in reference to (1) commerce; (2) pure arithmetic; and shew by duodecimals that the area of a room whose length is 29 feet $7\frac{1}{2}$ inches, and breadth is 33 feet $9\frac{1}{4}$ inches, is 704 feet $30\frac{3}{8}$ inches.

*Question 6.* In a given equilateral triangle a circle is inscribed, and then in the triangle formed by a tangent to that circle parallel to any side and the parts of the original triangle cut off by it, another circle is inscribed, and so on *ad infinitum.* Find the sum of the radii of these circles.

**Mary** *[reading] Christ Church, Oxford; December 13th 1854*

*My dear Sister,*

*I have just given my Scout a bottle of wine to drink to my First. We shall be made Bachelors on Monday. I have just been to Mr. Price to see how I did in the papers, and the result will I hope be gratifying to you. The following were the sum total of the marks for each in the 1st class, as nearly as I can remember:*

*Dodgson 279*
*Bosanquet 261*
*Cookson 254*
*Fowler 225*
*Ranken 213*

*All this is very satisfactory. I must also add (this is a very boastful letter) that I ought to get the Senior Scholarship next term. One thing more I will add, to crown all, and that is—I find I am the next 1st class Math. student to Faussett (with the exception of Kitchin, who has given up Mathematics) so that I stand next (as Bosanquet is going to leave) for the Lectureship.*

*Your very affectionate brother, Charles L. Dodgson.*

## Fit the Fifth: Lectures and Puzzles

**Mrs Liddell** Mr Dodgson did indeed get the Mathematics Lectureship at Christ Church. He became the College's Sub-librarian and some years later Curator of the Senior Common Room, moving into a sumptuous suite of rooms for which he commissioned the eminent artist William De Morgan, son of the mathematician Augustus De Morgan, to design the tiles around his fireplace.

But in 1855 his teaching experience was just beginning.

**Dodgson** My one pupil has begun his work with me, and I will give you a description how the lecture is conducted. It is the most important point, you know, that the tutor should be dignified, and at a distance from the pupil, and that the pupil should be as much as possible degraded—otherwise, you know, they are not humble enough.

So I sit at the further end of the room—outside the door (which is shut) sits the scout—outside the outer door (also shut) sits the sub-scout—half-way down stairs sits the sub-sub-scout—and down in the yard sits the pupil.

The questions are shouted from one to the other, and the answers come back in the same way—it is rather confusing till you are well used to it. The lecture goes on, something like this.

What is twice three?

**Scout** *[calls to Sub-scout]* What's a rice tree?

**Sub-scout** *[calls to Sub-sub-scout]* When is ice free?

**Sub-sub-scout** *[calls to Student]* What's a nice fee?

**Student** *[timidly—to Sub-sub-scout]* Half a guinea!

**Sub-sub-scout** *[calls to Sub-scout]* Can't forge any!

**Sub-scout** *[calls to scout]* Ho for Jinny!

**Scout** *[calls to Dodgson]* Don't be a ninny!

**Dodgson** *[looks offended, but tries another question]* Divide a hundred by twelve!

**Scout** *[calls to Sub-scout]* Provide wonderful bells!

**Sub-scout** *[calls to Sub-sub-scout]* Go ride under it yourself!

**Sub-sub-scout** *[calls to Student]* Deride the dunder-headed elf!

**Student** *[surprised—to Sub-sub-scout]* Who do you mean?

**Sub-sub-scout** *[calls to Sub-scout]* Doings between!

**Sub-scout** *[calls to scout]* Blue is the screen!

**Scout** *[calls to Dodgson]* Soup-tureen!

**Dodgson** And so the lecture proceeds.

**Mrs Liddell** It was during his early years as a lecturer that he started to teach a class of young children at the school across the road. He varied the lessons with stories and puzzles, and may have been the first to use recreational mathematics as a vehicle for teaching mathematical ideas.

**Dodgson** From my diary, February 5th 1856.

I worked for them the puzzle of writing the answer to an addition sum, when only one of the five rows has been written. Let me show you. Alice has already given me the first number:

**Alice** Twenty-one thousand, eight hundred and seventy-nine.

**Dodgson** Could I please have another five-digit number? *[writes it down]*

And another? . . . *[writes it down]*

And another? . . . *[writes it down]*

And finally one more . . . *[writes it down]*

$$\begin{array}{r} 21879 \\ +\ 62593 \\ +\ 17472 \\ 37406 \\ 82527 \\ \hline 221877 \\ \hline \end{array}$$

Now the gentleman in the front row will add up these numbers—and the answer is 221,877, as I predicted.

I enjoyed showing these puzzles to my young child-friends. Here's another one, involving pounds, shillings and pence—remember that there are twelve pence in a shilling and twenty shillings in a pound.

Put down any number of pounds not more than twelve, any number of shillings under twenty, and any number of pence under twelve. Under the pounds put the number of pence, under the shillings the number of shillings, and under the pence the number of pounds, thus reversing the line.

Subtract—reverse the line again—then add.

Answer, £12 18s. 11d., *whatever* numbers may have been selected.

| | | | |
|---|---|---|---|
| £ 9 | 16s | 5d | |
| £ 5 | 16s | 9d | (reverse) |
| £ 3 | 19s | 8d | (subtract) |
| £ 8 | 19s | 3d | (reverse) |
| £ 12 | 18s | 11d | (add) |

**Bat Price** Another problem, hotly debated in Carroll's day, was the *monkey on a rope puzzle.* A rope goes over a pulley—on one side is a monkey, and on the other is an equal weight. The monkey starts to climb the rope—what happens to the weight? I say that it goes *up*, with increasing velocity; Clifton says that it goes *up* uniformly, while Sampson says that it goes *down.*

**Dodgson** In later years I produced two books of my problems. One of them, *Pillow problems,* consists of seventy-two problems thought out during wakeful hours. All of these problems I thought up in bed, solving them completely in my head, and I never wrote anything down until the next morning. There were problems in algebra:

**Alice** *Prove that 3 times the sum of 3 squares is also the sum of 4 squares.*

**Dodgson** plane geometry:

**Student 1** *To double down part of a given triangle, making a crease parallel to the base, so that, when the lower corners are folded over it, their vertices may meet.*

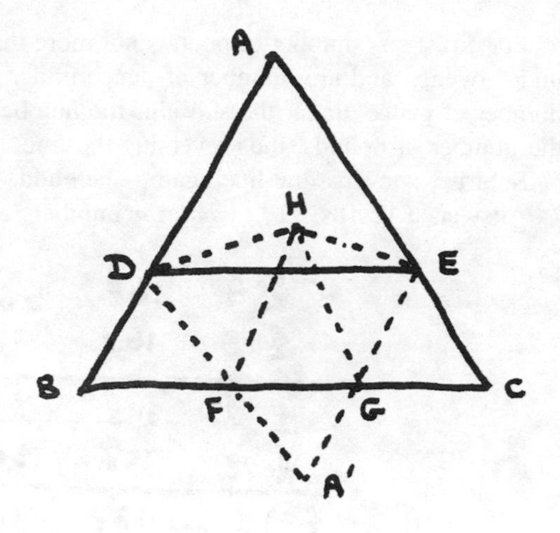


**Dodgson** solid geometry:

**Student 2** *If four equilateral triangles be made the sides of a square pyramid, find the ratio which its volume has to that of a tetrahedron made of the triangles.*

**Dodgson** and chances:

**Student 3** *A bag contains 2 counters, each of which is known to be black or white. 2 white and a black are put in, and 2 white and a black drawn out. Then a white is put in, and a white drawn out. What is the chance that it now contains 2 white?*

**Dodgson** My other puzzle book, *A tangled tale,* contains ten stories each involving a mathematical problem. Here's the preface—can you guess which of my child-friends I dedicated it to?

**Edith Rix**
*B**e**loved Pupil! Tamed by thee,*
*A**d**dish-, Subtrac-, Multiplica-tion,*
*D**i**vision, Fractions, Rule of Three,*
*A**t**test thy deft manipulation!*
*T**h**en onward! Let the voice of Fame*
*F**r**om Age to Age repeat thy story,*
*T**i**ll thou hast won thyself a name*
*E**x**ceeding even Euclid's glory!*

**Dodgson** The second letters of each line spell Edith Rix, the sister of Wilton Rix to whom I wrote the algebra letter earlier. Miss Rix went on to study mathematics at Cambridge, although I tried to persuade her to do so in Oxford.

**Mrs Liddell** And while we're on the subject of verse, here's another ingenious one by Mr Dodgson. It can be read both horizontally and vertically.

**Poet**

| | | | | | |
|---|---|---|---|---|---|
| *I* | *often* | *wondered* | *when* | *I* | *cursed,* |
| *Often* | *feared* | *where* | *I* | *would* | *be—* |
| *Wondered* | *where* | *she'd* | *yield* | *her* | *love* |
| *When* | *I* | *yield,* | *so* | *will* | *she,* |
| *I* | *would* | *her* | *will* | *be* | *pitied!* |
| *Cursed* | *be* | *love!* | *She* | *pitied* | *me . . .* |

## Fit the Sixth: Photographs and Postulates

**Mrs Liddell** In his early years as a lecturer at Christ Church Mr Dodgson took up the hobby of photography, using the new collodion process. He was one of the first to regard photography as an art, rather than just a means of recording images, and I've been told that if he were not known for his Alice books, he'd be primarily remembered as a pioneering photographer who took many hundreds of fine pictures.

My daughters used to love spending the afternoon with Mr Dodgson, watching him mix his chemicals, dressing up in costumes, and posing quite still for 45 seconds until the picture was done. One was a picture of my Alice, dressed as a beggar girl: Alfred Tennyson described it as the most beautiful photograph he had ever seen.

**Alice**

*From Hiawatha's photographing:*
*From his shoulder Hiawatha*
*Took the camera of rosewood*
*Made of sliding, folding rosewood;*
*Neatly put it all together.*
*In its case it lay compactly,*
*Folded into nearly nothing;*
*But he opened out the hinges,*
*Pushed and pulled the joints and hinges,*
*Till it looked all squares and oblongs,*
*Like a complicated figure*
*In the Second Book of Euclid...*

**Bat Price** And talking of Euclid brings us to Dodgson's enthusiasm for the writings of this great Greek author. Influenced by him, Dodgson produced for his pupils a *Syllabus of plane algebraic geometry,* described as the 'algebraic analogue' of Euclid's pure geometry, and systematically arranged with formal definitions, postulates, and axioms.

A few years later he gave an algebraic treatment of the *Fifth book of Euclid*—the one on proportion, possibly due to Eudoxus—taking the propositions in turn and recasting them in algebraic notation.

But he is best known for his celebrated book *Euclid and his modern rivals,* which appeared in 1879. Some years earlier, the Association for the Improvement of Geometrical Teaching had been formed, with the express purpose of replacing Euclid in schools by newly devised geometry books. Dodgson was bitterly opposed to these aims, and his book, dedicated to the memory of Euclid, is a detailed attempt to compare Euclid's *Elements,* favourably in every case, with the geometry texts of such as Legendre, J. M. Wilson, Benjamin Peirce, and others of the time.

It is written as a drama in four acts, with four characters—Minos and Radamanthus (two of the judges in Hades), Herr Niemand (the phantasm of a German professor), and Euclid himself. After demolishing each rival book in turn, Euclid approaches Minos to compare notes.

**Euclid** Are all gone?

**Minos** Be cheerful, sir: *Our revels now are ended: these our actors, as I foretold you, were all spirits, and are melted into air, into thin air!*

**Euclid** Good. Let us to business. And first, have you found any method of treating parallels to supersede mine?

**Minos** No! A thousand times, no! The infinitesimal method, so gracefully employed by M. Legendre, is unsuited to beginners: the method by transversals, and the method by revolving lines, have not yet been offered in a logical form: the 'equidistant' method is too cumbrous: and as for the 'method of direction', it is simply a rope of sand—it breaks to pieces wherever you touch it!

**Euclid** We may take it as a settled thing, then, that you have found no sufficient cause for abandoning either my sequence of Propositions or their numbering.

**Bat Price** And so they continued, comparing Euclid's constructions, demonstrations, style, and treatment of lines and angles, with those of his modern rivals.

Dodgson's love of geometry surfaced in other places, too.

His *Dynamics of a parti-cle* was a witty pamphlet concerning the parliamentary election for the Oxford University seat. Dodgson starts with his definitions:

**Dodgson** *Plain anger* is the inclination of two voters to one another, who meet together, but whose views are not in the same direction.

When a proctor, meeting another proctor, makes the votes on one side equal to those on the other, the feeling entertained by each side is called *right anger.*

*Obtuse anger* is that which is greater than *right anger.*

**Bat Price** He then introduced his postulates:

**Dodgson** 1. Let it be granted, that a speaker may digress from any one point to any other point.

2. That a finite argument (that is, one finished and disposed with) may be produced to any extent in subsequent debates.

3. That a controversy may be raised about any question, and at any distance from that question.

**Bat Price** And so he went on for several pages, leading up to the following geometrical construction. Here, *WEG* is the sitting candidate William Ewart Gladstone (too liberal for Dodgson), *GH* is Gathorne-Hardy (Dodgson's preferred choice), and *WH* is William Heathcote, the third candidate.

**Dodgson** Let *UNIV* be a large circle, and take a triangle, two of whose sides *WEG* and *WH* are in contact with the circle, while *GH,* the base, is not in contact with it. It is required to destroy the contact of *WEG* and to bring *GH* into contact instead.... When this is effected, it will be found most convenient to project *WEG* to infinity.

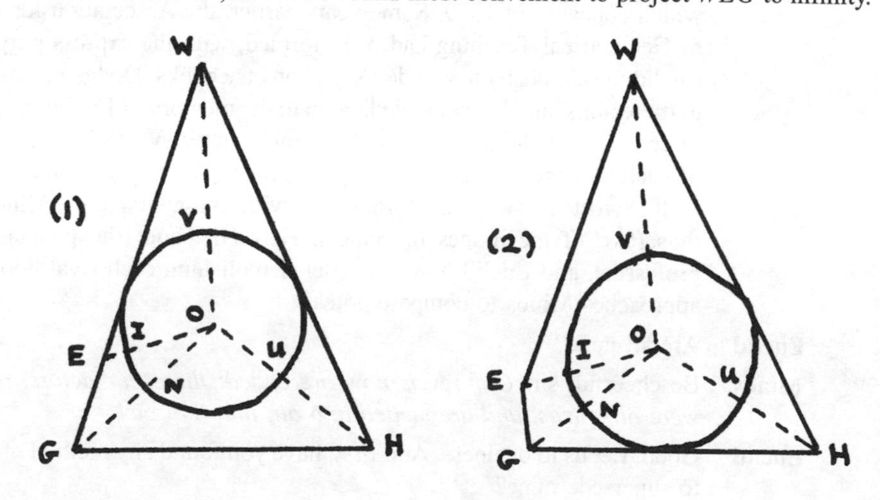

**Mrs Liddell** There were other occasions when Mr Dodgson applied his whimsy to university issues. In 1868 he wrote to the Senior Censor putting forward the case for a specially designed mathematical institute.

**Censor** *[reading] Dear Senior Censor,*

*In a desultory conversation on a point connected with the dinner at our high table, you incidentally remarked to me that lobster sauce, 'though a necessary adjunct to turbot, was not entirely wholesome.'*

*It is entirely unwholesome. I never ask for it without reluctance: I never take a second spoonful without a feeling of apprehension on the subject of possible nightmare. This naturally brings me on to the subject of Mathematics, and of the accommodation provided by the University for carrying on the calculations necessary in that important branch of Science.*

*It may be sufficient for the present to enumerate the following requisites: others might be added as funds permitted.*

*A. A very large room for calculating Greatest Common Measure. To this a small one might be attached for Least Common Multiple: this, however, might he dispensed with.*

*B. A piece of open ground for keeping roots and practising their extraction: it would be advisable to keep Square Roots by themselves, as their corners are apt to damage others.*

*C. A room for reducing Fractions to their Lowest Terms. This should be provided with a cellar for keeping the Lowest Terms when found, which might also be available to the general body of undergraduates, for the purpose of "keeping Terms".*

*D. A large room, which might be darkened, and fitted up with a magic lantern, for the purpose of exhibiting Circulating Decimals in the act of circulation.*

*E. A narrow strip of ground, railed off and carefully levelled, for investigating the properties of Asymptotes, and testing practically whether Parallel Lines meet or not; for this purpose it should reach, to use the expressive language of Euclid, 'ever so far.'*

*This last process, of 'continually producing the Lines,' may require centuries or more: but such a period, though long in the life of an individual, is as nothing in the life of the University.*

*May I trust that you will give your immediate attention to this most important subject?*

*Believe me, Sincerely yours, MATHEMATICUS*

## Fit the Seventh: Equations to Elections

**Bat Price** In 1865, Dodgson wrote his only algebra book, *An elementary treatise on determinants, with their application to simultaneous linear equations and algebraical geometry.* In later years the story went around, which Dodgson firmly denied, that Queen Victoria had been utterly charmed by *Alice's Adventures in Wonderland*—

**Q. Victoria** Send me the next book Mr Carroll produces—

**Bat Price** —the next book being the one on determinants—

**Q. Victoria** We are not amused.

**Bat Price** Unfortunately, Dodgson's book didn't catch on, because he used his own cumbersome terminology and notation instead of standard ones, but it did contain the first appearance in print of a well-known result involving the solutions of simultaneous linear equations. It also included a new method of his for evaluating large determinants in terms of small ones, a method that I presented on his behalf to the Royal Society of London, who subsequently published it in their *Proceedings*.

Another interest of his was the analysis of tennis tournaments.

**Dodgson** At a lawn tennis tournament where I chanced to be a spectator, the present method of assigning prizes was brought to my notice by the lamentations of one player who had been beaten early in the contest, and who had the mortification of seeing the second prize carried off by a player whom he knew to be quite inferior to himself.

Let us take sixteen players, for example, ranked in order of merit, and let us organise a tournament with 1 playing 2, 3 playing 4, and so on.

Then the winners of the first round will be 1, 3, 5, and so on; those of the second round will be 1, 5, 9 and 13; the final will then be won by player 1, defeating player 9 who wins the second prize but actually started in the lower half of the ranking.

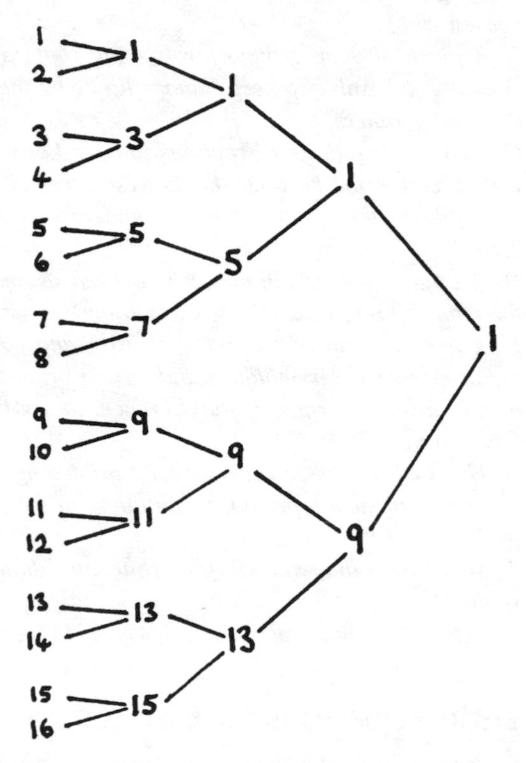


To avoid this difficulty, I have managed to devise a method for re-scheduling all the rounds so that the first three prizes will go to the best three players.

**Bat Price** Yet another interest of his was the study of voting patterns. Some of his recommendations were adopted in England, such as the rule that allows no results to be announced until all the voting booths have closed. Others, such as his various methods of proportional representation, were not.

As the philosopher Sir Michael Dummett has remarked:

**Dummett** It is a matter for the deepest regret that Dodgson never completed the book he planned to write on this subject. Such was the lucidity of his exposition and mastery of this topic that it seems possible that, had he published it, the political history of Britain would have been significantly different.

**Bat Price** The simplest example he gave of the failure of conventional methods is that of a simple majority.

**Dodgson** In this example there are eleven electors, each deciding among four candidates $a$, $b$, $c$, $d$. The first three electors rank them $a$, $c$, $d$, $b$; the next four rank them $b$, $a$, $c$, $d$; and so on. Which candidate, overall, is the best?

| 1 | 2 | 3 | 4 | 5 | 6 | 7 | 8 | 9 | 10 | 11 |
|---|---|---|---|---|---|---|---|---|---|---|
| a | a | a | b | b | b | b | c | c | c | d |
| c | c | c | a | a | a | a | a | a | a | a |
| d | d | d | c | c | c | c | d | d | d | c |
| b | b | b | d | d | d | d | b | b | b | b |

Candidate $a$ is considered best by three electors and second-best by the remaining eight electors.

In spite of this, candidate $b$ is the winner, although he is ranked worst by over half of the electors.

## Fit the Eighth: Symbolic Logic

**Mrs Liddell** Throughout his life, Mr Dodgson was interested in logic. In *Through the looking glass,* Tweedledum and Tweedledee are bickering as always:

**Tweedledum** *I know what you're thinking about—but it isn't so, nohow.*

**Tweedledee** *Contrariwise—if it was so, it might be; and if it were so, it would be; but as it isn't, it ain't. That's logic.*

**Mrs Liddell** Mr Dodgson believed that symbolic logic could be understood by his many child-friends, and devised *The game of logic* in order to help them sort out syllogisms. This contained a board and nine red and grey counters which are placed on sections of the board to represent true and false statements in order to sort out syllogisms.

**Student 1** *That story of yours, about your once meeting the sea-serpent, always sets me off yawning.*

**Student 2** *I never yawn, unless when I'm listening to something totally devoid of interest.*

**Alice** *Conclusion: That story of yours, about your once meeting the sea-serpent, is totally devoid of interest.*

**Dodgson** If, dear Reader, you will faithfully observe these Rules, and so give my book a really fair trial, I promise you most confidently that you will find *Symbolic logic* to be one of the most, if not *the* most, fascinating of mental recreations.

In this first part I have carefully avoided all difficulties which seemed to be beyond the grasp of an intelligent child of (say) twelve or fourteen years of age. I have myself taught most of its contents, *viva voce,* to many children, and have found them to take a real intelligent interest in the subject.

Some are very straightforward:

**Student 1** *Babies are illogical.*

**Student 2** *Nobody is despised who can manage a crocodile.*

**Student 3** *Illogical persons are despised.*

**Alice** *Conclusion: Babies cannot manage crocodiles.*

**Dodgson** Others need more thought, but can readily be sorted out using my counters. The following example contains five statements, but some of my examples go up to ten.

**Student 1** *No kitten that loves fish is unteachable.*

**Student 2** *No kitten without a tail will play with a gorilla.*

**Student 3** *Kittens with whiskers always love fish.*

**Student 1** *No teachable kitten has green eyes.*

**Student 2** *No kittens have tails unless they have whiskers.*

**Alice** *Conclusion: no kitten with green eyes will play with a gorilla.*

**Dodgson** And here's one that I once set as a challenge to logicians.

**Student 1** *If some a are b and some not, then some c are not d.*

**Student 2** *If some e are f, and if some g are h, then some j are k.*

**Student 3** *If all l are m, then no n are p.*

**Student 1** *If some c are d and some not, then some g are h.*

**Student 2** *If no e are f, and if some n are p, then some j are not k.*

**Student 3** *If some e are not f, and if some g are not h, then some n are p.*

**Student 1** *If some c are not d, and if some j are k, then no e are f.*

**Student 2** *If some g are not h, and if some j are not k, then some l are m.*

**Student 3** *If some e are not f, and if some n are p, then some a are not b.*

**Student 1** *If some a are b, and if some c are d, then some g are not h.*

**Student 2** *If some c are not d, and if some l are not m, then some e are f.*

**Alice** *Conclusion: if some a are b, and if some e are not f, then no c are d.*

**Mrs Liddell** Sadly, Mr Dodgson died just before Volume 2 of his *Symbolic Logic* was completed, and his manuscript version did not turn up until the 1970s. If it *had* appeared, then Charles Dodgson would almost certainly have been recognised as the greatest British logician between the time of George Boole and Augustus De Morgan and that of Bertrand Russell.

But let's leave the final word with his *alter ego,* Lewis Carroll. One night in

1857, while sitting alone in his room listening to the music from a Christ Church ball, he composed a double acrostic, one of whose lights has often been quoted as his own whimsical self-portrait:

**Dodgson** *Yet what are all such gaieties to me*
*Whose thoughts are full of indices and surds?*
$x^2 + 7x + 53 = \frac{11}{3}$.

## Notes and production suggestions

This production needs a minimum of six performers, with parts allocated as follows:

1. Charles Dodgson
2. Alice Liddell
3. Mrs Liddell, White Queen, and Queen Victoria
4. Robinson Duckworth, Gryphon, Butcher, Mein Herr, Hatter, Poet, Euclid, Senior Censor, Sir Michael Dummett, Tweedledum, Student 1
5. Humpty Dumpty, Mock Turtle, The Earl, Bartholomew Price, Finals Examiner, Tweedledee, Student 2
6. Red Queen, Lady Muriel, Louisa, Elizabeth and Mary Dodgson, Edith Rix, Minos, Student

The three college scouts can be played by members of the audience.

Like *The hunting of the snark,* the presentation is divided into eight 'fits', or scenes. If desired, these can be introduced by Dodgson.

An overhead projector can be used to display some of the mathematical items and the poems. In the original performances, slides of Oxford scenes, Dodgson's photographs, diagrams from the Alice books, and mathematical title pages were also projected.

The name 'Liddell' should be pronounced to rhyme with 'fiddle.'

### Fit 1

Dodgson's self-portrait, the division sum in the letter to Isabella Bowman, and the algebraic argument in the letter to Wilton Rix can all be displayed.

### Fit 2

The different 'scenes'—the trip on the river, the mock turtle scene, the scene with the two queens, and the Humpty Dumpty scene, should preferably take place in different parts of the performing area.

Alice's poem should preferably be displayed, so that the audience can see the initial letters of each line spelling out Alice's name.

Humpty Dumpty's subtraction sum can be displayed.

### Fit 3

The sum in the Butcher's poem can be displayed.

Fortunatus's purse is difficult to stage. Lady Muriel should be sitting down, with the Earl standing behind her. If handkerchieves are used, it is a good idea to stick velcro, or some other material, on to the correct sides so that they can more easily be joined. In any case, this does not have to be done accurately.

### Fit 4

Louisa should pronounce the final e at the end of each word: 'onn-e moon-e his day-e next-e,...'

The Hatter should be appropriately mad, and should exit from the side of the stage still singing 'Twinkle, twinkle,... '.

The examination questions can be displayed.

## Fit 5

For the lecture scene the three scouts can be members of the audience, placed some way from the others. Each should stand up to shout to the next one. The student (also standing in the audience) can be a member of the audience, or one of the three allocated students, or Alice. The addition sum can be very effective. After Alice's number, the next two numbers should be volunteered by random members of the audience. The fourth and fifth numbers should come from previously selected members of the audience—the fourth number should be 99,999 minus the second number, and the fifth number should be 99,999 minus the third number. The sum, 221,877 should have been written down previously, and revealed at the right moment. It is a good idea to find someone with a calculator to perform the addition sum, and shout out the answer.

The money sum can be projected, the plane geometry pillow problem, the Edith Rix poem (with the second letter of each line highlighted), and the horizontal-and-vertical poem can all be displayed.

## Fit 6

The geometrical diagram from *The dynamics of a parti-cle* should be displayed.

## Fit 7

The tennis tournament diagram and the voting example can be displayed.

## Fit 8

Each of the logic examples should preferably be displayed, with the conclusion revealed at the appropriate time; the final poem can also be displayed.

In the last logic example ('If some $a$ are $b$, ... '), the readers should gradually speed up.

## Acknowledgements

The cartoon of Dodgson is taken from Stuart Dodgson Collingwood's *The Life and Letters of Lewis Carroll,* Century, 1898. The drawings from *The Hunting of the Snark* and *Sylvie and Bruno Concluded* are by Harry Furniss, and those from *Alice's Adventures in Wonderland* and *Through the Looking Glass* are by John Tenniel. Lewis Carroll's monkey puzzle is taken from Sam Loyd's *Cyclopedia of Puzzles,* 1914.

# Biographical Notes

**Sheldon Axler** Sheldon Axler is Dean of the College of Science and Engineering at San Francisco State University. Former Editor-in-Chief of the *Mathematical Intelligencer* and Associate Editor of the *American Mathematical Monthly,* he has received the Lester R. Ford Award for expository writing from the Mathematical Association of America and the Distinguished Faculty Award from Michigan State University. One of his books has the semi-ungrammatical title of *Linear Algebra Done Right.*

**Arthur Benjamin** Arthur Benjamin earned his BS in Applied Mathematics from Carnegie Mellon University and his PhD in Mathematical Sciences from The John Hopkins University. Since 1989 he has taught at Harvey Mudd College, where he is currently Professor of Mathematics. He has served as Editor of the Spectrum book series for the Mathematical Association of America (MAA), and currently serves on the editorial boards of *Mathematics Magazine,* the *UMAP Journal,* and is co-editor (with Jennifer Quinn) of *Math Horizons.* Benjamin and Quinn recently wrote the book *Proofs That Really Count: The Art of Combinatorial Proof,* upon which their chapter is based. In 2000, he received the MAA's Haimo Award for Distinguished Teaching. Aside from his research interests in combinatorics and game theory, he enjoys tournament backgammon, racing calculators, and performing magic.

**Joe Buhler** A graduate of Reed College, Buhler received his PhD from Harvard University in 1977. Since 1980 he has been a member of the Reed faculty and is currently Professor of Mathematics. His fields of research are computational algebraic number theory, algebra, and cryptography. He shares with Ron Graham a serious interest in juggling, having learned to juggle from Professor Tom Brown at Simon Fraser University, who in turn had learned juggling from Ron Graham himself. Away from the Reed campus, Buhler has served as Deputy Director of the Mathematical Sciences Research Institute (MSRI) in Berkeley.

**Don Chakerian** Don Chakerian received his PhD from the University of California, Berkeley, and then started his career as an instructor of mathematics at the California Institute of Technology. For more than 30 years, he was a Professor of Mathematics at the University of California, Davis. During this time he received several awards and prizes both for his teaching achievements and his publications in the areas of expository mathematics and mathematics education. His scientific interests lie primarily in the theory of geometric inequalities and the geometry of convex sets.

**Dmitry Fuchs** Dmitry Fuchs is Professor of Mathematics at the University of California, Davis. He graduated from and was a Professor at Moscow State University in Russia. His research ranges from topology to homological algebra and representation theory that has applications in string theory and quantum field theory. The author of about 100 publications, including several mathematics textbooks, Fuchs is a brilliant expositor, actively participating in popularizing mathematics. He has been involved in all kinds of work with mathematically curious high school students both in Russia and in the United States.

**Joseph A. Gallian** Joseph A. Gallian is a Distinguished Professor of Teaching and Professor of Mathematics at the University of Minnesota, Duluth. Winner of the Haimo Award of the Mathematical Association of America (MAA) for distinguished teaching, he has also received the MAA's Allendoerfer and Evans Awards for Expository Writing. In addition he has served the MAA as a Pólya Lecturer, as Second Vice President, and as Co-Director of Project NExT. For more than 20 years he has directed an NSF sponsored Research Experience for Undergraduates program. Among the courses he teaches: one on the Beatles and one on Math and Sports.

**Ronald L. Graham** Ron Graham currently holds the Irwin and Joan Jacobs Chair of Computer and Information Science at the University of California at San Diego. Shortly after receiving his PhD from the University of California, Berkeley, he joined the staff of Bell Labs, Murray Hill, and he later became Chief Scientist at AT&T Laboratories. He has been President of the American Mathematical Society (AMS) and the International Jugglers Association and is currently president of the Mathematical Association of America (MAA). He is a fellow of the American Academy of Arts and Sciences and a member of the National Academy of Sciences (U.S.), which he serves as Treasurer. Among his many awards are the Pólya Prize in Combinatorics from the Society for Industrial and Applied Mathematics, the Euler Medal from the Institute of Combinatorics and Its Applications, and both the Carl B. Allendoerfer Award and the Lester R. Ford Award from the MAA and the Leroy Steele Award for Lifetime Achievement from the American Mathematical Society. In 1983 he gave an invited address to the International Congress of Mathematicians in Warsaw and in 2000 he was the AMS Gibbs Lecturer.

**Susan Holmes** Susan Holmes received her early education in France and England. She did graduate work in pure mathematics before moving to statistics and earning her PhD under the direction of Professor Yves Escoufier at l'Université Montpellier II in 1985. Her thesis was on "Computer-Intensive Methods for the Evaluation of Results after an Exploratory Analysis," studying the techniques of bootstrapping for multivariate analyses. She has been a tenured research scientist with INRA (Institut National de la Recherche Agronomique) in France for over ten years, a visiting scholar at Stanford, MIT, and Harvard, an associate professor of Biometry at Cornell, and is currently an Associate Professor in the Statistics Department at Stanford. Currently her main area of research is in the application of mathematics and statistics to Molecular Biology and Genetics. Reference to some of her many publications and talks may be found at `www-stat.stanford.edu/~susan/`.

**Helen Moore** Helen Moore is the Associate Director of the American Institute of Mathematics, which is located in Palo Alto, CA, and is one of several federally-supported mathematics institutes in the United States. She graduated from the North Carolina School of Science and Mathematics and the University of North Carolina at Chapel Hill. She received her PhD in mathematics in 1995 from the State University of New York, Stony Brook, where she won a university-wide teaching award. She taught and did research at Bowdoin College, and then Stanford University (where she won a departmental teaching award). She received a National Science Foundation grant for her research on the mathematics of soap films, and currently works on mathematical models of diseases such as leukemia and hepatitis. She loves hiking in the California hills and playing Ultimate Frisbee and acoustic guitar.

**Robert Osserman** Robert Osserman is Professor Emeritus of Mathematics at Stanford University and Special Projects Director at the Mathematical Sciences Research Institute in Berkeley, where he was previously Deputy Director. His research interests have always had a geometric component, including the theory of minimal surfaces, isoperimetric inequalities, and geometric function theory. He has been a Guggenheim Fellow at the University of Warwick, a Fulbright Lecturer at the University of Paris, and gave an invited address at the International Congress of Mathematicians in Helsinki in 1978. He has also written a book for the general reader on geometry and cosmology called *Poetry of the Universe: a Mathematical Exploration of the Cosmos.* He served as Pólya Lecturer for the MAA and is a Fellow of the AAAS. He has received the Lester R. Ford award for expository writing from the MAA, and in 2003 he was awarded the Communications Award from the Joint Policy Board for Mathematics, based partly on the success of *Poetry of the Universe,* but also for his interviews of playwrights and actors about plays with a connection to mathematics: David Auburn on his play, *Proof*; G. V. Coyne, S.J., and Michael Winters on Brecht's *Galileo*; Tom Stoppard on *Arcadia*; Michael Frayn on *Copenhagen*; and Steve Martin on his play *Picasso at the Lapin Agile* and his book, *The Pleasure of My Company.* Several of these public conversations are available on video and distributed by the AMS and MAA.

**Jean J. Pedersen** Jean Pedersen's main mathematical interests are polyhedral geometry, number theory and combinatorics. For the past 25 years she has worked with Peter Hilton doing research leading to over 80 joint research publications and six books. Among her books that may interest readers of this volume are *Geometric Playthings* (with Kent Pedersen), *Build Your Own Polyhedra* (with Peter Hilton), and two companion volumes, *Mathematical Reflections/In a Room of Many Mirrors* and *Mathematical Vistas/From a Room with Many Windows* (both with Peter Hilton and Derek Holton). Pedersen, a Professor of Mathematics and Computer Science at Santa Clara University, is a past member of the Board of Governors of the Mathematical Association of America (MAA) and, in 1997, she received an MAA Section Award for Distinguished Teaching.

**Carl Pomerance** Carl Pomerance received his BA from Brown University in 1966 and his PhD from Harvard University in 1972 under the direction of John Tate. During the period 1972–99 he was a professor at the University of Georgia, with visiting positions at the University of Illinois at Urbana-Champaign, the University of Limoges, Bell Communications Research, and the Institute for Advanced Study. Subsequently he has been at Bell Labs but has recently moved to Dartmouth College. A number theorist, Pomerance specializes in analytic, combinatorial, and computational number theory. He considers the late Paul Erdős to be his greatest influence. Pomerance was an invited speaker at the 1994 International Congress of Mathematicians in Zürich, and he was the Mathematical Association of America Pólya Lecturer for 1993–95, as well as the MAA Hedrick Lecturer in 1999. He has won the Chauvenet Prize (1985), the Haimo Award for Distinguished Teaching (1997), and the Conant Prize (2001). In addition he is the coauthor with Richard Crandall of the new book, *Prime Numbers: A Computational Perspective.*

**Jennifer Quinn** Jennifer Quinn grew up in Rhode Island. She graduated from Williams College where she studied mathematics, biology, and theater. In 1993, she received her PhD in mathematics from the University of Wisconsin. Since then she has taught at Occidental College in Los Angeles. In 2001, she received the Distinguished Teaching Award from the Southern California Section of the MAA. She and her coauthor, Arthur Benjamin, are constantly searching for proofs that really count! Their book on the subject, *Proofs That Really Count: The Art of Combinatorial Proof,* was published by the MAA in 2003. They are serving a 5-year term as co-editors of *Math Horizons.* Professor Quinn's other pursuits include avoiding committee work, home renovations, raising two beautiful boys, Anson and Zachary, and trying to get out of Los Angeles.

**Karl Rubin** Karl Rubin received his AB from Princeton University and his PhD from Harvard, after which he taught at The Ohio State University and Columbia University before joining the Stanford faculty in 1997, where he is currently Professor of Mathematics. He has spent most of his research career studying elliptic curves. He was the first PhD student of Andrew Wiles, who used elliptic curves to prove Fermat's Last Theorem. As an undergraduate he was a Putnam Fellow in 1975 and subsequently has won many honors, including the Cole Prize in Number Theory from the American Mathematical Society and Guggenheim and Sloan Fellowships. In 2002 he was an invited speaker at the International Congress of Mathematicians in Beijing. He likes to travel and has lectured in more than a dozen countries.

**Edward F. Schaefer** Ed Schaefer earned his PhD from the University of California, Berkeley, in 1992, where his advisor was Hendrik W. Lenstra, Jr., and since then has been at Santa Clara University, where he is currently Associate Professor of Mathematics and Computer Science. His main research interests are cryptography and arithmetic geometry, the subject of his MAA Invited Student Lecture in 1997. Arithmetic geometry uses geometry to solve problems from number theory, as in this article. He has given talks on mathematics in Guatemala and Peru, in addition to several European countries, and helped to establish the Department of Mathematics at Landivar University in Guatemala in 1997.

**Richard A. Scott** Richard Scott received his PhD from MIT in 1993. After holding postdoctoral positions at the Institute for Advanced Study in Princeton and The Ohio State University (the Zassenhaus Assistant Professorship), he came to Santa Clara University in 1997 where he is an Associate Professor of Mathematics and Computer Science. His research interests are in the topology of nonpositively curved spaces and the topology of algebraic varieties. He is currently a Councilor for the Council of Undergraduate Research, a national organization that promotes funding and programs for student/faculty research at undergraduate institutions.

**Zvezdelina Stankova** Zvezda Stankova was born and grew up in Bulgaria. Her mathematics career started quite unexpectedly in the 5th grade when, three months after joining her middle school Math Circle, she won the Regional Math Olympiad. Since then she has participated in and won many national and international mathematics contests. In 1997 she received her PhD in the area of algebraic geometry from Harvard University. After holding postdoctoral positions at the Mathematical Sciences Research Institute and the University of California, Berkeley, she joined (in 1999) the faculty at Mills College in Oakland, California, where she is currently an associate professor. She is one of the founders of the Bay Area Mathematical Olympiad (BAMO) for middle and high school students and also leads the Berkeley Math Circle at the University of California, Berkeley. Professor Stankova was one of the coaches of the US national high school mathematics team in preparation for the International Mathematical Olympiad (IMO) in 1998–2002 and 2004.

**Sherman Stein** After earning a PhD at Columbia University, Sherman Stein spent almost all of his teaching career at the University of California, Davis, from which he retired in 1993. In his own words, "During that time, and since, I have devoted my time to spreading the gospel of mathematics and to research, primarily in the application of algebra to geometry." Professor Stein has been a winner of several awards for his teaching and writing, for example, the Beckenbach Prize for *Algebra and Geometry: Homomorphisms in the Service of Geometry* (with Sándor Szabó), the Lester R. Ford Award for expository writing, and the Distinguished Teaching Award on his own campus. He is the author of several best-selling books, such as *The Strength in Numbers*; *Mathematics: The Man-made Universe*; *Archimedes: What Did He Do Besides Cry Eureka?*; and *How the Other Half Thinks*.

**Peter Stevenhagen** Peter Stevenhagen graduated from the University of Amsterdam in 1985 and received his PhD from the University of California, Berkeley, in 1988. He formed a number theory group in Amsterdam during the 1990s and in 2000 he was appointed Professor of Mathematics at the University of Leiden, the oldest university in The Netherlands, founded in 1575. Besides his research papers in algebraic number theory, he has written several highly acclaimed popular articles on number theoretic topics.

**Robin Wilson** Robin Wilson is head of the Pure Mathematics Department at the Open University, UK, Fellow of Keble College, Oxford University, and is Gresham Professor of Geometry, Gresham College, London. He is also a frequent visitor to The Colorado College. He has written and edited over twenty books on subjects ranging from graph theory, via the Gilbert and Sullivan operas, to the history of mathematics, and most recently has produced a book on the history and proof of the four color theorem. He is internationally known for his bright shirts and his awful puns.

**Paul Zeitz** Paul Zeitz received a PhD in 1992 from the University of California, Berkeley. He has been at the University of San Francisco since then, and is now professor and chair of the mathematics department. Before graduate school, he taught high school math in San Francisco and Colorado. He also had flings with history (his undergraduate major), journalism, and geology before finally settling down with mathematics.

He has worked extensively with various mathematical competitions, starting with high school, when he was a winner of the USA Mathematical Olympiad and a member of the first American team to enter the International Mathematical Olympiad. He has helped to coach American IMO teams, and, with Zvezdelina Stankova, co-founded the Bay Area Mathematical Olympiad in 1999. His book, *The Art and Craft of Problem Solving* (Wiley, 1999), is used widely in problem seminars. He received the Haimo Award in 2003.

## About the Editors

**David Hayes** David Hayes became interested in geometry in the seventh grade and was further encouraged at North High School in Oildale, California, by excellent math instructors and by Courant and Robbins' book *What is Mathematics?*. He went on to earn a PhD in Mathematics from the University of California at Davis where he worked with David Barnette in graph theory. His principal mathematical interests are in the areas of graph theory, combinatorics, and number theory. He is currently the Chair of the Computer Science Department at San Jose State University.

**Tatiana Shubin** Tatiana Shubin obtained her undergraduate mathematical education in the USSR, at Moscow State and Kazakh State Universities. In 1983 she earned her PhD from University of California at Santa Barbara, and after holding a visiting lecturer position at UC Davis, she joined the faculty of San Jose State University in 1985. She is a co-founder and one of the coordinators of the Bay Area Mathematical Adventures (BAMA), and the San Jose Math Circle, and is involved in various local and national math competitions for middle and high school students.

# Index

ASHLAND COMMUNITY COLLEGE
3 3631 1113290 X